T0301020

Chemotaxis, Reaction, Network

Mathematics for Self-Organization

Chemotaxis, Reaction, Network

Mathematics for Self-Organization

Takashi Suzuki

Osaka University, Japan

World Scientific

NEW JERSEY · LONDON · SINGAPORE · BEIJING · SHANGHAI · HONG KONG · TAIPEI · CHENNAI · TOKYO

Published by

World Scientific Publishing Co. Pte. Ltd.
5 Toh Tuck Link, Singapore 596224
USA office: 27 Warren Street, Suite 401-402, Hackensack, NJ 07601
UK office: 57 Shelton Street, Covent Garden, London WC2H 9HE

Library of Congress Cataloging-in-Publication Data
Names: Suzuki, Takashi, author.
Title: Chemotaxis, reaction, network : mathematics for self-organization / by
    Takashi Suzuki (Osaka University, Japan).
Description: New Jersey : World Scientific, 2018. | Includes bibliographical references and index.
Identifiers: LCCN 2018011304 | ISBN 9789813237735 (hardcover : alk. paper)
Subjects: LCSH: Self-organizing systems. | Chemotaxis. | Thermodynamics.
Classification: LCC Q325 .S89 2018 | DDC 003/.7--dc23
LC record available at https://lccn.loc.gov/2018011304

British Library Cataloguing-in-Publication Data
A catalogue record for this book is available from the British Library.

For any available supplementary material, please visit
https://www.worldscientific.com/worldscibooks/10.1142/10926#t=suppl

Printed in Singapore

Preface

What is the origin of self-organization? This question has attracted many scientists in physics, chemistry, biology, engineering, informatics, and medicine. Two aspects are pointed out; static and kinetic theories. Static theory is concerned with the equilibria, while kinetic theory is further divided into two parts; near from and far from equilibrium.

First, role of far from equilibrium is noted. Fluctuations of limit cycles, traveling waves with pulses, and self-similar autoreproduction are particularly of interest, which are observed in thermodynamically open systems provided with the entropy delivery. Then, near from equilibrium is noticed in accordance with the bottom up self-organization. This phenomenon is realized in thermo-dynamically closed system where non-stable stationary states control the transient dynamics. Importance of the mathematical study on equilibria is now well recognized in this context, which has induced the fusion of non-linear mathematics and non-equilibrium statistical mechanics. This is the theme of this monograph.

We begin with mathematical modeling and mathematical analysis of 2D Smoluchowski-Poisson equation. Based on macroscopic reduction of microscopic structure, we have observed the scaling invariance and the dual variation which results in the quantized blowup mechanism and the recursive hierarchy in both dynamic and static levels. In the static theory we reach Onsager's theory on point vortices, which leads to a natural extension of these structures to higher dimensions. This part is described in the context of self-gravitating fluids, reviewing basic theories on non-equilibrium statistical mechanics.

The second aim is to describe new mathematical techniques for thermo-dynamically isolated systems. Essentially, these systems are not associated with self-organization. Spatial homogenization is the leading principle for

v

their dynamics, that is, the asymptotic ODE control. A variety of references are devoted to the case where the ODE part admits a unique global attractor. We shall describe the basic tool of relative entropy, or diversity. Then we turn to more complicated dynamics, showing new insights to what emerges from the reaction network, that is, spatial homogenization, pathway grouping, nested periodic orbits derived from the structure of the Poisson manifold, break down of dynamical equilibrium revealed by the catastrophic theory, and the reproducible dumping oscillation as a result of a stable periodic orbit realized in transient dynamics.

This monograph is totally devoted to the recent mathematical theories on the bottom up self-organization, observed in closed and isolated thermo-dynamical systems. It is composed of five chapters. Chapter 1 deals with a simple system of chemotaxis. This system is provided with a striking feature of the solution, that is, the quantized blowup mechanism observed both in finite and in infinite time blowup solutions. Actually, this feature is due to the microscopic dynamics controlled by the total set of stationary solutions, revealed by several scaling transformations. Steady state of this system coincides with the mean field limit of many static point vortices derived from a Hamiltonian. Then Chapter 2 is devoted to this modeling and its extension to the kinetic theory. Mathematical analysis to the stationary states is also provided. Chapter 3 is a generalization to the higher dimension, noticed in several areas of mathematical physics. Similar structures to the 2D Smoluchowski-Poisson equation are observed in a degenerate parabolic equation with critical growth which arises in accordance with the mean field theory for self-interacting fluids associated with the Tsallis entropy. In contrast that the first three chapters are strongly associated with the canonical statistical mechanics provided with the quantized blowup mechanism and the recursive hierarchy, the last two chapters are connected with the micro-canonical statistical mechanics where the spatially homogenization is widely observed. Chapter 4 deals with the models directly related to the fundamental process arising in the theories of chemical reaction, population dynamics, and virus dynamics. The notion of weak solution is efficient to control global-in-time behavior of the solution, under the presense of the non-negativity of the solution. Several new estimates and mathematical techniques are presented there. Chapter 5, finally, is devoted to the recent developments in the theory of networks and their applications to mathematical oncology, a fusion of mathematics and biology, in accordance with the theories of integrable and dynamical systems.

The author took the efforts for this monograph to be self-contained.

Our previous book [Suzuki and Senba (2011)], basically addressed for undergraduate students, however, will be useful to understand motivations and technicalities. The reader thus can regard this monograph as a continuation of [Suzuki and Senba (2011)] to approach recent studies on physics, chemistry, and biology, using mathematical analysis for nonlinear partial and ordinary differential equations, from which really new concepts and notions are emerged.

The author expresses his sincere thanks to the collaborators in wide areas, nonlinear analysis, fluid dynamics, statistical mechanics, biology, chemistry, and medicine.

June, 2018
Takashi Suzuki

Acknowledgments

This work was supported by JSPS Core to Core Program, International Research Network and JSPS Kakenhi 16H06576 and 26247013.

Contents

Chapter 1

Chemotaxis

Chemotaxis is a feature that living things are attracted by special chemical sources, which results in, for example, the formation of spores in the case of cellular slime molds. Mathematical study on 2D Smoluchowski-Poisson equation is motivated by this phenomenon, and the quantized blowup mechanism is observed in three levels in the blowup of solutions; stationary, in finite time, and in infinite time. This equation, however, also describes the motion of mean field of many point vortices in relaxation time, that is, from quasi-equilibrium to equilibrium. This physical background is the origin of recursive hierarchy emerged from this model, of which details are stated in later sections, and then quantized blowup mechanism arises as its consequence. This chapter is concerned on a modeling from the view point of mathematical biology and several features observed for the blowup in finite time and in infinite time solutions to this model, particularly, mass quantization and collapse dynamics controlled by a Hamiltonian.

1.1 Keller-Segel Systems

The Keller-Segel system models the chemotactic features of cellular slime molds at the scale of several hours in time and cell populations in space, and is given by

$$u_t = \nabla \cdot (d_1(u,v)\nabla u) - \nabla \cdot (d_2(u,v)\nabla v)$$
$$v_t = d_v \Delta v - k_1 vw + k_{-1}p + f(v)u$$
$$w_t = d_w \Delta w - k_v w + (k_{-1} + k_2)p + g(v,w)u$$
$$p_t = d_p \Delta p + k_1 vw - (k_{-1} + k_2)p, \tag{1.1}$$

where $u = u(x,t)$, $v = v(x,t)$, $w = w(x,t)$, and $p = p(x,t)$ stand for the density of cellular slime molds, the concentrations of chemical substances,

enzymes, and complexes, respectively. This is a multi-scale model where
several factors are taken into account.

The first concept involved in this model is the gradient. Given the
scalar field φ, its gradient $\nabla\varphi$ stands for the vector field of which direction
maximizes the growth of ϕ and of which length is its rate. The gradient
operator ∇ is then defined, which leads to the divergence $\nabla\cdot j$ of the vector
field j. If $\rho = \rho(x,t)$ stands for the mass density varying in t, its flux
$j = j(x,t)$ is the vector field varying in t which satisfies

$$\frac{d}{dt}\int_\omega \rho\,dx = -\int_{\partial\omega} \nu\cdot j\,dS \tag{1.2}$$

for arbitrary domain ω, where ν and dS are the outer normal unit vector
and the surface element, respectively. Then the equation of conservation,

$$\rho_t = -\nabla\cdot j, \tag{1.3}$$

follows from the divergence theorem of Gauss because ω is arbitrary in
(1.2).

Diffusion equation then emerges as

$$\rho_t = d_\rho\Delta\rho \tag{1.4}$$

if j is proportional to $\nabla\rho$ in (1.3). Here and henceforth, $d.$ stands for the
physical constant associated with the material $\cdot$. Equation (1.4), however,
arises also as a limit of the distribution function of a randomly moving
particle. The diffusion coefficient $D = d_\rho > 0$ is associated with the mean
jump length Δx and the mean waiting time τ through the Einstein formula

$$D = \frac{(\Delta x)^2}{2N\tau},$$

where N denotes the space dimension.

In the transport theory, time variation of ρ is formulated in accordance
with the velocity of particles, denoted by $v = v(x,t)$. Under this flow,
position $x = x(t)$ of the particle at time t is subject to

$$\frac{dx}{dt} = v(x,t),$$

and if ω_t denotes the region made by the particles at time t which are in ω
at $t = 0$, Liouville's theorem implies

$$\frac{d}{dt}\int_{\omega_t}\rho\,dx = \int_{\omega_t}\frac{D\rho}{Dt}+\rho\nabla\cdot v\,dx = \int_{\omega_t}\rho+\nabla\cdot v\rho\,dx, \tag{1.5}$$

where

$$\frac{D}{Dt} = \frac{\partial}{\partial t}+v\cdot\nabla$$

denotes the material derivative. By (1.3) and (1.5) it follows that

$$j = v\rho,$$

which means that the flux of the density of particles is the product of mass and velocity.

We see that the terms in (1.1) associated with the material transport are composed of the diffusion and chemotaxis,

$$u_t = \nabla \cdot (d_1(u,v)\nabla u) - \nabla \cdot (d_2(u,v)\nabla v)$$
$$v_t = d_v \Delta v, \;\; w_t = d_w \Delta w, \;\; p_t = d_p \Delta p.$$

Concentrations of chemical material v, enzyme w, and complexes p are subject to the diffusion, while the density of slime molds u is under the influence of fluctuations of environments. Hence the diffusion coefficient $d_1 = d_1(u,v)$ and the chemotactic velocity $d_2/u = d_2(u,v)/u$ are under the control of (u,v).

The other part of (1.1) is an ODE model, composed of the productions of u and v of which rates depend on v and (v,w),

$$v_t = f(v)u, \quad w_t = g(v,w)u,$$

and the chemical reaction

$$V + W \to P \;(k_1), \quad P \to V + W \;(k_{-1}), \quad P \to W \;(k_2),$$

is described by the law of mass action,

$$v_t = -k_1 vw + k_{-1}p$$
$$w_t = -k_1 vw + (k_{-1} + k_2)p$$
$$p_t = k_1 vw - (k_{-1} + k_2)p. \tag{1.6}$$

A reduction is made in the quasi-stationary state of (1.6) where $p_t = 0$. Using the mass conservation $(w + p)_t = 0$, we reach

$$u_t = \nabla \cdot (d_1(u,v)\nabla u) - \nabla \cdot (d_2(u,v)\nabla v)$$
$$v_t = d_v \Delta v - k(v)v + f(v)u \tag{1.7}$$

by (1.1), where

$$k(v) = \frac{ck_1 k_2}{k_{-1} + k_2 + k_1 v}.$$

A number of mathematical studies is done for the linear approximation of (1.7),

$$u_t = d_u \Delta u - \chi \nabla \cdot (u \nabla v)$$
$$v_t = d_v \Delta v - b_1 v + b_2 u$$

and its reduction to the elliptic-parabolic system, e.g.,

$$u_t = \nabla \cdot (\nabla u - u \nabla v), \quad -\Delta v = u - \frac{1}{|\Omega|} \int_\Omega u \; dx \quad \text{in } \Omega \times (0, T)$$

$$\frac{\partial u}{\partial \nu} - u \frac{\partial v}{\partial \nu} = \frac{\partial v}{\partial \nu} = 0 \quad \text{on } \partial\Omega \times (0, T). \tag{1.8}$$

As is described in the beginning of this section, this system is nothing but the Smoluchowski-Poisson equation describing the motion of mean field of many self-gravitating Brownian particles.

1.2   Blowup in Finite Time

As is described in the previous section, there is a quantized blowup mechanism to (1.8) in stationary, in finite time, and in infinite time. We note that system (1.8) is divided into the Smoluchowski part

$$u_t = \nabla \cdot (\nabla u - u \nabla v) \text{ in } \Omega \times (0, T), \quad \left. \frac{\partial u}{\partial \nu} - u \frac{\partial v}{\partial \nu} \right|_{\partial\Omega} = 0 \tag{1.9}$$

and the Poisson part

$$-\Delta v = u - \frac{1}{|\Omega|} \int_\Omega u \; dx, \quad \left. \frac{\partial v}{\partial \nu} \right|_{\partial\Omega} = 0, \quad \int_\Omega v \; dx = 0. \tag{1.10}$$

Then the initial condition is provided as

$$u|_{t=0} = u_0(x). \tag{1.11}$$

To make it well-posed, here we added the last condition to (1.10), because otherwise, the solution is not definite up to an additional constant. This modification, however, effects nothing to (1.9) where only ∇v is involved in the first equation.

The above mentioned quantized blowup mechanism holds to the other form of the Smoluchowski-Poisson equation where (1.10) is replaced by

$$-\Delta v = u, \quad v|_{\partial\Omega} = 0. \tag{1.12}$$

In more details, boundary blowup points are excluded for both blowup in finite time and blowup in infinite time to (1.9) with (1.12), while boundary blowup points actually exist for (1.9) with (1.10) where the collapse masses are reduced to a half of those of interior blowup points.

The description of the kinetic blowup mechanism, that is, blowup in finite time and in infinite time, becomes simpler if we take the system on the whole space $\mathbf{R}^2$,

$$u_t = \Delta u - \nabla \cdot (u \nabla \Gamma * u), \quad u|_{t=0} = u_0(x) \geq 0 \quad \text{in } \mathbf{R}^2 \times (0, T) \tag{1.13}$$

where

$$\Gamma(x) = \frac{1}{2\pi} \log \frac{1}{|x|} \quad \text{and} \quad (\Gamma * u)(x, t) = \int_{\mathbf{R}^2} \Gamma(x - x') u(x', t) \, dx'.$$

If $u_0 \in X \equiv L^1(\mathbf{R}^2) \cap L^\infty(\mathbf{R}^2)$ the semi-group theory guarantees a unique local-in-time solution $0 \leq u \in C([0, T), X)$ smooth in (x, t) for $t > 0$, provided with the total mass conservation

$$\|u(\cdot, t)\|_1 = \lambda \equiv \|u_0\|_1. \tag{1.14}$$

If the maximal existence time denoted by $T = T_{\max}$ is finite, it follows that

$$\lim_{t \uparrow T} \|u(\cdot, t)\|_\infty = +\infty, \tag{1.15}$$

because the existence time of the solution is estimated below by $\|u_0\|_\infty$.

Given $\varphi \in C_b^2(\mathbf{R}^2)$, we have the weak form of (1.13),

$$\frac{d}{dt} \int_{\mathbf{R}^2} u(x, t) \varphi(x) \, dx = \int_{\mathbf{R}^2} u(x, t) \Delta \varphi(x) \, dx$$
$$+ \frac{1}{2} \iint_{\mathbf{R}^2 \times \mathbf{R}^2} \rho_\varphi(x, x') u(x, t) u(x', t) \, dx dx' \tag{1.16}$$

for

$$\rho_\varphi(x, x') = -\frac{x - x'}{2\pi |x - x'|^2} \cdot (\nabla \varphi(x) - \nabla \varphi(x')) \in L^\infty(\mathbf{R}^2 \times \mathbf{R}^2),$$

recalling that $C_b^2(\mathbf{R}^2)$ denotes the set of C^2 functions in $\mathbf{R}^2$ uniformly bounded up to its second derivatives. The process of deriving (1.16) is called *symmetrization*, because it uses the symmetry of the potential kernel for the Poisson part, of which origin is in the action-reaction law valid to self-interacting particles. Then, taking $\varphi = |x|^2$ is justified in (1.16) if we assume $|x|^2 u_0 \in L^1(\mathbf{R}^2)$ which results in

$$\frac{d}{dt} \int_{\mathbf{R}^2} |x|^2 u(x, t) \, dx = 4\lambda - \frac{\lambda^2}{2\pi}. \tag{1.17}$$

Equality (1.17) implies $T < +\infty$ if $\lambda > 8\pi$. The other property derived from this equality is the boundedness of the blowup set defined by

$$\mathcal{S} = \{ x_0 \in \mathbf{R}^2 \cup \{\infty\} \mid \text{there exist } x_k \to x_0 \text{ and } t_k \uparrow T \text{ such that}$$
$$\lim_{k \to \infty} u(x_k, t_k) = +\infty \}.$$

In fact, using the following lemma derived from the Gagliardo-Nirenberg inequality, we obtain $\mathcal{S} \subset \mathbf{R}^2$ by (1.14), (1.16), and Chebyshev's inequality.

Lemma 1.1. *There is $\varepsilon_0 > 0$ such that*

$$\lim_{R \downarrow 0} \limsup_{t \uparrow T} \|u(\cdot, t)\|_{L^1(B(x_0, R))} < \varepsilon_0 \quad \Rightarrow \quad x_0 \notin \mathcal{S}.$$

Lemma 1.1 is called ε-regularity.

The proof of this lemma is divided into two parts. First, we show the existence of $\varepsilon_0 > 0$ such that

$$\limsup_{t \uparrow T} \|u(\cdot,t)\|_{L^1(\Omega \cap B(x_0,4R))} < \varepsilon_0$$

$$\Rightarrow \quad \limsup_{t \uparrow T} \int_{\Omega \cap B(x_0,2R)} [u(\log u - 1)](x,t) \, dx < +\infty \qquad (1.18)$$

for $R > 0$. Then it follows that

$$\limsup_{t \uparrow T} \int_{\Omega \cap B(x_0,2R)} [u(\log u - 1)](x,t) \, dx < +\infty$$

$$\Rightarrow \quad \limsup_{t \uparrow T} \|u(\cdot,t)\|_{L^\infty(B(x_0,R))} < +\infty. \qquad (1.19)$$

Here, a nice cut-off function is used for the proof. It is $\varphi = \varphi_{x_0,R}(x) \in C^2(\overline{\Omega})$ with the support radius $0 < R \le 1$ satisfying

$$0 \le \varphi = \varphi_{x_0,R}(x) = \begin{cases} 1, & x \in B(x_0, R/2) \\ 0, & x \in B(x_0, R)^c, \end{cases} \qquad \left.\frac{\partial \varphi}{\partial \nu}\right|_{\partial \Omega} = 0 \qquad (1.20)$$

and

$$|\nabla \varphi_{x_0,R}| \le CR^{-1}\varphi^{5/6}, \quad |\nabla^2 \varphi_{x_0,R}| \le CR^{-2}\varphi^{2/3}. \qquad (1.21)$$

Free energy

$$\mathcal{F}(u) = \int_{\mathbf{R}^2} u(\log u - 1) \, dx - \frac{1}{2} \iint_{\mathbf{R}^2 \times \mathbf{R}^2} \Gamma(x - x')u \otimes u \, dxdx'$$

defined for $u \otimes u = u(x,t)u(x',t)$ satisfies formally that

$$\frac{d}{dt}\mathcal{F}(u) = -\int_{\mathbf{R}^2} u|\nabla(\log u - v)|^2 \, dx \le 0. \qquad (1.22)$$

Then the dual Trudinger-Moser inequality

$$\inf\{\mathcal{F}(u) \mid u \ge 0, \|u\|_1 = 8\pi\} > -\infty \qquad (1.23)$$

guarantees the global-in-time existence of uniformly bounded solution to (1.13) for the case of $\lambda < 8\pi$. Hence $\lambda = 8\pi$ is the threshold for the existence of the solution global-in-time.

Justifying (1.22) requires additional assumptions on u_0. This process is technical, but the following theorem, proven without free energy, assures that $T < +\infty$ occurs only if $\lambda > 8\pi$. Thus we obtain $T = +\infty$ even if $\lambda = 8\pi$, where blowup in infinite time may arise.

Theorem 1.1. *Given* $0 \leq u_0 \in L^1(\mathbf{R}^2) \cap L^\infty(\mathbf{R}^2)$ *with* $|x|^2 u_0 \in L^1(\mathbf{R}^2)$, *let* $u = u(\cdot, t)$ *be the solution to (1.13). Assume blowup in finite time indicated by* $T < +\infty$. *Then* $u(x,t)dx = \mu(dx,t)$ *is extended up to* $t = T$ *as*

$$\mu(dx, t) \in C_*([0, T], \mathcal{M}(\mathbf{R}^2 \cup \{\infty\}))$$

where $\mathcal{M}(\mathbf{R}^2 \cup \{\infty\}) = C(\mathbf{R}^2 \cup \{\infty\})'$ *with* $\mathbf{R}^2 \cup \{\infty\}$ *standing for the one-point compactification of* $\mathbf{R}^2$. *It holds that* $\sharp \mathcal{S} < +\infty$ *and*

$$\mu(dx, T) = \sum_{x_0 \in \mathcal{S}} m(x_0) \delta_{x_0}(dx) + f(x)dx \tag{1.24}$$

where $0 < f = f(x) \in L^1(\mathbf{R}^2) \cap C(\mathbf{R}^2 \setminus \mathcal{S})$ *and* $m(x_0) \in 8\pi\mathbf{N}$.

For the proof, first, the weak form (1.16) implies

$$\left| \frac{d}{dt} \int_{\mathbf{R}^2} u(x,t)\varphi(x) \, dx \right| \leq C\|\nabla\varphi\|_{C^1}(\lambda + \lambda^2) \tag{1.25}$$

for each $\varphi \in C_b^2(\mathbf{R}^2)$. Inequality (1.25) is called the *monotonicity formula*. Since this inequality implies

$$\int_0^T \left| \frac{d}{dt} \int_{\mathbf{R}^2} u(x,t)\varphi(x) \, dx \right| \, dt < +\infty,$$

the extension of $u(x,t)dx$ to

$$\mu(dx, t) \in C_*([0, T], \mathcal{M}(\mathbf{R}^2 \cup \{\infty\}))$$

is achieved. Then, the ε-regularity, Lemma 1.1, implies

$$\mu(\{x_0\}, T) \geq \varepsilon_0, \quad \forall x_0 \in \mathcal{S}.$$

Hence $\mathcal{S} < +\infty$ follows from the total mass conservation, $\mu(\mathbf{R}^2, T) = \lambda$. It holds also that (1.24) with

$$m(x_0) \geq \varepsilon_0 \quad \text{and} \quad 0 \leq f = f(x) \in L^1(\mathbf{R}^2).$$

The property $f \in L^\infty_{loc}(\mathbf{R}^2 \setminus \mathcal{S})$ follows from Lemma 1.1. Therefore, $f = f(x)$ is smooth in $\mathbf{R}^2 \setminus \mathcal{S}$ by the elliptic and parabolic regularity to

$$u_t = \Delta u - \nabla \cdot (u\nabla v) \quad \text{and} \quad -\Delta v = u \qquad \text{in } \mathbf{R}^2 \times (0, T)$$

derived from (1.13). Then the strong maximum principle guarantees $f = f(x) > 0$ for any $x \in \mathbf{R}^2 \setminus \mathcal{S}$ unless $f \equiv 0$. This case is not consistent to $T < +\infty$ because it implies $u_0 \equiv 0$ and hence $u \equiv 0$.

Thus we have only to show quantization of the collapse mass, $m(x_0) \in 8\pi\mathbf{N}$, for each $x_0 \in \mathcal{S}$ to complete the proof of Theorem 1.1. We call this property the *quantized blowup mechanism*.

It is not certain whether $f \in L^\infty(\mathbf{R}^2)$ holds or not. To approach this problem, the following lemma may be useful.

Lemma 1.2. *There is $\varepsilon_0 > 0$ such that if $\{u_k\}$ is a family of solutions to (1.13) satisfying $0 \le u_k \in C([0, T), X)$ and $u_{kt} \in L^\infty(0, T; X)$ for $X = L^1(\mathbf{R}^2) \cap L^\infty(\mathbf{R}^2)$, and if*

$$\|u_{k0}\|_{L^1(B(x_0, 2R))} < \varepsilon_0, \quad k = 1, 2, \cdots,$$

holds for $x_0 \in \mathbf{R}^2$ and $R > 0$, then it follows that

$$\sup_k \|u_k\|_{L^\infty(B(x_0, R) \times (\tau, t_0))} < +\infty$$

for any $0 < \tau < t_0$, where $0 < t_0 \ll 1$ is determined by R.

Lemma 1.2 is a refinement of Lemma 1.1 where local L^∞ norm of $u(\cdot, t)$ is estimated by that of u_0. The proof of Lemma 1.2 is, therefore, based on the smoothing effect of several norms of the solution.

To describe the essence of this proof, let $u_k = u_k(x, t) \in C^{1,2}(\mathbf{R}^2 \times (-T, T))$ be the solution to (1.13). By (1.25) there is $0 < t_1 < T$ such that

$$\|u_{k0}\|_{L^1(B(x_0, 4R))} < \varepsilon_0/2 \implies \sup_{t \in [-t_1, t_1]} \|u_k(\cdot, t)\|_{L^1(B(x_0, 2R))} < \varepsilon_0 \quad (1.26)$$

for $u_{k0} = u_k|_{t=0}$. Inequality (1.18) relies on the estimate derived from (1.13),

$$\frac{d}{dt} \int_{\mathbf{R}^2} u(\log u - 1)\varphi \, dx + \frac{1}{4} \int_{\mathbf{R}^2} u^{-1}|\nabla u|^2 \varphi \, dx \le 2 \int_{\mathbf{R}^2} u^2 \varphi \, dx + C_\varphi \quad (1.27)$$

applied to $\varphi = \varphi_{x_0, R}$. See Chapter 11 of [Suzuki (2005)]. Then the first term on the right-hand side of (1.27) is absorbed into the second term of the left-hand side by the Gagliardo-Nirenberg inequality, under the cost of the smallness of local L^1 of norm of the solution, say, the conclusion of (1.26). A direct application of (1.27), however, involves the norm of the initial value of the solution in (1.18). Hence the above described parabolic smoothing of several norms is necessary to proceed.

For this purpose, we derive

$$\frac{dJ}{dt} + 2 \int_{\mathbf{R}^2} u^2 \varphi \, dx \le C_R$$

from (1.27) for $\varphi = \varphi_{x_0, R}$ and

$$J = \int_{\mathbf{R}^2} (u \log u + e^{-1}) \varphi \, dx,$$

recalling $s \log s + e^{-1} \geq 0$ for $s \geq 0$. Then we use

$$\int_{\mathbf{R}^2} (u \log u + e^{-1}) \varphi \leq \left\{ \int_{\mathbf{R}^2} [(u \log u + e^{-1}) \varphi]^{3/2} \right\}^{2/3} |B_R|^{1/3}$$

to infer

$$\frac{dJ}{dt} + 3J^{3/2} \leq C_R + \int_{\mathbf{R}^2} \left[-2u^2 + C_R (u \log u + e^{-1})^{3/2} \right] \varphi \, dx.$$

An elementary to the right-hand side then guarantees

$$\frac{dJ}{dt} + 3J^{3/2} \leq C_R'. \tag{1.28}$$

Then we use

$$\frac{d}{dt} t^{-2} + 3(t^{-2})^{3/2} = t^{-3}$$

to derive

$$J(t) \leq t^{-2}, \quad 0 < t \leq \min\{t_1, t_0\} \tag{1.29}$$

where $t_0^{-3} = C_R'$.

In equality (1.29) indicates a parabolic smoothing and is the starting point of the proof of Lemma 1.2. We refer to Chapter 12 of [Suzuki (2005)] for the completion of the proof of Lemma 1.2.

There is a scaling invariance of (1.13) described by

$$u^\mu(x, t) = \mu^2 u(\mu x, \mu^2 t), \quad \mu > 0. \tag{1.30}$$

Thus if $u = u(x, t)$ is the solution to (1.13), then $u^\mu = u^\mu(x, t)$ defined by (1.30) satisfies its first equation. The monotonicity formula (1.25) is also applicable to the reverse direction of t. Then we can make Lemma 1.2 to the following scaling invariant form:

Lemma 1.3. *There are positive constants ε_0, σ_0, and C_1 such that*

$$\|u_0\|_{L^1(B(x_0, 2R))} < \varepsilon_0$$

$$\Rightarrow \sup_{t \in [-\sigma_0 R^2, \sigma_0 R^2]} \|u(\cdot, t)\|_{L^\infty(B(x_0, R))} \leq C_1 R^{-2} \tag{1.31}$$

for any $x_0 \in \mathbf{R}^2$ and $0 < R \leq (T/\sigma_0)^{1/2}$, if $u = u(x, t) \geq 0$ is a classical solution to

$$u_t = \Delta u - \nabla \cdot (u \nabla \Gamma * u) \quad \text{in } \mathbf{R}^2 \times (-T, T), \quad T > 0.$$

Given $x_0 \in \mathcal{S}$, we take $0 < R \ll 1$ satisfying $\|f\|_{L^1(B(x_0, 4R))} < \varepsilon_0/2$ to apply Lemma 1.3 for $B = B(x_1, R)$ for each $x_1 \in \partial B(x_0, 2R)$. It then follows that

$$\sup_{x \in \partial B(x_0, R)} u(x, t) \leq CR^{-2}, \quad t \in [0, T). \tag{1.32}$$

Hence $f \in L^\infty_{loc}(\mathbf{R}^2)$ is not obtained by ε-regularity, Lemmas 1.1, 1.2, or 1.3. In fact, inequality (1.32) does not even imply $f \in L^1_{loc}(\mathbf{R}^2)$ although $f \in L^\infty(\mathbf{R}^2 \setminus B(0, R))$ follows for $R \gg 1$.

1.3   Scaling Limit

To complete the proof of Theorem 1.1 we use the backward self-similar transformation

$$y = (x - x_0)/(T - t)^{1/2}, \quad s = -\log(T - t), \quad z(y, s) = (T - t)u(x, t)$$

to reach

$$z_s = \nabla \cdot (\nabla z - z\nabla(\Gamma * z + |y|^2/4)) \quad \text{in } \mathbf{R}^2 \times (-\log T, +\infty) \tag{1.33}$$

together with $z(y, s) \geq 0$ and $\|z(\cdot, s)\|_1 = \lambda$. Similarly to (1.25) it holds also that

$$\left| \frac{d}{dt} \int_{\mathbf{R}^2} z(y, s)\varphi(y) \, dy \right| \leq C_\varphi, \quad \varphi \in C_0^2(\mathbf{R}^2). \tag{1.34}$$

Here, we cannot take $\varphi \in C_b^2(\mathbf{R}^2)$ in (1.34) because of the $|y|^2/4$ term in (1.25). Inequality (1.34), however, is sufficient to guarantee the weak scaling limit. Thus any $s_k \uparrow +\infty$ admits a subsequence denoted by the same symbol such that

$$z(y, s + s_k)dy \rightharpoonup \zeta(dy, s) \quad \text{in } C_*(-\infty, +\infty; \mathcal{M}(\mathbf{R}^2)) \tag{1.35}$$

where $\mathcal{M}(\mathbf{R}^2) = C_0(\mathbf{R}^2)'$ for $C_0(\mathbf{R}^2) = \{\varphi \in C(\mathbf{R}^2 \cup \{\infty\}) \mid \varphi(\infty) = 0\}$. This $\zeta(dy, s) \geq 0$ defined by (1.35) is a *weak solution* to

$$z_s = \nabla \cdot (\nabla z - z\nabla(\Gamma * z + |y|^2/4)) \quad \text{in } \mathbf{R}^2 \times (-\infty, +\infty). \tag{1.36}$$

To define this notion, let $\mathcal{E}$ be the closure in $L^\infty(\mathbf{R}^2 \times \mathbf{R}^2)$ of

$$\{\rho_\varphi + \psi \mid \varphi \in C_0^2(\mathbf{R}^2), \ \psi \in C_0(\mathbf{R}^2 \times \mathbf{R}^2)\}$$

defined for

$$\rho_\varphi(y, y') = -\frac{y - y'}{2\pi|y - y'|^2} \cdot (\nabla\varphi(y) - \nabla\varphi(y')) \in L^\infty(\mathbf{R}^2 \times \mathbf{R}^2)$$

where $C_0(\mathbf{R}^2 \times \mathbf{R}^2)$ denotes the set of continuous functions in $\mathbf{R}^2 \times \mathbf{R}^2$ with compact supports. We note that $\rho_\varphi(y, y')$ is not continuous at $y = y'$, while $\mathcal{E}$ is separable. Thanks to this property, there is $0 \leq \mathcal{K} \in L^\infty(0, T; \mathcal{E}')$ in accordance with $\zeta(dy, s)$ generated by (1.35) which satisfies $\|\mathcal{K}(\cdot, s)\|_{\mathcal{E}'} \leq \lambda^2$ such that

$$\mathcal{K}(\cdot, s)|_{C_0(\mathbf{R}^2 \times \mathbf{R}^2)} = \zeta(dy, s) \otimes \zeta(dy', s) \quad \text{a.e. } s$$

and

$$\frac{d}{ds}\langle\varphi, \zeta(dy, s)\rangle = \left\langle \Delta\varphi + \frac{y}{2} \cdot \nabla\varphi, \zeta(dy, s) \right\rangle + \frac{1}{2}\langle\rho_\varphi, \mathcal{K}(\cdot, s)\rangle \tag{1.37}$$

in the sense of measures in $s \in (-\infty, +\infty)$ for each $\varphi \in C_0^2(\mathbf{R}^2)$. Generally, given a family weak solutions of which initial values take uniformly bounded total variations, and the associate operators denoted by $\mathcal{K}$ above are uniformly bounded in $L_*^\infty(0, T; \mathcal{E}')$, then we can subtract a sequence converging weakly to a weak solution. These conditions are always satisfied when we generate a weak solution from their sequences, although we do not necessarily mention explicitly.

Let $\varphi = \varphi_{x_0,R}(x) \in C^2(\overline{\Omega})$ be the cut-off function satisfying (1.20)–(1.21). Since

$$\|\nabla \varphi_{x_0,R}\|_{C^1} \leq CR^{-2}, \quad 0 < R \ll 1$$

it holds that

$$\left| \int_{\mathbf{R}^2} \varphi_{x_0,R}(x) u(x,t) \, dx - \langle \varphi_{x_0,R}, \mu(dx, T) \rangle \right| \leq C_\lambda R^{-2}(T-t)$$

which implies

$$\lim_{b\uparrow+\infty} \limsup_{t\uparrow T} \left| \langle \varphi_{x_0, b(T-t)^{1/2}}, \mu(\cdot, t) \rangle - m(x_0) \right| = 0 \qquad (1.38)$$

by (1.24). We thus obtain

$$m(x_0) = \zeta(\mathbf{R}^2, s), \quad -\infty < s < +\infty. \qquad (1.39)$$

Equality (1.39) is called the *parabolic envelope*. Similarly, there is the second parabolic envelope described by

$$\langle |y|^2, \zeta(dy, s) \rangle \leq C, \quad -\infty < s < +\infty \qquad (1.40)$$

which implies the tightness of the measures $\{\zeta(dy, s)\}$.

The *scaling back* $A(dy, s)$ of $\zeta(dy, s)$ is defined by

$$\zeta(dy, s) = e^{-s} A(dy', s'), \quad y' = e^{-s/2} y, \quad s' = -e^{-s}. \qquad (1.41)$$

It is a weak solution to

$$A_s = \nabla \cdot (\nabla A - A \nabla \Gamma * A), \quad A \geq 0 \quad \text{in } \mathbf{R}^2 \times (-\infty, 0) \qquad (1.42)$$

defined similarly as in the weak solution to (1.36), satisfying

$$A(\mathbf{R}^2, s) = m(x_0), \quad -\infty < s < 0.$$

The following lemma is concerned on the full orbit of (1.42) and is called the weak Liouville property.

Lemma 1.4. *If $a = a(dy, s) \in C_*(-\infty, +\infty; \mathcal{M}(\mathbf{R}^2))$ is a weak solution to*

$$a_s = \nabla \cdot (\nabla a - a \nabla \Gamma * a), \quad a \geq 0 \quad \text{in } \mathbf{R}^2 \times (-\infty, +\infty) \qquad (1.43)$$

satisfying $a(\mathbf{R}^2, s) = m$, $-\infty < s < +\infty$, then it follows that either $m = 0$ or $m = 8\pi$.

The proof is similar to the case of strong solution, using local second moment and then scaling invariance of (1.43):

$$a^\mu(dy, s) = \mu^2 a(dy', s'), \quad y' = \mu y, \quad s' = \mu^2 s. \qquad (1.44)$$

Concerning (1.42) on the half orbit $-\infty < s < 0$, however, this argument assures only $m(x_0) \geq 8\pi$. It is also worth mentioning that by the definition, the total measure $a(\mathbf{R}^2, s)$ of the weak solution is constant if it is uniformly bounded in s.

Here we take $\tilde{s}_\ell \uparrow +\infty$ and apply the concentration compactness principle to the sequence of probability measures $\{A_\ell(dy)\}$ defined by

$$\tilde{A}_\ell(dy) = A(dy, -\tilde{s}_\ell)/m(x_0).$$

Passing to a subsequence, we have the following alternatives:

(1) **compact.** Any $0 < \varepsilon < 1$ admits $y_\ell \in \mathbf{R}^2$ and $R > 0$ such that

$$\tilde{A}_\ell(B(y_\ell, R)) > 1 - \varepsilon, \quad \forall \ell \gg 1. \qquad (1.45)$$

(2) **vanishing.** For any $R > 0$ there holds that

$$\lim_{\ell \to \infty} \sup_x \tilde{A}_\ell(B(x, R)) = 0. \qquad (1.46)$$

(3) **dichotomy.** There is $0 < \lambda < 1$ such that any $\varepsilon > 0$ admits $x_\ell \in \mathbf{R}^2$ and $R > 0$ such that

$$\liminf_{\ell \to \infty} \tilde{A}_\ell(B(y_\ell, R)) \geq \lambda - \varepsilon$$
$$\lim_{R' \uparrow +\infty} \liminf_{\ell \to \infty} \tilde{A}_\ell(\mathbf{R}^2 \setminus B(y_\ell, R')) \geq \lambda - \varepsilon. \qquad (1.47)$$

Assuming the compactness to $\{\tilde{A}_\ell(dy)\}$, let $A_\ell(dy, s) = A(dy', s - \tilde{s}_\ell)$, $y' = y + y_\ell$. Then we obtain the convergence

$$A_\ell(dy, s) \rightharpoonup a(dy, s) \quad \text{in } C_*(-\infty, +\infty; \mathcal{M}(\mathbf{R}^2)),$$

up to a subsequence, with $a = a(dy, s)$ standing for a weak solution to (1.42). Since $\tilde{A}_\ell(dy) = A_\ell(dy', 0)/m(x_0)$, it follows that $a(\mathbf{R}^2, 0)/m(x_0) = 1$. Hence we obtain

$$a(\mathbf{R}^2, 0) = m(x_0) = 8\pi$$

by Lemma 1.2, and the proof is complete.

Assuming the vanishing of $\{\tilde{A}_\ell(dy)\}$, we shall show that each $\tilde{A}_\ell(dy)$, $\ell \geq 1$, has a density which converges uniformly to zero as $\ell \to \infty$. For this purpose, first, we use Lemma 1.3. Given $R > 0$, we assume (1.45), to obtain

$$\sup_x A(B(x, R), -\tilde{s}_\ell) = \sup_y \zeta(B(y, \tilde{s}_\ell^{-1/2} R), -\log \tilde{s}_\ell) < \varepsilon_0/2$$

for $\ell \gg 1$. We fix such ℓ and put $c_\ell = \tilde{s}_\ell^{-1/2} R$. From (1.35), any $b > 0$ admits k_0 such that

$$\sup_{|y| \leq b} \|z(\cdot, -\log \tilde{s}_\ell + s_k)\|_{L^1(B(y,c_\ell))} < \varepsilon_0, \quad \forall k \geq k_0$$

which means

$$\sup_{x \in B(x_0, b(T-t_k)^{1/2})} \|u(\cdot, t_k)\|_{L^1(B(x,c_\ell(T-t_k)^{1/2}))} < \varepsilon_0$$

for $t_k \uparrow T$ defined by $T - t_k = e^{-s_k}\tilde{s}_\ell$. Then Lemma 1.3 guarantees

$$\sup_{x \in B(x_0,b(T-t_k)^{1/2})} \sup_{|t-t_k| \leq \sigma_0 c_\ell^2 (T-t_k)} \|u(\cdot, t)\|_{L^\infty(B(x_0,b(T-t_k)^{1/2}))}$$
$$\leq C_1 c_\ell^{-2} (T - t_k)^{-1}. \tag{1.48}$$

We may assume $\sigma_0 c_\ell^2 \leq 1/2$, taking $\ell \gg 1$. Then it follows that

$$|t - t_k| \leq \sigma_0 c_\ell^2 (T - t_k) \quad \Rightarrow \quad \frac{1}{2} \leq \frac{T-t}{T-t_k} \leq \frac{3}{2}$$

which implies

$$\sup_{|t-t_k| \leq \sigma_0 c_\ell^2 (T-t_k)} (T-t)\|u(\cdot,t)\|_{L^\infty(B(x_0,b'(T-t)))} \leq 3C_1 c_\ell^{-2}/2 \tag{1.49}$$

for $b' = \sqrt{2/3}b$. Since

$$|t - t_k| \leq \sigma_0 c_\ell^2 (T - t_k) \quad \Leftrightarrow \quad \left|1 - e^{s_k - s}\right| \leq \sigma_0 c_\ell^2,$$

inequality (1.49) is reduced to

$$\sup_{|s-s_k'| \leq s_0 c_\ell^2} \|z(\cdot, s)\|_{L^\infty(B(0,b'))} \leq 3C_1 c_\ell^{-2}/2$$

for $s_k' = -\log(T - t_k) = s_k - \log \tilde{s}_\ell$, where $s_0 > 0$ is an absolute constant. Therefore, we obtain

$$\sup_{s \in [s_k - \log \tilde{s}_\ell - s_0 c_\ell^2, s_k - \log \tilde{s}_\ell + s_0 c_\ell^2]} \|z(\cdot, s)\|_{L^\infty(B(0,b'))} \leq 3C_1 c_\ell^{-2}/2. \tag{1.50}$$

Sending $k \to \infty$ and then making $b \uparrow +\infty$ in (1.50), we get

$$\sup_{|s+\log \tilde{s}_\ell| \leq s_0 c_\ell^2} \|\zeta(\cdot, s)\|_\infty \leq 3C_1 c_\ell^{-2}/2 = 3C_1 R^{-2}\tilde{s}_\ell/2$$

by the parabolic regularity to (1.33). Then we put $s = -\log \tilde{s}_\ell$ to obtain

$$\|A(\cdot, -\tilde{s}_\ell)\|_\infty \leq 3C_1 R^{-2}/2$$

by (1.41). Here we make $\ell \to \infty$ and then $R \uparrow +\infty$. Consequently, there arises

$$\lim_{\ell \to \infty} \|\tilde{A}_\ell(dy)\|_\infty = 0. \tag{1.51}$$

We note that (1.51) implies (1.46), and hence vanishing of $\{\tilde{A}_\ell(dy)\}$.

If dichotomy occurs to $\{\tilde{A}_\ell(dy)\}$, there is $0 < \lambda_1 < 1$ such that any $\varepsilon > 0$ admits $y_\ell^1 \in \mathbf{R}^2$ and $R_1 > 0$ such that

$$\liminf_{\ell \to \infty} \tilde{A}_\ell^1(B(y_\ell^1, R_1)) \geq \lambda_1 - \varepsilon$$
$$\lim_{R' \uparrow +\infty} \liminf_{\ell \to \infty} \tilde{A}_\ell^1(\mathbf{R}^2 \setminus B(y_\ell^1, R')) \geq (1 - \lambda_1) - \varepsilon \qquad (1.52)$$

for $\tilde{A}_\ell^1 = \tilde{A}_\ell$. The first bubble is detected as in the compact case from the first inequality of (1.52). Thus, any $\varepsilon > 0$ admits $b_1 > 0$ such that

$$\limsup_{\ell \to \infty} \left| m(x_0) \cdot \tilde{A}_\ell(B_\ell^1) - 8\pi \right| < \varepsilon, \quad B_\ell^1 = B(y_\ell^1, b_1)$$

and hence $\lambda_1 \cdot m(x_0) = 8\pi$. Furthermore, the second parabolic envelope (1.40) implies

$$A(B_R^c, -\tilde{s}) \leq C_2 \tilde{s}/R^2$$

by Chebyshev's inequality and hence

$$|y_\ell^1| \leq C_3(1 + b_1)\tilde{s}_\ell^{1/2}.$$

Then we proceed to the residual part: $A_\ell^2 = \tilde{A}_\ell^1 \big|_{B(y_\ell, R_1)^c}$. We introduce the probability measure $\tilde{A}_\ell^2 = A_\ell^2/A_\ell^2(\mathbf{R}^2)$ and apply the concentration compactness principle to $\{\tilde{A}_\ell^2(dy)\}$. If compactness occurs to this sequence, any $\varepsilon > 0$ admits $y_\ell^2 \in \mathbf{R}^2$ and $R_2 > 0$ such that

$$\tilde{A}_\ell^2(B(y_\ell^2, R_2)) > 1 - \varepsilon, \quad \forall \ell \gg 1.$$

By the inequalities in (1.51) we obtain

$$\lim_{\ell \to \infty} \left| y_\ell^1 - y_\ell^2 \right| = +\infty.$$

Similarly to the compact case of $\{\tilde{A}_\ell^1(dy)\}$ it follows also that $m_2 = m(x_0) - 8\pi = (1 - \lambda_1)m(x_0) = 8\pi$ from the weak Liouville property, and hence $m(x_0) = 16\pi$. In the vanishing case of $\{\tilde{A}_\ell^2(dy)\}$, we obtain

$$\lim_{\ell \to \infty} \|\tilde{A}_\ell\|_{L^\infty_{loc}((\mathbf{R}^2 \bigcup\{\infty\}) \setminus B_\ell^1)} = 0.$$

Here, uniform bound touching ∂B_ℓ^1 is impossible, while the uniform L^∞ control is valid similarly as in (1.32). We take the residual part of $\tilde{A}_\ell^2(dy)$ to continue this argument if $\{\tilde{A}_\ell^2(dy)\}$ is dichotomy, and finally reach the following lemma [Suzuki (2015a)].

Lemma 1.5. *Given $\tilde{s}_\ell \uparrow +\infty$, let $A_\ell(dy) = A(\cdot, -\tilde{s}_\ell)$. Then, passing to a subsequence we have $m \in \mathbf{N} \cup \{0\}$ satisfying the following property. First, any $\varepsilon > 0$ admits $y_\ell^j \in \mathbf{R}^2$ and $b_j > 0$ for $1 \leq j \leq m$ such that*

$$\lim_{\ell \to \infty} |y_\ell^i - y_\ell^j| = +\infty, \ i \neq j, \quad |y_\ell^j| \leq C_3(1 + \max_j b_j)\tilde{s}_\ell^{1/2}$$

$$\limsup_{\ell \to \infty} |A_\ell(B_\ell^j) - 8\pi| < \varepsilon, \quad \forall j$$

$$\lim_{\ell \to \infty} \sup_{x \in \mathbf{R}^2} A_\ell(B(x,R) \setminus \bigcup_{j=1}^m B_\ell^j) = 0, \quad \forall R > 0,$$

where $B_\ell^j = B(y_\ell^j, b_j)$. Second, there arises one of the following alternatives.

(1) $m(x_0) - 8\pi m - \varepsilon > 0$ and $A_\ell(dy)$ is smooth in $\mathbf{R}^2 \setminus \bigcup_{j=1}^m B_\ell^j$ and it holds that

$$\liminf_{\ell \to \infty} A_\ell(\mathbf{R}^2 \setminus \bigcup_{j=1}^m B_\ell^j) \geq m(x_0) - 8\pi - \varepsilon$$

$$\lim_{\ell \to \infty} \|A_\ell\|_{L_{loc}^\infty((\mathbf{R}^2 \cup \{\infty\}) \setminus \bigcup_{j=1}^m B_\ell^j)} = 0.$$

(2) $m(x_0) = 8\pi m$ and

$$\limsup_{\ell \to \infty} A_\ell(\mathbf{R}^2 \setminus \bigcup_{j=1}^m B_\ell^j) < \varepsilon.$$

The above $\tilde{m}$ is estimated above by $[m(x_0)/\lambda]+1$. By the diagonal argument, furthermore, we can assume $b_0 > 0$ such that $b_j \leq b_0$, $\forall j$, regardless of $\tilde{s}_\ell \to \infty$ or its subsequence. Then, again by an argument of contradiction, any $\varepsilon > 0$ admits $\tilde{s}_1$, $m(\tilde{s}) \in \{0, 1, 2, \cdots, \tilde{m}\}$, and $y_j(\tilde{s}) \in \mathbf{R}^2$ and $0 < b_j(\tilde{s}) \leq b_0$ defined for $\tilde{s} \geq \tilde{s}_1$ and $1 \leq j \leq m(\tilde{s})$ such that

$$\lim_{\tilde{s} \uparrow +\infty} |y_i(\tilde{s}) - y_j(\tilde{s})| = +\infty, \ i \neq j, \quad |y_j(\tilde{s})| \leq C_3(1 + b_0)\tilde{s}^{1/2}$$

$$\limsup_{\tilde{s} \uparrow +\infty} \sup_{1 \leq j \leq m(\tilde{s})} |A(B_j(\tilde{s}), -\tilde{s}) - 8\pi| < \varepsilon$$

$$\lim_{\tilde{s} \uparrow +\infty} \sup_{x \in \mathbf{R}^2} A(B(x,R) \setminus \bigcup_{j=1}^{m(\tilde{s})} B_j(\tilde{s}), -\tilde{s}) = 0, \quad \forall R > 0, \tag{1.53}$$

where $B_j(\tilde{s}) = B(y_j(\tilde{s}), b_j(\tilde{s}))$. From (1.41), relation (1.53) means

$$\limsup_{\tilde{s} \uparrow +\infty} \max_{1 \leq j \leq m(\tilde{s})} |\zeta(\tilde{B}_j(\tilde{s}), -\log \tilde{s}) - 8\pi| < \varphi$$

$$\limsup_{\tilde{s} \uparrow +\infty} \sup_{y \in \mathbf{R}^2} \zeta(B(y,R) \setminus \bigcup_{j=1}^{m(\tilde{s})} \tilde{B}_j(\tilde{s}), -\log \tilde{s}) = 0, \quad \forall R > 0 \tag{1.54}$$

for

$$B_j(\tilde{s}) = B(\tilde{s}^{-1/2}y_j(\tilde{s}), \tilde{s}^{-1/2}b_j(\tilde{s})) = B(\tilde{y}_j(\tilde{s}), \tilde{b}_j(\tilde{s})),$$

which satisfies

$$\bigcup_{j=1}^{m(\tilde{s})} \tilde{B}_j(\tilde{s}) \subset B(0, R_1), \quad \forall \tilde{s} \geq \tilde{s}_1$$

with some $R_1 > 0$. By (1.35), then we obtain $\tilde{s}_2 \geq \tilde{s}_1$ such that

$$\limsup_{k\to\infty} \max_{1\leq j\leq m(\tilde{s})} \left| \|z(\cdot, -\log \tilde{s} + s_k)\|_{L^1(\tilde{B}_j(\tilde{s}))} - 8\pi \right| > \varepsilon$$

$$\limsup_{k\to\infty} \sup_{y\in\mathbf{R}^2} \|z(\cdot, -\log \tilde{s} + s_k)\|_{L^1(B(y,R)\backslash \tilde{F}(\tilde{s}))} < \varepsilon_0, \quad \forall R > 0 \quad (1.55)$$

for any $\tilde{s} \geq \tilde{s}_2$, where $\tilde{F}(\tilde{s}) = \displaystyle\bigcup_{j=1}^{m(\tilde{s})} \tilde{B}_j(\tilde{s})$.

Let $R(t) = (T - t)^{1/2}$ and define $t_k \uparrow T$ by $s_k = -\log(T - t_k)$. Then relation (1.55) means

$$\limsup_{k\to\infty} \max_{1\leq j\leq m(\tilde{s})} \left| \|u(\cdot, T - \tilde{s}R(t_k)^2)\|_{L^1(B_j^k(\tilde{s}))} - 8\pi \right| < \varepsilon$$

$$\lim_{b\uparrow+\infty} \limsup_{k\to\infty} \|u(\cdot, T - \tilde{s}R(t_k))\|_{L^1(B(x_0,bR(t_k))\backslash F_k(\tilde{s}))} < \varepsilon_0, \quad (1.56)$$

where

$$F_k(\tilde{s}) = \bigcup_{j=1}^{m(\tilde{s})} B_j^k(\tilde{s}), \quad B_j^k(\tilde{s}) = B(x_j^k(\tilde{s}), r_j^k(\tilde{s}))$$

and

$$x_j^k(\tilde{s}) = x_0 + R(t_k)\tilde{y}_j(\tilde{s}), \quad r_j^k(\tilde{s}) = R(t_k)\tilde{b}_j(\tilde{s}).$$

By the diagonal argument, this $\tilde{s}_2$ can be independent of the choice of the subsequence of $t_k \uparrow T$. Thus given $t_k \uparrow T$, we have $\tilde{s}_2$ satisfying (1.56) for any $\tilde{s} \geq \tilde{s}_2$. Then the improved ε regularity, Lemma 1.3 guarantees

$$\lim_{b\uparrow+\infty} \limsup_{k\to\infty} R(t_k)^2\|u(\cdot, T - \tilde{R}(t_k))\|_{L^\infty(B(x_0,bR(t_k))\backslash \tilde{F}_k^r(\tilde{s}))} \leq C_4 r^{-2}, \quad (1.57)$$

where $\tilde{F}_k^r(\tilde{s}) = \bigcup_{x_1\in F_k(\tilde{s})} B(x_1, rR(t_k))$ for $0 < r \ll 1$.

Now we take $t_k \uparrow T$ by $R(t_k) = 2^{-k/2}$ for $k \geq k_0$ with $k_0 \gg 1$. Then there is $t_0 > 0$ in $0 < T - t_0 \ll 1$, such that

$$[t_0, T) \subset \bigcup_{k=k_0}^{\infty} [T - 2\tilde{s}_2 R(t_k)^2, T - \tilde{s}_2 R(t_k)^2].$$

Applying (1.57) for $\tilde{s} = \tilde{s}_2$, therefore, we obtain the following lemma.

Lemma 1.6. *Given* $0 < \varepsilon \ll 1$, *any* $t \in [t_0, T)$ *admits* $m(t) \in \{0, 1, \cdots, \tilde{m}\}$ *and disjoint balls* $B_j(t)$, $1 \le j \le m(t)$, *satisfying*

$$F(t) = \bigcup_{j=1}^{m(t)} B_j(t) \subset B(x_0, C_4 R(t))$$

and

$$\lim_{b\uparrow+\infty} \limsup_{t\uparrow T} \max_{1\le j\le m(t)} \left| \|u(\cdot,t)\|_{L^1(B_j(t))} - 8\pi \right| < \varepsilon$$

$$\lim_{b\uparrow+\infty} \limsup_{t\uparrow T} R(t)^2 \|u(\cdot,t)\|_{L^\infty(B(x_0,bR(t))\setminus F_r(t))} \le C_5 r^{-2}, \qquad (1.58)$$

where $F_r(t) = \bigcup_{x_i\in F(t)} B(x_1, rR(t))$ *for* $0 < r \ll 1$.

Relation (1.58) implies the existence of $s_1 \gg 1$, $R_2 > 0$, $m(s) \in \{0, 1, \cdots, \tilde{m}\}$, and distinct balls $\tilde{B}_j(s)$ defined for $s \ge s_1$ and $1 \le j \le m(s)$, such that

$$\bigcup_{j=1}^{m(s)} \tilde{B}_j(s) \subset B(0, R_2)$$

and

$$\limsup_{s\uparrow+\infty} \max_{1\le s\le m(s)} \left| \|z(\cdot,s)\|_{L^1(\tilde{B}_j(s))} - 8\pi \right| < \varepsilon$$

$$\lim_{b\uparrow+\infty} \limsup_{s\uparrow+\infty} \|z(\cdot,s)\|_{L^\infty(B(0,b)\setminus G_r(s))} \le C_6 r^{-2}, \qquad (1.59)$$

where

$$G_r(s) = \bigcup_{y\in G(s)} B(y, r), \quad G(s) = \bigcup_{j=1}^{m(s)} \tilde{B}_j(s).$$

Using (1.59), we examine (1.35) again. In fact, analogous result to Lemma 1.2, the improved ε-regularity, is valid to (1.33) locally uniformly in $\mathbf{R}^2$, and therefore, the singular part of $\zeta(dy, s)$ denoted by $\zeta^s(dy, s)$ is composed of a finite sum of 8π times δ functions:

$$\zeta^s(dy, s) = 8\pi \sum_{y_0\in\mathcal{S}(s)} \delta_{y_0}(dy), \quad \sharp\mathcal{S}(s) \le \tilde{m}, \quad \mathcal{S}(s) \subset B(0, R_2). \qquad (1.60)$$

To show that the coefficient of each $\delta_{y_0}(dy)$ is so quantized as 8π, one may use the scaling argument applied for $A(dy, s)$ defined by (1.41) and the Liouville property Lemma 1.4. See the proof of Lemma 1.8 below.

By (1.59), there is $R > 0$ such that

$$\sup_{s \in \mathbf{R}} \|\zeta(dy, s)\|_{L^\infty(\mathbf{R}^2 \setminus B(0,R))} \leq C_7. \tag{1.61}$$

Combining (1.60)–(1.61), therefore, we obtain

$$\zeta(dy, s) = 8\pi \sum_{j=1}^{m(s)} \delta_{y_j(s)}(dy) + g(y, s)dy \tag{1.62}$$

in (1.35), where $m_j(s) \in \mathbf{N} \bigcup \{0\}$, $y_j(s) \in B(0, R/2)$, and $g = g(y, s) \geq 0$ such that

$$\|g(\cdot, s)\|_1 + \|g(\cdot, s)\|_{L^\infty(\mathbf{R}^2 \setminus B(0,R))} \leq C_8. \tag{1.63}$$

It thus holds that $\mathcal{S}(s) = \{y_j(s) \mid 1 \leq j \leq m(s)\} \subset B(0, R/2)$ for any $s \in \mathbf{R}$.

1.4 Residual Vanishing

We complete the following proof.

Proof of Theorem 1.1. If $m(x_0) \notin 8\pi\mathbf{N}$ in (1.39) there arises $g(\cdot, s) \not\equiv 0$ and $m(x_0) > 8\pi m(s)$ for any $s \in \mathbf{R}$. Hence we have $\delta > 0$ such that

$$\|g(\cdot, s)\|_1 \geq \delta \quad \text{a.e. } s. \tag{1.64}$$

We show that (1.64) gives a contradiction because the regular part of $\zeta(dy, s)$ is swept away to $y = \infty$ as $s \uparrow +\infty$ by the $|y|^2/4$ term in (1.36), while the second parabolic envelope (1.40) implies that $\zeta(dy, s)$ in $\mathbf{R}^2 \setminus B(0, R)$ with $R \gg 1$ is uniformly thin. For this purpose we use

$$\int_{\mathbf{R}^2} |y|^2 g(y, s) \, dy \leq C_9 \tag{1.65}$$

derived from this inequality. It holds also that

$$g > 0 \quad \text{in } H = \bigcup_s (\mathbf{R}^2 \setminus \mathcal{S}(s)) \times \{s\} \tag{1.66}$$

from the maximum principle for parabolic equations. Here we note that from ε-regularity to (1.33) the set H is open in $\mathbf{R}^2 \times (-\infty, +\infty)$.

Elliptic and parabolic regularity implies the smoothness of $g = g(y, s)$ in H. Then there arises

$$g_s = \Delta g - \nabla \cdot g \nabla w \quad \text{in } H$$

$$w(y, s) = \frac{|y|^2}{4} + 4 \sum_{j=1}^{m(s)} \log \frac{1}{|y - y_j(s)|} + (\Gamma * g)(y, s). \tag{1.67}$$

Given $s_0 \in \mathbf{R}$, let $\mathcal{S}_\varepsilon(s_0) = \bigcup_{j=1}^{m} B(y_j(s_0), \varepsilon)$ be the ε-neighborhood of $\mathcal{S}(s_0)$. Therefore, we obtain

$$y_i(s) \in B(y_j(s_0), \varepsilon) \Rightarrow (y - y_i(s)) \cdot \nu_y \geq 0, \ \forall y \in \partial B(y_j(s_0), \varepsilon) \quad (1.68)$$

for $1 \leq j \leq m(s_0)$ and $|s - s_0| \ll 1$, by $\zeta(dy, s) \in C_*(-\infty, +\infty : \mathcal{M}(\mathbf{R}^2))$, where ν denotes the outer unit normal vector. By (1.67), then it holds that

$$\frac{d}{ds} \int_{B(0,r) \setminus \mathcal{S}_\varepsilon(s_0)} g \, dy = \int_{\partial(B(0,r) \setminus \mathcal{S}_\varepsilon(s_0))} \frac{\partial g}{\partial \nu} - g \frac{\partial w}{\partial \nu} \, dS$$

$$\leq \int_{\partial(B(0,r) \setminus \mathcal{S}_\varepsilon(s_0))} \frac{\partial g}{\partial \nu} - g \frac{\partial v}{\partial \nu} \, dS \quad (1.69)$$

for $r > 2R$, $|s - s_0| \ll 1$, and

$$v = \frac{|y|^2}{4} + (\Gamma * g)(y, s).$$

Here we take $0 < h \ll 1$ and operate $\frac{1}{h} \int_s^{s+h} \cdot \, ds'$ to both sides of (1.67), and then make $\varepsilon \downarrow 0$ which results in

$$\frac{1}{h} \left[\int_{B(0,r)} g(y, s') \, dy \right]_{s'=s}^{s'=s+h} \leq \frac{1}{h} \int_s^{s+h} ds' \int_{\partial B(0,r)} (g_r - g v_r)(\cdot, s') \, dS$$

by (1.68). Since inequality (1.63) implies

$$\|\nabla(\Gamma * g)(\cdot, s)\|_{L^\infty(\mathbf{R}^2 \setminus B(0, 2R))} \leq C_{10},$$

it holds that

$$\int_{\partial B(0,r)} (g_r - g v_r) \, dS \leq \int_{\partial B(0,r)} \left(g_r - \frac{r}{2} g + C_{10} g \right) \, dS$$

$$= \frac{d}{dr} \left(r \int_0^{2\pi} g \, d\theta \right) - \int_0^{2\pi} \left(1 + \frac{r^2}{2} \right) g \, d\theta + C_{10} r \int_0^{2\pi} g \, d\theta.$$

Then we obtain

$$\frac{1}{h} \left[\int_0^r (r\bar{g})(r, s') \, dr' \right]_{s'=s}^{s'=s+h}$$

$$\leq \frac{1}{h} \int_s^{s+h} \left[\frac{d}{dr}(r\bar{g}) - \left(\frac{r^2}{2} - C_{11} r + 1 \right) \bar{g} \right] (r, s') \, ds' \quad (1.70)$$

for

$$\bar{g} = \frac{1}{2\pi} \int_0^{2\pi} g \, d\theta.$$

Inequality (1.70) means

$$\frac{1}{h}[B(r,s')]_{s'=s}^{s'=s+h} \leq \frac{1}{h}\int_s^{s+h} (B_{rr} - a(r)B_r)(r,s')\,ds'$$

$$= \frac{1}{h}\int_s^{s+h} [e^A(e^{-A}B_r)_r](r,s')\,ds' \quad \text{in } (\mathbf{R}^2 \setminus B(0,2R)) \times (-\infty,+\infty),$$

where

$$B = B(r,s) = \int_0^r r\bar{g}(r,s)\,dr$$

$$a = a(r) = \frac{r^2}{2} - C_{10}r + \frac{1}{r}$$

$$A = A(r) = \frac{r^2}{4} - C_{10}r + \log r.$$

Hence it follows that

$$\frac{1}{h}\left[(e^{-A}B)(r,s')\right]_{s'=s}^{s'=s+h} \leq \frac{1}{h}\int_s^{s+h} (e^{-A}B_r)_r(r,s')\,ds' \tag{1.71}$$

for $r > 2R$ and $-\infty < s < +\infty$.

By (1.64)–(1.65) we obtain

$$B(r,s) = \int_0^r r\bar{g}(r,s)\,dr \geq \frac{\delta}{2} - \frac{C_{11}}{r^2+1}, \quad r > 0,\ s \in \mathbf{R}. \tag{1.72}$$

It holds also that

$$B(r,s) > 0, \quad r > 0,\ s \in \mathbf{R}$$

by (1.66). Now we take $r_1 > 2R$ such that

$$a(r) \geq \frac{r}{4} + 1 \quad \text{and} \quad \frac{\delta}{2} - \frac{C_{11}}{r^2+1} \geq \frac{\delta}{4} \quad \text{for all } r \geq r_1, \tag{1.73}$$

to put

$$\varphi(r) = \begin{cases} c_1 r + c_2, & r \geq r_2 \\ \sin\beta(r - r_1), & r_1 \leq r_2 = r_1 + \frac{\pi}{4\beta}, \end{cases} \tag{1.74}$$

using

$$c_2 = \frac{1}{\sqrt{2}}\left(1 - \frac{\pi}{4}\right) - r_1 c_1 \quad \text{and} \quad \beta = \sqrt{2}c_1$$

defined for $0 < c_1 \ll 1$.

It holds that

$$0 \leq \varphi = \varphi(r) \in C^1[r_1,\infty), \quad \varphi(r_1) = 0, \quad \varphi_r > 0 \text{ in } (r_1,+\infty) \tag{1.75}$$

and

$$\int_{r_1}^{\infty} e^{-A}\varphi \, dr < +\infty, \qquad \lim_{r\uparrow+\infty} e^{-A}(\varphi + \varphi_1) = 0. \tag{1.76}$$

We have also $r_2 \uparrow +\infty$ as $c_1 \downarrow 0$, and then

$$\varphi_{rr} - a\varphi_r \le -\mu\varphi \tag{1.77}$$

is valid for $r \ge r_2$ with $0 < \mu \ll 1$ by (1.72)–(1.73). Since $\varphi_r > 0$ and $\varphi_{rr} = -\beta^2\varphi$ for $r_1 < r < r_2$, inequality (1.77) is extended to $r \ge r_1$ for $0 < \mu \ll 1$.

Now we apply (1.76)–(1.77) to (1.71). First, this inequality holds for $r \ge r_1$, $-\infty < s < +\infty$, and therefore, we obtain

$$\frac{1}{h}\int_{r_1}^{\infty} \left[(e^{-A}B)(r,s')\varphi(r)\right]_{s'=s}^{s'=s+h} dr \le \int_s^{s+h} \int_{r_1}^{\infty} \left[(e^{-A}B_r)_r\varphi\right](r,s') \, dr.$$

Second, by

$$B(\cdot,s') > 0, \quad B_r(\cdot,s') > 0, \quad B(\infty,s') \le 2\pi C_9$$

there is $r_j \uparrow +\infty$ such that $B_r(s',r_j) \to 0$, which guarantees

$$\int_{r_1}^{\infty} (e^{-A}B_r)_r(s',r) \, dr = -\int_{r_1}^{\infty} (e^{-A}B_r)(r,s')\varphi_r(r) \, dr$$

$$\le \int_{r_1}^{\infty} \left[(e^{-A}\varphi_r)_r B\right](r,s') \, dr = \int_{r_1}^{\infty} \left[(\varphi_{rr} - a\varphi_r)e^{-A}B\right](r,s') \, dr$$

by (1.75). Then it follows that

$$\frac{1}{h}\left[\int_{r_1}^{\infty} (e^{-A}B\varphi)(r,s') \, dr\right]_{s'=s}^{s'=s+h} \le -\mu \cdot \frac{1}{h}\int_s^{s+h} ds' \int_{r_1}^{\infty} (e^{-A}B\varphi)(r,s') \, dr$$

and hence

$$\frac{d^+}{ds}\int_{r_1}^{\infty} e^{-A}B\varphi \, dr \le -\mu \int_{r_1}^{\infty} e^{-A}B\varphi \, dr, \quad -\infty < s < +\infty, \tag{1.78}$$

where

$$\frac{d^+}{ds}H(s) = \limsup_{h\downarrow 0} \frac{1}{h}\left(H(s+h) - H(s)\right).$$

Since (1.78) implies

$$\frac{d^+}{ds}\left(e^{\mu s/2}\int_{r_1}^{\infty} e^{-A}B\varphi \, dr\right) \le -\frac{\mu}{2}e^{\mu s/2}\int_{r_1}^{\infty} e^{-A}B\varphi \, dr < 0$$

the mapping $s \mapsto e^{\mu s/2} \int_{r_1}^{\infty} e^{-A} B\varphi \, dr$ is monotone decreasing. It holds that

$$e^{\mu s/2} \int_{r_1}^{\infty} [(e^{-A}B\varphi](r,s) \, dr \leq \int_{r_1}^{\infty} [(e^{-A}B\varphi](r,0) \, dr, \quad s > 0$$

and hence

$$\lim_{s\uparrow+\infty} \int_{r_1}^{\infty} (e^{-A}B\varphi)(r,s) \, dr = 0.$$

This is a contradiction because

$$\int_{r_1}^{\infty} (e^{-A}B\varphi)(r,s)dr \geq \frac{\delta}{4} \int_{r_1}^{\infty} (e^{-A}\varphi)(r) \, dr > 0$$

follows from (1.70) and (1.72). Hence we have $g \equiv 0$, which implies $m(x_0) \in 8\pi \mathbf{N}$ and

$$\zeta(dy, s) = 8\pi \sum_{j=1}^{m} \delta_{y_j(s)}(dy) \tag{1.79}$$

with $m = m(x_0)/(8\pi)$. The proof of Theorem 1.1 is complete. $\square$

Putting $\varphi = |x|^2$ in (1.37) is justified by (1.39)–(1.40), which results in

$$\frac{dI}{ds} = m(x_0) - \frac{m(x_0)^2}{2\pi} + I, \quad \text{a.e. } s \in \mathbf{R}$$

for $I = \langle |y|^2, \zeta(dy, s) \rangle$. See [Suzuki (2013)] for a detailed proof. If $m(x_0) = 8\pi$, therefore, it holds that $I(s) \equiv 0$ by (1.40), and hence

$$\zeta(dy, s) = 8\pi\delta_0(dy), \quad -\infty < s < +\infty.$$

We say that the blowup point $x_0 \in \mathcal{S}$ is *simple* if $m(x_0) = 8\pi$. Then it holds that

$$A(dy, s) = 8\pi\delta_0(dy), \quad -\infty < s < 0 \tag{1.80}$$

by (1.41).

Otherwise, we say that *collision* of collapses occurs at $x_0 \in \mathcal{S}$. Equality (1.79) is then transformed into

$$A(dy, s) = 8\pi \sum_{j=1}^{m} \delta_{y'_j(s)}(dy), \quad -\infty < s < 0 \tag{1.81}$$

with $m \geq 2$. In this case we obtain

$$\frac{dy'_j}{ds} = 8\pi \nabla_{y_j} H_m(y'_1, \cdots, y'_m), \ 1 \leq j \leq m, \quad \text{a.e. } s < 0 \tag{1.82}$$

by (1.42), where

$$H_m(y_1, \cdots, y_m) = \sum_{1 \leq i < j \leq m} \Gamma(y_i - y_j), \quad \Gamma(y) = \frac{1}{2\pi} \log \frac{1}{|y|}. \tag{1.83}$$

1.5 Bounded Domains

For the case of bounded domains, free energy is easier to handle with. If $\Omega \subset \mathbf{R}^2$ is a bounded domain with smooth boundary $\partial\Omega$, the Poisson equation (1.10) or (1.12) takes the Green function denoted by $G = G(x, x')$. Then it holds that

$$v(x,t) = \int_\Omega G(x,x')u(x',t)\,dx'.$$

This $G = G(x,x')$ is smooth in $\overline{\Omega} \times \overline{\Omega} \setminus \overline{D}$ for $D = \{(x,x) \mid x \in \Omega\}$. The singularity on the diagonal D is distinguished according to the interior and the boundary. First, interior singularity is given by

$$G(x,x') = \Gamma(x - x') + K(x,x'), \qquad (1.84)$$

where

$$K = K(x,x') \in C^{1+\theta,\theta}(\Omega \times \overline{\Omega}) \cap C^{\theta,1+\theta}(\overline{\Omega} \times \Omega)$$

for $0 < \theta < 1$. Second, boundary singularity around $x_0 \in \partial\Omega$ is described by the conformal diffeomorphism

$$X : \overline{\Omega \times B(x_0, 2R)} \;\rightarrow\; \overline{R_+^2}$$

defined for $0 < R \ll 1$, where $\mathbf{R}_+^2 = \{(X_1, X_2) \mid X_2 > 0\}$. Using the reflection

$$\begin{pmatrix} X_1 \\ X_2 \end{pmatrix} \;\mapsto\; X_* = \begin{pmatrix} X_1 \\ -X_2 \end{pmatrix},$$

the parametrices of (1.10) and (1.12) are given by

$$E(x,x') = \Gamma(X - X') + \Gamma(X - X'_*)$$

and

$$E(x,x') = \Gamma(X - X') - \Gamma(X - X'_*), \qquad (1.85)$$

respectively. Then it follows that

$$G(x,x') = E(X,X') + K(x,x') \qquad (1.86)$$

with

$$K = K(x,x') \in C^{1+\theta,\theta} \cap C^{\theta,1+\theta}(\overline{\Omega \cap B(x_0, R)} \times \overline{\Omega \cap B(x_0, R)}).$$

Using these properties of Green's function, we obtain the following theorems analogous to Theorem 1.1. Below, equalities (1.84) and (1.86) are used locally around the blowup point, without any confusions of the notation.

Theorem 1.2. *Let $u = u(x,t)$ be the classical solution to (1.9) with (1.10) and (1.11). Assume the blowup in finite time, $T < +\infty$, and define the blowup set by*

$$\mathcal{S} = \{x_0 \in \overline{\Omega} \mid \text{there exist } x_k \to x_0 \text{ and } t_k \uparrow T \text{ such that}$$

$$\lim_{k \to \infty} u(x_k, t_k) = +\infty\}. \tag{1.87}$$

Then $\sharp \mathcal{S} < +\infty$ and it holds that

$$u(x,t)dx \rightharpoonup \sum_{x_0 \in \mathcal{S}} m(x_0)\delta_{x_0}(dx) + f(x)\, dx \quad \text{in } \mathcal{M}(\overline{\Omega}) \tag{1.88}$$

as $t \uparrow T$, where $0 < f = f(x) \in L^1(\Omega) \cap C(\overline{\Omega} \setminus \mathcal{S})$ and $m(x_0) \in m_*(x_0)\mathbf{N}$ for

$$m_*(x_0) = \begin{cases} 8\pi, & x_0 \in \Omega \\ 4\pi, & x_0 \in \partial\Omega. \end{cases}$$

If $x_0 \in \Omega \cap \mathcal{S}$, we define $A(dy, s)$ by (1.35) and (1.41). It holds that (1.80) when this x_0 is simple; $m(x_0) = m_*(x_0)$. In the other case of $m = m(x_0)/m_*(x_0) \geq 2$, we have the collapse dynamics (1.81)–(1.82).

If $x_0 \in \partial\Omega \cap \mathcal{S}$, we take zero extension of $z = z(y, s)$ to the region where it is not defined. Then we get (1.35) and this $\zeta(dy, s)$ has the support included in a closed half space independent of s. Then we take its even reflection denoted by $\tilde{\zeta}(dy, s)$ to define $\tilde{A}(dy, s)$ by (1.41). Then the same property as in (1.80) or (1.81)–(1.82) arises, according to $m = 1$ and $m \geq 2$, respectively.

Theorem 1.3. *Let $u = u(x,t)$ be the classical solution to (1.9) with (1.10) and (1.11). Assume the blowup in finite time, $T < +\infty$, and define the blowup set by (1.82). Then we obtain $\mathcal{S} \subset \Omega$ with $\sharp \mathcal{S} < +\infty$. There holds also that (1.87) as $t \uparrow T$, where $0 < f = f(x) \in L^1(\Omega) \cap C(\overline{\Omega} \setminus \mathcal{S})$ and $m(x_0) \in 8\pi\mathbf{N}$. Then the same conclusion as in the previous Theorem for interior blowup points occurs to each $x_0 \in \mathcal{S}$ for $A(dy, s)$ defined by (1.35) and (1.41).*

Free energy associated with system (1.9) with (1.10) or (1.9) with (1.12) is defined with the Green function $G = G(x, x')$ in the previous section and $u \otimes u = u(x,t)u(x',t)$ by

$$\mathcal{F}(u) = \int_\Omega u(\log u - 1)\, dx - \frac{1}{2} \iint_{\Omega \times \Omega} G(x,x')u \otimes u\, dxdx'. \tag{1.89}$$

Then it follows that

$$\frac{d}{dt}\mathcal{F}(u) = -\int_\Omega u|\nabla(\log u - v)|^2\, dx \leq 0 \tag{1.90}$$

for the classical solution $u = u(\cdot, t)$. In fact we have $u = u(x,t) > 0$ in $\overline{\Omega} \times (0, T)$ unless $u_0 \equiv 0$ by the strong maximum principle.

1.6 Simple Blowup Points

If the collapse on $x_0 \in S$ is simple, $m(x_0) = m_*(x_0)$, in (1.9) with (1.10) or (1.9) with (1.12), we obtain

$$\zeta(dy, s) = m_*(x_0)\delta_0(dy), \quad -\infty < s < +\infty$$

in (1.35), regardless of $s_k \uparrow +\infty$. Then it follows that

$$z(y, s)dy \rightharpoonup m_*(x_0)\delta_0(dy) \quad \text{in } \mathcal{M}(\mathbf{R}^2) \tag{1.91}$$

as $s \uparrow +\infty$.

In this case there arises type II blowup rate

$$\lim_{t \uparrow T}(T - t)\|u(\cdot, t)\|_{L^\infty(\Omega \cap B(x_0, b(T-t)^{1/2})} = +\infty, \quad \forall b > 0 \tag{1.92}$$

and free energy transmission

$$\lim_{t \uparrow T} \mathcal{F}_{x_0, b(T-t)^{1/2}}(u(\cdot, t)) = +\infty, \quad \forall b > 0 \tag{1.93}$$

where

$$\mathcal{F}_{x_0, R}(u) = \int_{\Omega \cap B(x_0, R))} u(\log u - 1) \, dx$$
$$-\frac{1}{2} \iint_{(\Omega \cap B(x_0, R)) \times (\Omega \cap B(x_0, R)))} u \otimes u \, dx dx'$$

stands for the local free energy.

This phenomenon is associated with the boundedness of free energy.

Theorem 1.4. *Let $T < +\infty$ in (1.9) with (1.10) or (1.9) with (1.12). Then any $x_0 \in S$ is simple if*

$$\lim_{t \uparrow T} \mathcal{F}(u(\cdot, t)) > -\infty. \tag{1.94}$$

The following comments may be useful. First, a short description of the Zygmund space $L \log L(\Omega)$ is in Section 3.4 of [Suzuki (2015)]. There is a duality between BMO and this space, and hence

$$\|(-\Delta)^{-1}u\|_\infty \le C_{12}\|u\|_{L \log L} \tag{1.95}$$

for

$$[(-\Delta)^{-1}u](x) = \int_\Omega G(x, x')u(x') \, dx'. \tag{1.96}$$

Second, by this duality and Theorem 1.4, any $x_0 \in S$ is simple if

$$\liminf_{t \uparrow T} \|u(\cdot, t)\|_{L \log L(\Omega)} < +\infty. \tag{1.97}$$

In fact, (1.97) implies the existence of $t_k \uparrow T$ such that

$$\|u(\cdot, t_k)\|_{L \log L(\Omega)} \le C_{13}.$$

Then we obtain

$$\|(-\Delta)^{-1} u(\cdot, t_k)\|_\infty \le C_{12} \cdot C_{13}$$

by (1.95), which guarantees

$$\mathcal{F}(u(\cdot, t_k)) \ge -|\Omega| + \frac{1}{2} \cdot \lambda \cdot C_{12} \cdot C_{13}$$

by

$$s(\log s - 1) + 1 \ge 0, \quad s \ge 0, \tag{1.98}$$

recalling $\lambda = \|u\|_1$. Then (1.94) follows from (1.90). Finally, (1.97) implies $f \in L \log L(\Omega)$ by Fatou's lemma and

$$u(x, t) \to f(x) \quad \text{locally uniformly in } x \in \overline{\Omega} \setminus \mathcal{S} \tag{1.99}$$

as $t \uparrow T$ by elliptic and parabolic regularity

We show Theorem 1.4 for the case of (1.9) with (1.12). Once the boundary blowup is excluded, this case is easier to handle with. In fact, the if part follows from (1.95). To show the only if part, we use an inequality analogous to (1.23), that is,

$$\inf\{\mathcal{F}(u) \mid u \ge 0, \ \|u\|_1 = 8\pi\} > -\infty, \tag{1.100}$$

where $\mathcal{F} = \mathcal{F}(u)$ is the free energy defined by (1.89). Inequality (1.100) is actually the dual form of the usual Trudinger-Moser inequality

$$\inf\{\mathcal{J}_{8\pi}(v) \mid v \in H_0^1(\Omega)\} > -\infty, \tag{1.101}$$

where

$$\mathcal{J}_\lambda(v) = \frac{1}{2} \|\nabla v\|_2^2 - \lambda \log \left(\int_\Omega e^v \, dx \right) + \lambda(\log \lambda - 1). \tag{1.102}$$

See Section 3.1 of [Suzuki (2015)] for the equivalence of (1.100) and (1.101).

Proof of Theorem 1.4. First, we assume (1.94) which implies

$$\int_0^T \int_\Omega u |\nabla(\log u - v)|^2 \, dx dt < +\infty. \tag{1.103}$$

Using $u > 0$, we rewrite (1.9) to

$$u_t = \nabla \cdot u \nabla(\log u - v), \quad u \frac{\partial}{\partial \nu}(\log u - v) \bigg|_{\partial \Omega} = 0.$$

Then it follows that

$$\left| \frac{d}{dt} \int_\Omega \varphi u \, dx \right| \le \left| \int_\Omega u \nabla (\log u - v) \cdot \nabla \varphi \, dx \right|$$

$$\le \|\nabla \varphi\|_\infty \lambda^{1/2} \|u^{1/2} \nabla (\log u - v)\|_2$$

for $\lambda = \|u_0\|_1$. With $\varphi = \varphi_{x_0,R}$ satisfying (1.20)–(1.21), therefore, it follows that

$$\left| \langle \varphi_{x_0,R}, \mu(dx,t) \rangle - \langle \varphi_{x_0,R}, \mu(dx,T) \rangle \right| \le \int_t^T \left| \frac{d}{dt} \int_\Omega \varphi_{x_0,R} \cdot u \, dx \right| dt$$

$$\le \|\nabla \varphi_{x_0,R}\|_\infty \lambda^{1/2} \int_t^T \|u^{1/2} \nabla (\log u - v)\|_2 \, dt$$

$$\le C_{14} \lambda^{1/2} R^{-1} (T-t)^{1/2} \left\{ \int_t^T \int_\Omega u |\nabla (\log u - v)|^2 \, dx dt \right\}^{1/2}$$

for $x_0 \in \mathcal{S}$.

Putting $R = b(T-t)^{1/2}$, we thus obtain

$$\left| \langle \varphi_{x_0,b(T-t)^{1/2}}, \mu(dx,t) \rangle - \langle \varphi_{x_0,b(T-t)^{1/2}}, \mu(dx,T) \rangle \right|$$

$$\le C_{14} \lambda^{1/2} b^{-1} \left\{ \int_t^T \int_\Omega u |\nabla (\log u - v)|^2 \, dx dt \right\}^{1/2}$$

for any $b > 0$. Equality (1.38) is thus improved as

$$\limsup_{t \uparrow T} \left| \langle \varphi_{x_0,b(T-t)^{1/2}}, \mu(dx,t) \rangle - m(x_0) \right| = 0, \quad \forall b > 0 \qquad (1.104)$$

by (1.103). Then it follows that

$$\zeta(dy,s) = m(x_0) \delta_0(dy), \quad -\infty < s < +\infty$$

in (1.35), and hence

$$A(dy,s) = m(x_0) \delta_0(dy), \quad s < 0. \qquad (1.105)$$

We thus obtain $m(x_0) = 8\pi$ by Theorem 1.3. $\qquad \square$

1.7 Blowup in Infinite Time

Blowup in infinite time indicates

$$T = +\infty, \quad \limsup_{t \uparrow +\infty} \|u(\cdot,t)\|_\infty = +\infty. \qquad (1.106)$$

In the other case of

$$T = +\infty, \quad \sup_{t \ge 0} \|u(\cdot,t)\|_\infty < +\infty, \qquad (1.107)$$

the orbit $\mathcal{O} = \{u(\cdot, t)\}$ is compact in $C(\overline{\Omega})$, and then, the ω-limit set defined by

$$\omega(u_0) = \{u_\infty \in C^2(\overline{\Omega}) \mid \text{there is } t_k \uparrow +\infty \text{ such that}$$
$$\lim_{k \to \infty} \|u(\cdot, t_k) - u_\infty\|_\infty = 0\} \tag{1.108}$$

is non-empty, compact, and connected from the parabolic regularity. See [Henry (1981)] for this classical theory of dynamical systems. In accordance with the free energy, acting as a Lyapunov function, $\omega(u_0)$ is contained in the set of stationary solutions.

To describe this fact more precisely, we assume the contrary, and define the ω-limit set by (1.108). This set is invariant under the flow, which means that the solution $\tilde{u} = \tilde{u}(\cdot, t)$ to (1.9) with (1.10) or (1.9) with (1.12), taking the initial value $u_\infty \in \omega(u_0)$ keeps to stay there, $\tilde{u}(\cdot, t) \in \omega(u_0)$, as far as it exists. Since

$$\|u_\infty\|_1 = \|\tilde{u}(\cdot, t)\|_1 = \|u_0\|_1 = \lambda > 0, \tag{1.109}$$

we obtain $\tilde{u}(x, t) > 0$ on $\overline{\Omega}$ for $t > 0$ by the strong maximum principle. Then $\mathcal{F}(\tilde{u}(\cdot, t))$ is well-defined by (1.89), and is independent of t from the LaSalle principle. See Section 3.3 of [Suzuki (2015)].

It thus follows that

$$\frac{d}{dt}\mathcal{F}(\tilde{u}(\cdot, t)) = 0, \quad t > 0 \tag{1.110}$$

in (1.90), and equalities (1.109)–(1.110) imply

$$\log \tilde{u}(\cdot, t) - \tilde{v}(\cdot, t) = \text{constant}, \quad \tilde{v}(\cdot, t) = (-\Delta)^{-1}\tilde{u}(\cdot, t)$$

for $(-\Delta)^{-1}$ defined by (1.96) and $0 < t \ll 1$. Then we obtain

$$\tilde{u}(\cdot, t) = \frac{\lambda e^{\tilde{v}(\cdot, t)}}{\int_\Omega e^{\tilde{v}(x, t)} dx}$$

by $\|\tilde{u}(\cdot, t)\|_1 = \lambda$. Sending $t \downarrow 0$, we thus get $v_\infty \equiv (-\Delta)^{-1}u_\infty$, where E_λ denotes the set of stationary solutions. Thus $v \in E_\lambda$ if and only if

$$-\Delta v = \lambda \left(\frac{e^v}{\int_\Omega e^v dx} - \frac{1}{|\Omega|} \right), \quad \left. \frac{\partial v}{\partial \nu} \right|_{\partial \Omega} = 0, \quad \int_\Omega v \, dx = 0 \tag{1.111}$$

for the case of (1.9) with (1.10) and

$$-\Delta v = \frac{\lambda e^v}{\int_\Omega e^v \, dx}, \quad v|_{\partial \Omega} = 0 \tag{1.112}$$

for the case of (1.9) with (1.12), respectively, where $\lambda = \|u_0\|_1$.

By Theorems 1.2 and 1.3, blowup in finite time $T < +\infty$ does not arise in the case of $\lambda = 4\pi$ and $\lambda = 8\pi$, for (1.9) with (1.10) and (1.9) with (1.12), respectively, where $\lambda = \|u_0\|_1$. Therefore, if there is no classical stationary solution for this value of λ, the orbit $\mathcal{O}$ cannot be bounded. For example, blowup in infinite time, (1.106), occurs to (1.9) with (1.12) for $\lambda = 8\pi$ if Ω is close to a ball because $E_{8\pi} = \emptyset$ in this case. Problem (1.111), on the other hand, admits the trivial solution $v = 0$ for any λ. Hence there is a possibility of the compact orbit $\mathcal{O}$ global-in-time for any Ω and λ.

Problem (1.111) or (1.112) is called the Boltzmann-Poisson equation. There is a quantized blowup mechanism for the family $\{\lambda_k, v_k\}$ of their solutions. Concerning (1.111), if

$$\lambda_k \to \lambda_0, \quad \|v_k\|_\infty \to +\infty \qquad (1.113)$$

then $\lambda_0 \in 4\pi\mathbf{N}$, and passing to a subsequence, there holds that

$$v_k \to \sum_{x_0 \in \mathcal{S}} m_*(x_0) G(\cdot, x_0) \quad \text{locally uniformly in } \overline{\Omega} \setminus \mathcal{S} \qquad (1.114)$$

with the blowup set defined by

$$\mathcal{S} = \{x_0 \in \overline{\Omega} \mid \text{there is } x_k \to x_0 \text{ such that } \lim_{k \to \infty} v_k(x_k) = +\infty\}. \qquad (1.115)$$

This set is finite denoted by $\mathcal{S} = \{x_1^*, \cdots, x_\ell^*\}$. Then $(x_1^*, \cdots, x_\ell^*)$ is a critical point of the Hamiltonian concerning the motion of ℓ-point vortices,

$$H_\ell(x_1, \cdots, x_\ell) = \frac{1}{2} \sum_{j=1}^{\ell} R(x_j) + \sum_{1 \le i < j \le \ell} G(x_i, x_j), \qquad (1.116)$$

where $R(x) = K(x, x)$ denotes the Robin function defined by $K = K(x, x')$ in (1.86).

Similar results holds to (1.112), where $\mathcal{S} \subset \Omega$ always occurs to the blowup set $\mathcal{S}$ defined by (1.115). Hence we obtain $\lambda_0 \in 8\pi\mathbf{N}$ in (1.113), and (1.114) is simplified to

$$v_k \to \sum_{x_0 \in \mathcal{S}} 8\pi G(\cdot, x_0) \quad \text{locally uniformly in } \overline{\Omega} \setminus \mathcal{S}.$$

The Robin function $R(x) = K(x, x)$ is now defined through $K = K(x, x')$ in (1.84). Finally, if $\mathcal{S} = \{x_1^*, \cdots, x_\ell^*\}$, then $(x_1^*, \cdots, x_\ell^*)$ is a critical point of the ℓ-th Hamiltonian defined by (1.116). Furthermore, the limit function in (1.114),

$$v_0(x) = \sum_{x_0 \in \mathcal{S}} m_*(x_0) G(x, x_0),$$

is called the singular limit of the stationary solutions.

Henceforth, we assume blowup in infinite time, (1.106), and

$$t_k \uparrow +\infty, \quad \lim_{k\to\infty} \|u(\cdot,t_k)\|_\infty = +\infty. \tag{1.117}$$

Then, the principle of the generation of weak solution works to $u_k = u_k(\cdot, t + t_k)$, $k = 1, 2, \cdots$, and passing to a subsequence we obtain

$$u(x, t + t_k)dx \;\rightharpoonup\; \mu(dx, t) \quad \text{in } C_*(-\infty, +\infty; \mathcal{M}(\overline{\Omega})) \tag{1.118}$$

where $\mu = \mu(dx, t)$ is a weak solution. First, we prove the following theorem concerning the singular part of $\mu^s(dx, t)$.

Theorem 1.5. *It holds that*

$$\mu^s(dx, t) = \sum_{x_0 \in \mathcal{S}_t} m_*(x_0)\delta_{x_0}(dx), \quad -\infty < t < +\infty \tag{1.119}$$

in (1.118), where $\mathcal{S}_t$ is the blowup set defined by

$$\mathcal{S}_t = \{x_0 \in \overline{\Omega} \mid \text{there exists } x_k \to x_0 \text{ such that}$$
$$\lim_{k\to\infty} u(x_k, t + t_k) = +\infty\}.$$

The numbers $\sharp(\partial\Omega \bigcap \mathcal{S}_t)$ and $\sharp(\Omega \bigcap \mathcal{S}_t)$ are independent of t. Letting $\partial\Omega \bigcap \mathcal{S}_t = \{x_1(t), \cdots, x_{\ell_1}(t)\}$ and $\Omega \bigcap \mathcal{S}_t = \{x_{\ell_1+1}(t), \cdots, x_{\ell_1+\ell_2}(t)\}$, there arises

$$\tau \cdot \frac{dx_i}{dt} = \tau \cdot \nabla_{x_i} H_\ell(x_1, \cdots, x_\ell), \quad 1 \le i \le \ell_1$$
$$\frac{dx_i}{dt} = \nabla_{x_i} H_\ell(x_1, \cdots, x_\ell), \quad \ell_1 + 1 \le i \le \ell = \ell_1 + \ell_2 \tag{1.120}$$

where $H_\ell = H_\ell(x_1, \cdots, x_\ell)$ denotes the ℓ-th Hamiltonian defined by (1.116),

$$\ell = 2 \cdot \sharp(\partial\Omega \cap \mathcal{S}) + \sharp(\Omega \cap \mathcal{S}), \tag{1.121}$$

and τ is the unit tangential vector. The solution $x = (x_i(t))$, $-\infty < t < +\infty$, makes a pre-compact orbit $\{x(t)\}$ in $\overline{\Omega}^\ell \setminus \overline{D}$, where

$$D = \{x = (x_1, \cdots, x_\ell) \in \Omega^\ell \mid \exists i \ne j \text{ such that } x_i = x_j\}, \tag{1.122}$$

and therefore, there arises a critical point of $H_\ell = H_\ell(x_1, \cdots, x_\ell)$. Finally, it holds that $\partial\Omega \cap \mathcal{S} = \emptyset$ for (1.9) with (1.12).

1.8 A Recursive Hierarchy

The proof of Theorem 1.5 is divided into several steps. First, from the improved ε-regularity, Lemma 1.2, the singular part of $\mu(dx, t)$ in (1.118), denoted by $\mu^s(dx, t)$, is a finite sum of delta functions:

$$\mu^s(dx, t) = \sum_{x_0 \in \mathcal{S}_t} m(x_0) \delta_{x_0}(dx) \tag{1.123}$$

with $m(x_0) \geq \varepsilon_0$. Second, we have the following lemma.

Lemma 1.7. *It holds that $m(x_0) = m_*(x_0)$ in (1.123).*

Proof. The fact $\partial \Omega \cap \mathcal{S}_t = \emptyset$ for (1.9) with (1.12) is due to the form of parametrix $E(x, x')$ in (1.85), and the proof is similar to the case of blowup in finite time.

In fact, assuming $x_0 \in \partial \Omega \cap \mathcal{S}_T$ for this case, we take

$$\zeta(dy, s) = (T - t)\mu(dx, t), \quad y = (x - x_0)/(T - t)^{1/2}, \quad s = -\log(T - t)$$

and put it to be zero where it is not defined. Then we can confirm the generation of the weak solution. Any $s_k \uparrow +\infty$ thus admits a subsequence denoted by the same symbol and the limit measure $\tilde{\zeta}(dy, s)$ such that

$$\zeta(dy, s + s_k) \rightharpoonup \tilde{\zeta}(dy, s) \quad \text{in } C_*(-\infty, +\infty; \mathcal{M}(\mathbf{R}^2)).$$

This $\tilde{\zeta}(dy, s)$ has the support included in a closed subspace independent of s, denoted by $\overline{L}$, and is a weak solution to

$$z_s = \nabla \cdot (\nabla z - z\nabla(E * z + |y|^2/4)) \text{ in } L \times (-\infty, +\infty), \quad z|_{\partial L} = 0.$$

It also holds that $\tilde{\zeta}(\mathbf{R}^2, s) = m(x_0) > 0$ similarly to the blowup in finite time of the classical solution. Then we get a contradiction, from the proof of exclusion of its boundary blowup.

Given $x_0 \in \Omega \cap \mathcal{S}_0$ in (1.9) with (1.12), we show $m(x_0) = 8\pi$ in (1.123). Then we get the conclusion by translating the time variable t. For this purpose we assume $x_0 = 0 \in \Omega$ to take

$$\mu_\beta(dx, t) = \beta^2 \mu(dx', dt'), \quad x' = \beta x, \ t' = \beta^2 t$$

for $\beta > 0$, which is a weak solution in $\beta^{-1}\Omega \times (-\infty, +\infty)$. Then each $\beta_k \downarrow 0$ admits a subsequence denoted by the same symbol and the limit measure $\hat{\mu}(dx, t)$ such that

$$\mu_{\beta_k}(dx, t) \rightharpoonup \tilde{\mu}(dx, t) \quad \text{in } C_*(-\infty, +\infty; \mathcal{M}(\mathbf{R}^2)).$$

This $\tilde{\mu}(dx, t)$ is a weak solution to (1.43). Then we obtain $\tilde{\mu}(\mathbf{R}^2, t) \leq \lambda$ and $\mu(\mathbf{R}^2, 0) = m(x_0)$, which implies $m(x_0) = 8\pi$ by Lemma 1.4.

The fact $m(x_0) = 4\pi$ in (1.123) for $x_0 \in \partial\Omega \cap \mathcal{S}_t$ to (1.9) with (1.10) is also similarly treated as in the blowup in finite time. Thus, using the parametrix (1.84), we take even reflection at the scaling limit to complete the proof. □

Henceforth, we are concentrated on (1.9) with (1.12). This case is simpler than the (1.9) with (1.10), once $\partial\Omega \cap \mathcal{S}_t = \emptyset$ is proven. Thus we obtain

$$\mu(dx, t) = \sum_{x_0 \in \mathcal{S}_t} 8\pi\delta_{x_0}(dx) + f(x, t)dx \qquad (1.124)$$

in (1.118) with $0 \le f = f(\cdot, t) \in L^1(\Omega)$. Since

$$\mu(dx, t) \in C_*(-\infty, ; \infty; \mathcal{M}(\overline{\Omega})),$$

the set $Q = \bigcup_t (\overline{\Omega} \setminus \mathcal{S}_t) \times \{t\}$ is relatively open in $\overline{\Omega} \times (-\infty, +\infty)$. From the elliptic and parabolic regularity this $f = f(x, t)$ is smooth in Q, and satisfies (1.9) there with smooth v. Hence it is also positive everywhere unless identically zero by the strong maximum principle.

Now we show the following lemma.

Lemma 1.8. *The number* $\sharp\mathcal{S}_t = \ell$ *in (1.124) is independent of* t. *Writing*

$$\mu^s(dx, t) = \sum_{i=1}^{\ell} 8\pi\delta_{x_i(t)}(dx),$$

we have

$$\frac{dx_i}{dt} = 8\pi\nabla_{x_i} H_\ell(x_1, \cdots, x_\ell), \quad 1 \le i \le \ell \qquad (1.125)$$

in the sense of distributions in $t \in \mathbf{R}$ *with* $H_\ell = H_\ell(x_1, \cdots, x_\ell)$ *defined by (1.116).*

Proof. Since $\mu(dx, t) \in C_*(-\infty, +\infty; \mathcal{M}(\overline{\Omega}))$ defined by (1.118) is a weak solution, there is $0 \le \nu \in L_*^\infty(-\infty, +\infty; \mathcal{E}')$ satisfying

$$\nu|_{C(\overline{\Omega}\times\overline{\Omega})} = \mu \otimes \mu$$

and

$$\frac{d}{dt}\langle\xi, \mu(dx, t)\rangle = \langle\Delta\xi, \mu(dx, t)\rangle + \frac{1}{2}\langle\rho_\xi, \nu(t)\rangle$$

in the sense of distributions for each $\xi \in C^2(\overline{\Omega})$ with $\frac{\partial\xi}{\partial\nu}\Big|_{\partial\Omega} = 0$, where

$$\rho_\xi(x, x') = \nabla\xi(x) \cdot \nabla_x G(x, x') + \nabla\xi(x') \cdot \nabla_{x'} G(x, x')$$

and $\mathcal{E} \subset L^\infty(\Omega \times \Omega)$ is the closure of

$$\{\rho_\xi + \psi \mid \xi \in C^2(\overline{\Omega}), \; \left.\frac{\partial \xi}{\partial \nu}\right|_{\partial\Omega} = 0, \; \psi \in C(\overline{\Omega} \times \overline{\Omega})\}.$$

Thus we obtain

$$-\int_{\mathbf{R}} \eta'(t)\langle \xi, \mu(dx, t)\rangle dt = \int_{\mathbf{R}} \eta(t)\left[\langle \Delta\xi, \mu(dx, t)\rangle + \frac{1}{2}\langle \rho_\xi, \nu(t)\rangle\right] dt$$

(1.126)

for any $\eta = \eta(t) \in C_0^1(\mathbf{R})$.

Henceforth, we put

$$m = 8\pi \quad \text{and} \quad \mathcal{S}_t = \{x_i(t) \mid 1 \le i \le \ell(t)\} \tag{1.127}$$

in (1.124). Then there arises the decomposition

$$\nu(t) = \nu_0(t) + \nu_1(t) + \nu_2(t) + f \otimes f \, dxdx'$$
$$0 \le \nu_0(t), \; \nu_1(t), \nu_2(t) \in L_*^\infty(-\infty, +\infty; \mathcal{E}') \tag{1.128}$$

with

$$\nu_0(t)|_{C(\overline{\Omega}\times\overline{\Omega})} = m^2 \sum_{j,k=1}^{\ell(t)} \delta_{x_j(t)}(dx) \otimes \delta_{x_k(t)}(dx')$$

$$\nu_1(t)|_{C(\overline{\Omega}\times\overline{\Omega})} = f(x,t)dx \otimes \sum_{k=1}^{\ell(t)} m\delta_{x_k(t)}(dx')$$

$$\nu_2(t)|_{C(\overline{\Omega}\times\overline{\Omega})} = \sum_{j=1}^{\ell(t)} m\delta_{x_j(t)}(dx) \otimes f(x',t)dx'. \tag{1.129}$$

Putting $x_0 = x_i(t_0)$, we have $0 < \delta \ll 1$ and $0 < r \ll 1$ such that $\sharp(\mathcal{S}_t \cap B(x_0, r)) \le 1$ for $|t - t_0| < \delta$ by $\mu(dx, t) \in C_*(-\infty, +\infty; \mathcal{M}(\overline{\Omega}))$. We may assume $x_0 = 0$ without loss of generality.

Let

$$m(t) = \begin{cases} 8\pi, & \sharp(\mathcal{S}_t \cap B(x_0, r)) = 1 \\ 0, & \text{otherwise} \end{cases}$$

and

$$\xi = |x|^2 \varphi(x),$$

where $0 \le \varphi = \varphi(x) \in C_0^\infty(\Omega)$ is a cut-off function taking the value 1 in $B(x_0, r/2)$ and 0 outside $B(x_0, r)$. We have also $0 < c_0 \ll 1$ such that

$$|x_j(t) - x_i(t)| \ge c_0, \quad j \ne i, \; |t - t_0| < \delta. \tag{1.130}$$

Then we obtain

$$\rho_\xi(x, x') = -\frac{1}{\pi} + 2x \cdot \nabla_x K(x, x') + 2x' \cdot \nabla_{x'} K(x, x')$$

in $B(x_0, r/2) \times B(x_0, r/2)$, using $K = K(x, x')$ defined by (1.84). Therefore, since

$$4m(t) - \frac{m(t)^2}{2\pi} = 0$$

it follows that

$$\langle \rho_\xi, \nu_0(t) \rangle = m(t)^2 x_i(t) \cdot (\nabla_x K(x_i(t), x_i(t)) + \nabla_{x'} K(x_i(t), x_i(t))$$
$$+ m(t)^2 x_i(t) \cdot \sum_{j \neq i} (\nabla_x G(x_i(t), x_j(t)) + \nabla_{x'} G(x_i(t), x_j(t))$$
$$= 2m(t)^2 x_i(t) \cdot \nabla_{x_i} H_{\ell(t)}(x_1(t), \cdots, x_{\ell(t)}). \qquad (1.131)$$

By (1.126), (1.128), and (1.131) we end up with

$$- \int_{\mathbf{R}} \eta'(t) \left\{ m(t) |x_i(t)|^2 + \int_\Omega \xi(x) f(x, t) \, dx \right\} \, dt$$
$$= \int_{\mathbf{R}} \eta(t) \{ 2m(t)^2 x_i(t) \cdot \nabla_{x_i} H_{\ell(t)}(x_1(t), \cdots, x_{\ell(t)}(t))$$
$$+ \frac{1}{2} \langle \rho_\xi, f \otimes f \rangle + \frac{1}{2} \langle \rho_\xi, \nu_1(t) + \nu_2(t) \rangle \} \, dt \qquad (1.132)$$

if $\eta = \eta(t) \in C_0^1(\mathbf{R})$ is supported in $|t - t_0| < \delta$.

Now we let $r \downarrow 0$ in (1.132). In fact we have

$$\rho_\xi(x, x') = \rho_\xi^0(x, x') + \rho_\xi^1(x, x')$$

with

$$\rho_\xi^0(x, x') = -\frac{1}{2\pi} \frac{\nabla\xi(x) - \nabla\xi(x')}{|x - x'|^2} \cdot (x - x')$$
$$\rho_\xi^1(x, x') = \nabla\xi(x) \cdot \nabla_x K(x, x') + \nabla\xi(x') \cdot \nabla_{x'} K(x, x').$$

It holds also that

$$|\nabla\xi(x)| \leq C_{15}|x|\chi_B(x), \quad |\nabla^2\xi(x)| \leq C_{15}\chi_B(x)$$

for $B = B(x_0, r)$ by

$$r|\nabla\varphi(x)| + r^2|\nabla^2\varphi(x)| \leq C_{16}, \qquad (1.133)$$

which implies

$$|\rho_\xi^0(x, x')| \leq C_{17}(\chi_B(x) + \chi_B(x'))$$
$$|\rho_\xi^1(x, x')| \leq C_{17}(|x|\chi_B(x) + |x'|\chi_B(x'))$$

for $B = B(x_0, r)$.

First, we have

$$\langle \rho_\xi, f \otimes f \rangle = \langle \rho_\xi^0, f \otimes f \rangle + \langle \rho_\xi^1, f \otimes f \rangle \leq C_{18} \int_B f \, dx$$

and hence

$$\int_{\mathbf{R}} \eta(t) \langle \rho_\xi, f \otimes f \rangle \, dt = o(1) \qquad (1.134)$$

as $r \downarrow 0$. It holds also that

$$\int_{\mathbf{R}} \eta(t) \langle \rho_\xi^1, \nu_1(t) + \nu_2(t) \rangle \, dt = o(1).$$

Second, we show

$$\int_{\mathbf{R}} \eta(t) \langle \rho_\xi^0, \nu_1(t) \rangle \, dt = o(1). \qquad (1.135)$$

In fact, the other case

$$\int_{\mathbf{R}} \eta(t) \langle \rho_\xi^0, \nu_2(t) \rangle \, dt = o(1)$$

is proven similarly, and then it follows that

$$\int_{\mathbf{R}} \eta(t) \langle \rho_\xi^0, \nu_1(t) + \nu_2(t) \rangle \, dt = o(1).$$

To this end we note

$$\nabla \xi(x) = 2x\varphi(x) + |x|^2 \nabla\varphi(x)$$

to infer

$$\nabla \xi(x) - \nabla \xi(x') = 2(x - x')\varphi(x) + 2x'(\varphi(x) - \varphi(x'))$$
$$+ (|x|^2 - |x'|^2)\nabla\varphi(x) + |x' - a|^2(\nabla\varphi(x) - \nabla\varphi(x')). \qquad (1.136)$$

Hence it follows that

$$\rho_\xi^0(x, x') = -\frac{1}{\pi}\varphi(x) - \frac{1}{\pi}\frac{x - x'}{|x - x'|^2} \cdot x'(\varphi(x) - \varphi(x'))$$
$$- \frac{1}{2\pi}(|x|^2 - |x'|^2) \cdot \frac{\nabla\varphi(x)}{|x - x'|^2} \cdot (x - x')$$
$$- \frac{1}{2\pi}\frac{|x'|^2}{|x - x'|^2}(\nabla\varphi(x) - \nabla\varphi(x')) \cdot (x - x'). \qquad (1.137)$$

Here we use the decomposition

$$\rho_\xi^0(x, x') = \rho_\xi^{0,0}(x, x') + \rho_\xi^{0,1}(x, x') \qquad (1.138)$$

for

$$\rho_\xi^{0,0}(x,x') = \varphi(x')\rho_\xi^0(x,x')$$
$$\rho_\xi^{0,1}(x,x') = (1 - \varphi(x'))\rho_\xi^0(x,x').$$

First, equality (1.136) implies

$$\rho_\varphi^{0,0}(x,x') + \frac{1}{\pi}\varphi(x)\varphi(x') = -I - II - III$$

with

$$I = \frac{1}{\pi}\frac{x - x'}{|x - x'|^2}\cdot x'(\varphi(x) - \varphi(x'))\varphi(x')$$

$$II = \frac{1}{2\pi}(x - x')\cdot(x + x')\frac{\nabla\varphi(x)}{|x - x'|^2}\varphi(x')\cdot(x - x')$$

$$III = \frac{1}{2\pi}\frac{|x'|^2}{|x - x'|^2}(\nabla\varphi(x) - \nabla\varphi(x'))\cdot(x - x')\varphi(x').$$

Using (1.133), we obtain

$$|I| \le \frac{C_{19}}{r}|x'\varphi(x')| \le C\chi_B(x')$$

by the mean value theorem. Similarly, we have

$$|II| \le |x\cdot\nabla\varphi(x)|\varphi(x') + |x'||\nabla\varphi(x)|\varphi(x')$$
$$\le C_{20}(\varphi(x') + \frac{1}{r}|x'\varphi(x')|) \le C_{21}\chi_B(x')$$

and

$$|III| \le C_{22}|x'|^2r^{-2}\varphi(x') \le C_{23}\chi_B(x').$$

It thus follows that

$$\left|\rho_\xi^{0,0}(x,x') + \frac{1}{\pi}\varphi(x)\varphi(x')\right| \le C_{24}\chi_{\hat{B}}(x').$$

and hence

$$\int_{\mathbf{R}}\eta(t)\langle\rho_\xi^{0,0},\nu_1(t)\rangle\,dt = -\frac{1}{\pi}\cdot 8\pi\iint_{\Omega\times\mathbf{R}}\eta(t)\varphi(x)f(x,t)dxdt + O(r)$$
$$= o(1).$$

We have also

$$\langle\rho_\xi^{0,1}(t),\nu_1\rangle = -\frac{1}{\pi}\cdot 8\pi\sum_{j\ne i}\int_\Omega\left[\frac{x - x_j(t)}{|x - x_j(t)|^2}\cdot x_j(t)\varphi(x)\right.$$

$$\left. +\frac{1}{2}\frac{|x|^2}{|x - x_j(t)|^2}\nabla\varphi(x)\cdot(x - x_j(t))\right]f(x,t)\,dx$$

using

$$\rho_\xi^0(x, x') = -\frac{1}{\pi}\varphi(x) - \frac{1}{\pi}\frac{x - x'}{|x - x'|^2} \cdot x'(\varphi(x) - \varphi(x'))$$
$$-\frac{|x|^2}{2\pi}\frac{\nabla\varphi(x)}{|x - x'|^2} \cdot (x - x') + \frac{|x'|^2}{2\pi}\frac{\nabla\varphi(x')}{|x - x'|^2} \cdot (x - x')$$

for (1.137). Then we obtain

$$\int_{\mathbf{R}} \eta(t)\left|\langle\rho_\xi^{0,1}, \nu_1(t)\rangle\right| dt \le C_{25} \cdot c_0^{-1} \iint_{B \times \mathbf{R}} \eta(t)f(x, t) \, dx dt = o(1)$$

by (1.130) and (1.133).

It thus follows that (1.135) and then equality (1.132) is reduced to

$$-\int_{\mathbf{R}} \eta'(t)m(t)|x_i(t)|^2 \, dt$$
$$= 2\int_{\mathbf{R}} \eta(t)m(t)^2 x_i(t) \cdot \nabla_{x_i} H_{\ell(t)}(x_1(t), \cdots, x_{\ell(t)}(t)). \qquad (1.139)$$

Therefore, we have $\frac{d}{dt}(m(t)|x_i(t)|^2) \in L_{loc}^\infty(\mathbf{R})$ in sense of distributions, which implies $m(t) \equiv m = 8\pi$. Since this property holds for each $x_i = x_i(t)$, the number $\ell(t) = \ell$ is locally constant in t, and hence is independent of t. Furthermore, in the sense of distributions, there arises

$$\frac{1}{2}\frac{d}{dt}|x_i|^2 = 8\pi x_i \cdot \nabla_{x_i} H_\ell(x_1, \ldots, x_\ell).$$

Replacing the above $B(x_0, r)$ with $x_0 = 0$ to $B(a, r)$ with $|a| \ll 1$, we obtain

$$\frac{1}{2}\frac{d}{dt}|x_i - a|^2 = 8\pi(x_i - a) \cdot \nabla_{x_i} H_\ell(x_1, \cdots, x_\ell).$$

From the arbitrariness of $a \in \mathbf{R}^2$ in $|a| \ll 1$, it follows that (1.125). $\qquad \square$

We conclude this section with several lemmas.

Lemma 1.9. *The orbit $\{x(t)\} \subset \Omega^\ell \backslash D$ made by (1.164) is pre-compact for D defined by (1.122). The ω-limit and α-limit sets of this orbit is contained in the critical point of H_ℓ.*

Proof. Recalling (1.127), we apply the scaling argument used for the proof of $\partial\Omega \cap S_t = \emptyset$ in Lemma 1.7. Then it follows that

$$\liminf_{t\uparrow\pm\infty} \text{dist}(x_j(t), \partial\Omega) > 0, \quad 1 \le j \le \ell. \qquad (1.140)$$

Similarly, assuming the existence $t_k \uparrow \pm\infty$ and $i \ne j$ such that

$$\lim_{k\to\infty} |x_i(t_k) - x_j(t_k)| = 0,$$

we take a subsequence satisfying

$$\mu(dx, t + t_k) \rightharpoonup \tilde{\mu}(dx, t) \quad \text{in } C_*(-\infty, +\infty; \mathcal{M}(\overline{\Omega}))$$

and $\lim_{k\to\infty} x_k = x_0 \in \Omega$. This $\tilde{\mu}(dx, t)$, however, is a weak solution to (1.9) with (1.12) satisfying $\tilde{\mu}(\{x_0\}, 0) \geq 16\pi$, a contradiction to Lemma 1.7.

Since the orbit $\{x(t)\} \subset \Omega^\ell \setminus D$ is pre-compact, the result follows from the LaSalle principle. □

By Lemma 1.9 we have $t_0 \gg 1$, $0 < r_0 \ll 1$, and $x_1^0, \cdots, x_\ell^0 \in \Omega$ such that

$$B(x_i^0, 4r_0) \cap \mathcal{S}_t = \{x_i(t)\}, \quad t \geq t_0, \ 1 \leq i \leq \ell \tag{1.141}$$

which implies

$$\inf_{t \geq t_0, \ i \neq j} |x_i(t) - x_j(t)| \geq c_0 \tag{1.142}$$

with $c_0 > 0$. Concerning (1.140) in Lemma 1.9 we can prove

$$\liminf_{t\uparrow+\infty} \text{dist}(x_j(t), \partial\Omega) > 0, \quad 1 \leq j \leq \ell$$

directly, using

$$\frac{d}{dt} H_\ell(x_1, \cdots, x_\ell) = 8\pi |\nabla H_\ell(x_1, \cdots, x_\ell)|^2 \geq 0, \quad -\infty < t < +\infty$$

derived from (1.164).

Since $H_\ell = H_\ell(x)$ is real-analytic in $x = (x_1, \cdots, x_\ell) \in \Omega^\ell \setminus D$, there are $x_\pm = (x_1^\pm, \cdots, x_\ell^\pm) \in \Omega^\ell \setminus D$ such that

$$\lim_{t\to\pm\infty} x_i(t) = x_i^\pm, \quad \nabla_{x_i} H_\ell(x_\pm) = 0, \quad 1 \leq i \leq \ell.$$

The other consequence is the following lemma.

Lemma 1.10. *Any $\tilde{t}_k \uparrow +\infty$ admits a subsequence denoted by the same symbol such that*

$$\mu^s(dx, t + \tilde{t}_k) \rightharpoonup \mu_0^s(dx) \quad \text{in } C_*(-\infty, +\infty; \mathcal{M}(\overline{\Omega})), \tag{1.143}$$

where

$$\mu_0^s(dx) = 8\pi \sum_{j=1}^\ell \delta_{x_j^*}(dx), \quad \nabla_{x_j} H(x_1^*, \cdots, x_\ell^*) = 0, \ 1 \leq j \leq \ell. \tag{1.144}$$

Here we show the following lemma.

Lemma 1.11. *There are $t_1 \geq t_0$, $0 < r_1 \leq r_0$, and k_0 such that*

$$\sup_{x \in \partial B(x_i(t), R)} u(x, t + t_k) \leq C_{26} R^{-2} \tag{1.145}$$

for any $1 \leq i \leq \ell$, $t \geq t_1$, $0 < R \leq 2r_1$, and $k \geq k_0$.

Proof. We show the existence of $t_1 \geq t_0$ and $0 < r_1 \leq r_0$ such that

$$\|f(\cdot, t)\|_{L^1(B(x_i(t), 4r_1))} < \varepsilon_0/2, \quad t \geq t_1 \tag{1.146}$$

in (1.118) with (1.124). Then (1.145) follows similarly to (1.32) for k sufficiently large.

If (1.146) is not the case, there arise $\tilde{t}_k \uparrow +\infty$ and $r_k \downarrow 0$ such that

$$\|f(\cdot, \tilde{t}_k)\|_{L^1(B(x_i^k, 4r_k))} \geq \varepsilon_0/2, \quad k = 1, 2, \cdots \tag{1.147}$$

for $x_i^k = x_i(\tilde{x}_k)$. We have a subsequence, denoted by the same symbol, satisfying (1.143)–(1.144). It holds also that

$$\mu(dx, t + \tilde{t}_k) \rightharpoonup \tilde{\mu}(dx, t) \quad \text{in } C_*(-\infty, +\infty; \mathcal{M}(\overline{\Omega})), \tag{1.148}$$

where $\tilde{\mu}(dx, t)$ is a weak solution to (1.9) with (1.12). This $\tilde{\mu}(dx, t)$ takes the form

$$\tilde{\mu}(dx, t) = 8\pi \sum_{j=1}^{\tilde{\ell}} \delta_{\tilde{x}_j(t)}(dx) + \tilde{f}(x, t) dx \tag{1.149}$$

with $\tilde{\ell} \geq \ell$ and $0 \leq \tilde{f}(\cdot, t) \in L^1(\Omega)$. We have $\tilde{x}_j(t) = x_j^*$ for $1 \leq j \leq \ell$,

$$\frac{d\tilde{x}_j}{dt} = 8\pi H_\ell(\tilde{x}_1, \cdots, \tilde{x}_{\tilde{\ell}}), \quad 1 \leq j \leq \tilde{\ell}$$

and

$$\liminf_{t \uparrow +\infty} \text{dist}(\tilde{x}_j(t), \partial\Omega) > 0, \quad 1 \leq j \leq \tilde{\ell}$$

$$\liminf_{t \uparrow +\infty} |\tilde{x}_i(t) - \tilde{x}_j(t)| > 0, \quad 1 \leq i \neq j \leq \tilde{\ell}$$

with

$$\lim_{k \to \infty} x_i^k = x_i^*.$$

Here, there is $0 < \tilde{r}_1 \ll 1$ such that $B(x_i^*, 5\tilde{r}_1) \bigcap \mathcal{S}_* = \{x_i^*\}$ for $\mathcal{S}_* = \{x_1^*, \cdots, x_\ell^*, \tilde{x}_{\ell+1}(0), \cdots, \tilde{x}_{\tilde{\ell}}(0)\}$ and also

$$\tilde{\mu}(B(x_i^*, 4\tilde{r}_1), 0) < \frac{\varepsilon_0}{4} + 8\pi$$

which implies

$$\|\tilde{f}(\cdot,0)\|_{L^1(B(x_i^*,4\tilde{r}_1))} < \varepsilon_0/4. \tag{1.150}$$

Here we have

$$f(x,\tilde{t}_k)dx \ \rightharpoonup \ 8\pi \sum_{j=\ell+1}^{\tilde{\ell}} \delta_{\tilde{x}_j(0)}(dx) + \tilde{f}(x,0)dx \quad \text{in } \mathcal{M}(\overline{\Omega}),$$

by (1.143) and (1.148)–(1.149). Hence inequality (1.150) implies the existence of $0 < \tilde{r}_2 \le \tilde{r}_1$ such that

$$\|f(\cdot,\tilde{t}_k)\|_{L^1(B(x_i^k,4\tilde{r}_2))} < \varepsilon_0/2$$

for k sufficiently large, a contradiction to (1.147).   □

1.9 Initial Mass Quantization

Comparing Theorems 1.3–1.4 to Theorem 1.5, we see that blowup in infinite time obeys rather more similar features than those of the family of stationary solutions:

(1) quantized blowup mechanism without collision.
(2) Hamiltonian control (1.120) of the location of blowup points.

The residual vanishing indicates $\mu^{ac}(dx,t) = 0$ in (1.119), where $\mu^{ac}(dy,t)$ denotes the absolutely continuous part of $\mu(dx,t)$ generated by (1.118). If this property is valid always, blowup in infinite time, (1.107), does not occur to (1.9) with (1.10) if the initial mass is so dis-quantized as $\lambda \equiv \|u_0\|_1 \notin 4\pi\mathbf{N}$. Blowup in infinite time does not occur even for the quantized initial mass, $\lambda \in 4\pi\mathbf{N}$, if $H_\ell(x_1,\cdots,x_\ell)$ does not take the critical point for any $\ell \in \mathbf{N}$ in the form of

$$\ell = \ell_1 + \ell_2, \quad (\ell_1,\ell_2) \in (\mathbf{N}\bigcup\{0\})^2, \quad 4\pi\ell_1 + 8\pi\ell_2 = \lambda. \tag{1.151}$$

In particular, if the domain Ω does not admit any singular limit of stationary solutions to (1.111) for this $\lambda \in 4\pi\mathbf{N}$, blowup in infinite time does not occur.

Similar but simpler properties are valid to (1.9) with (1.12). Thus, blowup in infinite time does not occur in the following cases, where $\lambda = \|u_0\|_1$, that is, either $\lambda \notin 8\pi\mathbf{N}$, or $\lambda = 8\pi\ell$, $\ell \in \mathbf{N}$ and there is no critical point of H_ℓ.

When Ω is convex, there is no critical point of $H_\ell = H_\ell(x_1,\cdots,x_\ell)$ for $\ell \ge 2$, and, furthermore, $H_1 = R(x)$ takes a unique critical point.

Therefore, blowup in infinite time occurs only to $\lambda = 8\pi$ for such domains, and furthermore,

$$u(x,t)dx \rightharpoonup 8\pi\delta_{x_0}(dx) \quad \text{in } \mathcal{M}(\overline{\Omega})$$

as $t \uparrow +\infty$, where $x_0 \in \Omega$ is the critical point of $R = R(x)$. It should be noted here that a class of convex domains admits stationary solutions even for $\lambda \geq 8\pi$. If Ω is a ball and the initial value is radially symmetric, however, $\lambda > 8\pi$ implies blowup in finite time of the solution.

So far, the residual vanishing is shown only for radially symmetric solutions by the above references. The proof is based on the use of the defect measures. Below we follow the argument to establish the property for the general case. To this end we put

$$x_i = x_i(t), \ u_k(x,t) = u(x,t+t_k), \ v_k(x,t) = v(x,t+t_k) \quad (1.152)$$

in Lemma 1.8.

Lemma 1.12. *It holds that*

$$\frac{d}{dt}\int_{B(x_i,r)} (|x-x_i|^2 - r^2)u_k dx$$

$$\leq 4\int_{B(x_i,r)} u_k \ dx + 2\int_{B(x_i,r)} (x-x_i)\cdot u_k\nabla v_k dx$$

$$-2\int_{B(x_i,r)} (x-x_i)\cdot \dot{x}_i u_k \ dx \quad (1.153)$$

for $0 < r \leq r_1$.

Proof. First, by Liouville's theorem, we have

$$\frac{d}{dt}\int_{B(x_i,r)} |x-x_i|^2 u_k \ dx$$

$$= \int_{B(x_i,r)} \frac{\partial}{\partial t}(|x-x_i|^2 u_k) + \dot{x}_i \cdot \nabla(|x-x_i|^2 u_k) \ dx$$

$$= \int_{B(x_i,r)} |x-x_i|^2 u_{kt} - 2(x-x_i)\cdot \dot{x}_i u_k + 2\dot{x}_i \cdot (x-x_i)u_k$$

$$+\dot{x}_i|x-x_i|^2 \cdot \nabla u_k \ dx$$

$$= \int_{B(x_i,r)} |x-x_i|^2 u_{kt} + \dot{x}_i \cdot |x-x_i|^2 \nabla u_k \ dx. \quad (1.154)$$

Here, it holds that

$$\int_{B(x_i,r)} |x - x_i|^2 u_{kt} \, dx = \int_{B(x_i,r)} |x - x_i|^2 \nabla \cdot (\nabla u_k - u_k \nabla v_k) \, dx$$

$$= \int_{\partial B(x_i,r)} |x - x_i|^2 \left(\frac{\partial u_k}{\partial \nu} - u_k \frac{\partial v_k}{\partial \nu} \right) dS$$

$$- \int_{B(x_i,r)} 2(x - x_i) \cdot (\nabla u_k - u_k \nabla v_k) \, dx$$

$$= r^2 \int_{\partial B(x_i,r)} \left(\frac{\partial u_k}{\partial \nu} - u_k \frac{\partial v_k}{\partial \nu} \right) dS - 2 \int_{\partial B(x_i,r)} (x - x_i) \cdot \nu \, u_k \, dS$$

$$+ 4 \int_{B(x_i,r)} u_k \, dx + 2 \int_{B(x_i,r)} (x - x_i) \cdot u_k \nabla v_k \, dx$$

and hence the first term of the right-hand side on (1.154) is estimated as

$$\int_{B(x_i,r)} |x - x_i|^2 u_{kt} \, dx \le r^2 \int_{B(x_i,r)} \nabla \cdot (\nabla u_k - u_k \nabla v_k) \, dx$$

$$+ 4 \int_{B(x_i,r)} u_k \, dx + \int_{B(x_i,r)} 2(x - x_i) \cdot u_k \nabla v_k \, dx$$

$$= r^2 \int_{B(x_i,r)} u_{kt} \, dx + 4 \int_{B(x_i,r)} u_k \, dx$$

$$+ 2 \int_{B(x_i,r)} (x - x_i) \cdot u_k \nabla v_k \, dx. \tag{1.155}$$

For the second term of the right-hand side on (1.154), we note

$$\int_{B(x_i,r)} \dot{x}_i \cdot |x - x_i|^2 \nabla u_k \, dx = \int_{\partial B(x_i,r)} (\dot{x}_i \cdot \nu) |x - x_i|^2 u_k \, dS$$

$$- \int_{B(x_i,r)} 2(x - x_i) \cdot \dot{x}_i u_k \, dx$$

$$= r^2 \int_{\partial B(x_i,r)} (\dot{x}_i \cdot \nu) u_k \, dx - \int_{B(x_i,r)} 2(x - x_i) \cdot \dot{x}_i u_k \, dx$$

$$= r^2 \int_{B(x_i,r)} \dot{x}_i \cdot \nabla u_k \, dx - \int_{B(x_i,r)} 2(x - x_i) \cdot \dot{x}_i u_k \, dx. \tag{1.156}$$

By (1.154), (1.155), and (1.156), it holds that (1.153) because

$$\frac{d}{dt} \int_{B(x_i,r)} u_k \, dx = \int_{B(x_i,r)} u_{kt} + \dot{x}_i \cdot \nabla u_k \, dx$$

by Liouville's theorem again.   □

The last term on the right-hand side of (1.153) we use Lemma 1.9 to deduce $|\dot{x}| \leq C_1$. For the second term, we divide v_k as

$$v_k(x,t) = \int_{B(x_i,r)} \Gamma(x-x')u_k(x',t)\,dx' + \int_{B(x_i,r)} K(x,x')u_k(x',t)\,dx'$$
$$+ \int_{\Omega \setminus B(x_i,r)} G(x,x')u_k(x',t)\,dx' \equiv v_k^0(x,t) + v_k^1(x,t) + v_k^2(x,t).$$

We use the method of symmetrization for the first term, to obtain

$$2\int_{B(x_i,r)} (x-x_i) \cdot u_k \nabla v_k^0\,dx = -\frac{1}{2\pi} \left(\int_{B(x_i,r)} u_k\,dx \right)^2. \qquad (1.157)$$

For the second term, we have $\|\nabla v_k^1(\cdot,t)\|_{L^\infty(B(x_i,r))} \leq C_1$, because $K = K(x,x')$ is smooth in $\Omega \times \Omega$. We have also $\|\nabla v_k^2(\cdot,t)\|_{L^\infty(B(x_i,r))} \leq C_2$ for t and k sufficiently large. In fact, $u_k(\cdot,t)$ is locally uniformly bounded in $\Omega \setminus S_t$ with respect to $k \gg 1$, while $B(x_i(t),4r_i) \cap S_t = \{x_i(t)\}$ holds for t sufficiently large. Since

$$v_k^2(x,t) = \int_{\Omega \setminus S_t^{2r}} G(x,x')u_k(x',t)dx' + \int_{S_t^{2r} \setminus B(x_i,r)} G(x,x')u_k(x',t)dx',$$

the result follows from $\|u_k(\cdot,t)\|_1 = \lambda$ and the smoothness of $G = G(x,x')$ in $\Omega \times \Omega \setminus D$ for the first term on the right-hand side, and

$$\sup_x \int_\Omega |\nabla_x G(x,x')|\,dx' < +\infty$$

for the second term on the right-hand side by Lemma 1.9. We thus end up with

$$\frac{d}{dt} \int_{B(x_i,r)} (|x-x_i|^2 - r^2)u_k dx \leq 4\int_{B(x_i,r)} u_k\,dx - \frac{1}{2\pi} \left(\int_{B(x_i,r)} u_k\,dx \right)^2$$
$$+ C_3 \int_{B(x_i,r)} |x-x_i|u_k\,dx \qquad (1.158)$$

for $0 < r \leq r_2$, $t \geq t_0$, and $k \geq k_0$ with some r_2, t_0, and k_0.

We are ready to complete the proof of the residual vanishing for general (non-radially symmetric) case.

Theorem 1.6. *It holds that $f \equiv 0$ in (1.124).*

Proof. We continue to write $x_i = x_i(t)$. Making $k \to \infty$ in (1.158), we obtain

$$\frac{d}{dt} \int_{B(x_i,r)} (|x - x_i|^2 - r^2) f \, dx \leq 4 \left(8\pi + \int_{B(x_i,r)} f \, dx \right)$$

$$- \frac{1}{2\pi} \left(8\pi + \int_{B(x_i,r)} f \, dx \right)^2 + C_3 \int_{B(x_i,r)} |x - x_i| f \, dx$$

$$\leq -4 \int_{B(x_i,r)} f \, dx + C_3 \int_{B(x_i,r)} |x - x_i| f \, dx$$

in the sense of distributions in t, and hence

$$\frac{d}{dt} \int_{B(x_i,r)} (|x - x_i|^2 - r^2) f \, dx$$

$$\leq -2 \int_{B(x_i,r)} f \, dx + C_4 \int_{B(x_i,r)} |x - x_i|^2 f \, dx$$

$$= \frac{2}{r^2} \int_{B(x_i,r)} (|x - x_i|^2 - r^2) f \, dx + \left(C_4 - \frac{2}{r^2} \right) \int_{B(x_i,r)} |x - x_i|^2 f \, dx$$

$$\leq \frac{3}{r^2} \int_{B(x_i,r)} (|x - x_i|^2 - r^2) f \, dx \qquad (1.159)$$

for $0 < r \leq \min\{r_2, (2/C_4)^{1/2}\}$.

Let

$$I(t) = \int_{B(x_i,r)} (|x - x_i|^2 - r^2) f \, dx \leq 0.$$

By (1.159) if there is $t_1 \geq t_0$ such that $I(t_1) < 0$ then it holds that

$$\lim_{t \uparrow +\infty} I(t) = -\infty,$$

a contradiction. Hence we have $I(t) \equiv 0$ which implies $f(\cdot, t) = 0$ in $B(x_i(t), r)$. Then we obtain $f \equiv 0$ from the strong maximum principle for the parabolic equation. $\qquad \square$

1.10 Bounded Free Energy

Here we show the following theorem.

Theorem 1.7. *If*

$$\lim_{t \uparrow +\infty} \mathcal{F}(u(\cdot, t)) > -\infty \qquad (1.160)$$

then (1.119) in Theorem 1.5 is independent of t.

Proof. We have

$$\mathcal{F}(u(\cdot, t + t_k)) \geq -C_{27}.$$

Then it holds that (1.160). Using

$$\|\nabla u - u\nabla v\|_1^2 \leq \lambda \int_\Omega u^{-1}|\nabla u - u\nabla v|^2 \, dx = \lambda \int_\Omega u|\nabla(\log u - v)|^2 \, dx,$$

we obtain

$$\int_0^\infty \|\nabla u - u\nabla v\|_1^2 \, dt < +\infty \qquad (1.161)$$

by (1.160).

Now we show the residual vanishing, $f \equiv 0$. In fact, if this property is not the case we have $f > 0$ in

$$Q = \bigcup_t (\overline{\Omega} \setminus \mathcal{S}_t) \times \{t\}$$

which is relatively open in $\overline{\Omega} \times (-\infty, +\infty)$ by

$$\mu(dx, t) \in C_*(-\infty, +\infty; \mathcal{M}(\overline{\Omega})).$$

Given $x_0 \in \mathcal{S}_0$ and $0 < r \ll 1$, we have $0 < \varepsilon \ll 1$ and $R > r$ such that $A \bigcap \mathcal{S}_t = \emptyset$ for $|t| < \varepsilon$, where $A = B(R, x_0) \setminus B(x_0, r) \subset \Omega$. Passing to a subsequence, we may assume $t_{k+1} > t_k + 2\varepsilon$, and then it follows that

$$\lim_{k \to \infty} \int_{t_k - \varepsilon}^{t_k + \varepsilon} dt \int_A u|\nabla(\log u - v)|^2 \, dx = 0$$

from

$$\sum_k \int_{t_k - \varepsilon}^{t_k + \varepsilon} dt \int_A u|\nabla(\log u - v)|^2 dx \leq \int_0^\infty u|\nabla(\log u - v)|^2 dx < +\infty.$$

By the elliptic and parabolic regularity there is smooth $g = g(x, t)$ such that

$$\int_{-\varepsilon}^\varepsilon \int_K f|\nabla(\log f - g)|^2 \, dxdt = 0. \qquad (1.162)$$

In accordance with (1.118) we have

$$v(x, t + t_k) \rightharpoonup v_*(x, t) \quad \text{in } C_*(-\infty, +\infty; \mathcal{M}(\overline{\Omega})$$

which implies

$$\log f_* - g_* = \text{constant} \quad \text{in } A$$

for $f_* = f(\cdot, 0$ and $g_* = g(\cdot, 0)$ by (1.162). Hence it holds that

$$f_* = \sigma e^{g_*} \quad \text{in } A$$

for some $\sigma > 0$, and therefore,

$$\sigma \int_A e^{g_*} \, dx \le \int_A f_* \, dx \le \lambda. \tag{1.163}$$

We have, however,

$$g_*(x) \ge 8\pi G(x, x_0)$$

and then the left-hand side on (1.163) gets to $+\infty$ as $r \downarrow 0$, a contradiction.

Equation $-\Delta v = u$, on the other hand, is equivalent to

$$v_{z\bar{z}} = -\frac{u}{4}, \quad z = x_1 + \imath x_2, \ \bar{z} = x_1 - \imath x_2, \ x = (x_1, x_2)$$

and hence it holds that

$$s_{\bar{z}} = -\frac{1}{4}(u_z - uv_z) \tag{1.164}$$

for $s = v_{zz} - \frac{1}{2}v_z^2$. Thus we obtain

$$\int_0^\infty \|s_{\bar{z}}\|_1^2 \, dt < +\infty$$

by (1.161).

Elliptic L^1 estimate implies

$$v(\cdot, t + t_k) \rightharpoonup \tilde{v}(\cdot, t) \quad \text{in } W^{1,q}(\Omega)$$

for $1 < q < 2$ in accordance with (1.118), where $1 < q < 2$. Then it holds that

$$\tilde{v}(\cdot, t) = \sum_{x_0 \in \mathcal{S}_t} m_*(x_0) G(\cdot, x_0) \tag{1.165}$$

by $f \equiv 0$.

Given $t_k \uparrow +\infty$, we have a subsequence such that $t_{k+1} > t_k + 2$. Then it follows that

$$\lim_{k \to \infty} \int_{t_k - 1}^{t_k + 1} \|s_{\bar{z}}\|_1^2 \, dt = 0,$$

from

$$\sum_k \int_{t_k - 1}^{t_k + 1} \|s_{\bar{z}}\|_1^2 \, dt \le \int_0^\infty \|s_{\bar{z}}\|_1^2 \, dt < +\infty.$$

Then we obtain

$$\tilde{s}(\cdot,t)_{\overline{z}} = 0 \quad \text{in } \overline{\Omega} \setminus \mathcal{S}_t \tag{1.166}$$

for $\tilde{s} = \tilde{s}(\cdot,t)$ defined by

$$\tilde{s} = \tilde{v}_{zz} - \frac{1}{2}\tilde{v}_z^2. \tag{1.167}$$

Then the complex analysis developed for the stationary solutions guarantees

$$\nabla_{x_i} H_\ell(x_1(t), \cdot, x_\ell(t)) = 0, \quad 1 \le i \le \ell$$

for $\mathcal{S}_t = \{x_1(t), \cdots, x_\ell(t)\}$, and hence $(x_1(t), \cdots, x_\ell(t))$ is a stationary point of (1.164). It is thus independent of t and a critical point of $H_\ell = H_\ell(x_1, \cdots, x_\ell)$. $\qquad \square$

1.11 Notes

The Keller-Segel system [Keller and Segel (1970)] is described in Section 8.1 of [Suzuki and Senba (2011)]. See Section 2.1 of [Suzuki and Senba (2011)] for the derivation of the gradient operator ∇. Einstein's formula on diffusion coefficient is described in Section 6.1 of [Suzuki and Senba (2011)]. The transport theory is described in Section 2.1 of [Suzuki and Senba (2011)]. See Section 1.1 of [Suzuki and Senba (2011)] for the law of mass action. Equality (1.15) is well-known. See Section 8.2 of [Suzuki and Senba (2011)].

Among a huge number of references on the chemotaxis system, this chapter is focused on the quantized blowup mechanism observed in $2D$ Smoluchowski-Poisson equation. Quantized blowup mechanism was first noticed in [Suzuki (2005)], and then several arguments were added for the proof by [Suzuki (2015)]. Derivation of the Smoluchowski equation from the random walk of the particle is described in Section 6.1.4 of [Suzuki and Senba (2011)]. The other derivations and applications to physics and chemistry are given in Section 6.2.3 of [Suzuki and Senba (2011)], Section 10.1 of [Suzuki (2015)], and Chapter 5 of [Doi and Edwars (1986)]. Effective modeling is a combination of the master equation and Langevin equation. Kramers-Moyal theory uses moment expansion to reach the Kramers equation

$$\frac{\partial}{\partial t}P(x,v,t \mid x_1,v_1,t_1) = \left[-\frac{\partial}{\partial x}v + \frac{\partial}{\partial v}(-F(x)+\gamma v) + \frac{D}{m^2}\frac{\partial^2}{\partial v^2}\right]$$
$$\cdot P(x,v,t \mid x_1,v_1,t), \tag{1.168}$$

where the position, velocity, and time are independent variables. This Kramers equation is reduced to the Fokker-Planck and Smoluchowski equations, independent of the position and the velocity, respecively, as is described in Section 10.1 of [Suzuki (2015)]. This modeling is used in the theory of polymer dynamics [Doi and Edwars (1986)]. Moment expansion is used also in semi-conductor physics, where the transport equation, the Boltzmann equation, and the drift-diffusion model are induced in turn. See Section 6.2 of [Suzuki and Senba (2011)]. The other modeling relies on the principle of maximum entropy production to the transport equation. These notions are described in Section 10.3 of [Suzuki (2015)] and Section 3.3 of the present monograph. This modeling is used for several problems of fluid motion in the context of astrophysics. See Sections 2.3 and 3.3 below.

System (1.8) was introduced by [Jäger and Luckhaus (1992)]. There, global-in-time existence of the solution to (1.9)–(1.10) is proven for $\|u_0\|_1 \ll 1$. This model is also derived by [Sire and Chavanis (2002)] in the context of statistical mechanics in astro-physics, concerning the motion of mean field of many self-gravitating Brownian particles.

Blowup in finite time was noticed by [Biler, Hilhorst, and Nadzieja (1994)], using (1.6). Then [Nagai (1995)] detected the blowup threshold $\|u_0\|_1$ for the existence of the global-in-time radially symmetric classical solution, using second moment as in (1.22) and free energy associated with the Trudinger-Moser inequality for an analogous system of (1.9)–(1.10).

Local-in-time well-posedness and global-in-time existence of the solutions to (1.13), (1.8) with (1.9), and (1.8) with (1.12), or, related parabolic-parabolic systems were studied by [Biler (1998)], [Gajewski and Zacharias (1999)], and [Nagai, Senba, and Yoshida (1997)]. Here we show Theorem 1.1 first, supplementing the final part of Section 8.4.5 of [Suzuki and Senba (2011)] and also Sections 1.9 and 1.10 of [Suzuki (2015)].

The ε-regularity, Lemma 1.1 and construction of the nice cut-off function $\varphi = \varphi_{x_0,R}$ satisfying (1.20)–(1.21) were established by [Senba and Suzuki (2001)]. How to construct this function is described in Chapter 5 of [Suzuki (2005)]. See also Chapter 11 of [Suzuki (2005)] for details of the proof of (1.18) and (1.19). This function is used, for example, also in [Suzuki and Yamada (2015)]. To define this function, first, we take $0 \leq \psi = \psi_{x_0,R} \in C^\infty(\overline{\Omega})$ satisfying

$$\psi_{x_0,R}(x) = \begin{cases} 1, & x \in \Omega \cap B(x_0, R/2) \\ 0, & x \in \Omega \setminus B(x_0, R), \end{cases} \qquad \left.\frac{\partial \psi}{\partial \nu}\right|_{\partial\Omega} = 0. \qquad (1.169)$$

Then, setting $\varphi = \psi_{x_0,R}^6$, we obtain (4.178)–(4.179). Second, to define $\psi = \psi_{x_0,R}$ satisfying (1.169) we distinguish two cases, $x_0 \in \Omega$ and $x_0 \in \partial\Omega$.

If $x_0 \in \Omega$, we take $\psi_{x_0,R}$ as the standard radially symmetric cut-off function, assuming $0 < R \ll 1$. If $x_0 \in \partial\Omega$, on the other hand, this $\psi = \psi_{x_0,R}$ is constructed by a composition of the standard radially symmetric cut-off function and the conformal diffeomorphism $X : \overline{\Omega \cap B(x_0, 2R)} \to \overline{\mathbf{R}_+^2}$. See p. 91 of [Suzuki (2005)].

Free energy is easy to handle with for the classical solution defined on bounded domains. Then the global-in-time behavior of the solution to the Smoluchowski-Poisson equation becomes clearer as is described in later sections. The proof of (1.23) is provided in Chapter 15 of [Suzuki (2005)]. Monotonicity formula (1.25) is also noticed there for the proof of the finiteness of blowup points. The improved ε-regularity, Lemma 1.4) is presented by [Senba and Suzuki (2002b)] in the study of blowup in infinite time. Notion of the weak solution and its generation from their bounded family are used by [Senba and Suzuki (2002a)] for pre-scaled classical solutions. Generation of the weak scaling limit indicated as (1.40) is shown in Section 1.6 of [Suzuki (2015)]. The first parabolic envelope, (1.39), fundamental notion for the proof of Theorem 1.1, was presented in Chapter 15 of [Suzuki (2005)], using nice cut-off function $\varphi = \varphi_{x_0,R}$ defined by [Senba and Suzuki (2001)]. Then (1.40), the second parabolic envelope, was noted by [Senba (2007)].

Scaling back (1.41) and the translation limit are introduced in [Suzuki and Senba (2011)]. Alternatives (1.45), (1.46), and (1.47) are called the concentration principle [Lions (1984)]. This principle is very efficient to detect all collapses. This fact is suspected in formal blowup solutions with collapse collision due to [Luckhaus, Sugiyama, and Velázquez (2012)], and actual process is executed by [Suzuki (2015a)]. There, improved ε regularity and weak Liouville property established in [Senba and Suzuki (2002b)] and [Kurokiba and Ogawa (2003)] for classical and strong solutions, respectively, are used. See also Section 8.4.5 of [Suzuki and Senba (2011)] for the detailed proof of Lemma 1.4. Lemma 1.3 to control the vanishing was provided by [Suzuki (2015a)]. Equality (1.60) is obtained by [Senba and Suzuki (2003)], where Schwarz symmetrization is used instead of the Liouville property of the scaling limit. Finally, the residual vanishing was shown by [Suzuki (2015c)].

Equality (1.62) indicates the collapse mass normalization in the backward scaling limit. This property was first shown in [Senba and Suzuki (2003)] using Schwarz symmetrization. Equality (1.62), however, is derived also by the trique of scaling back, the translation limit, and the Liouville property as in the study of blowup in infinite time. Once (1.62) is established, the second parabolic envelope (1.40) guarantees the existence of $R > 0$ such that $|y_j(s)| \leq R$ and $\zeta(\mathbf{R}^2 \setminus B(0, R), s) < \varepsilon_0$ for any $s \in \mathbf{R}$. Then the improved ε regularity, Lemma 1.3, applied to $u = u(x, t)$ implies (1.61). Thus we have an alternative proof of (1.63) without using concentration compactness principle or Schwarz symmetrization.

The collapse dynamics (1.82)–(1.83) was derived by [Suzuki (2015b)] in the context of blowup in infinite time. These relations are regarded as recursive hierarchy, observed in blowup in finite, in infinite, and in stationary states as we have seen above. Exclusion of the boundary blowup for (1.9) with (1.13), on the other hand, was shown in [Suzuki (2013)]. See Chapter 5 of [Suzuki (2005)] for the boundary behavior of the Green function. Physical background of (1.89)–(1.90) in the context of thermodynamics was noticed by [Suzuki (2005)]. See also Section 4.1 of [Suzuki (2015)] for its mathematical framework. The case of (1.9)–(1.10) in Theorem 1.4 is studied in Section 1.10 of [Suzuki (2015)].

As a consequence of Theorems 1.2 and 1.3, radially symmetric blowup solution takes the simple blowup point at the origin. Such a solution was constructed in [Herrero and Velázquez (1996)] by the method of matched asymptotic expansion. Type II blowup rate (1.92) and free energy transmission (1.93) under (1.91) were noticed by [Senba (2007); Naito and Suzuki (2008)] and [Suzuki (2005)], respectively. For the uniform BMO estimate of the Green function, see [Chanillo and Li (1992)]. The improved Trudinger-Moser inequality for (1.101) was noticed by [Chen and Li (1991)], in the context of several geometric problems as in [Chang and Yang (1987)], [Moser (1971)].

Problem (1.111) is the Euler-Lagrange equation of $\mathcal{J}_\lambda = \mathcal{J}_\lambda(v)$, $v \in E$, defined by (1.102) and

$$ E = \{v \in H^1(\Omega) \mid \int_\Omega v = 0\}, $$

while problem (1.112) is that for $v \in H_0^1(\Omega)$. Trudinger-Moser inequality arises in this context, and thus $\mathcal{J}_{4\pi}$ and $\mathcal{J}_{8\pi}$ are bounded below for the former and the latter, respectively. These problems, (1.111) and (1.112) are called the Boltzmann-Poisson equation, as we mentioned, arising in the statistical mechanics for the point vortices initiated by [Onsager (1949)].

Quantized blowup mechanism for the family of Boltzmann-Poisson equation, and recursive hierarchy concerning the location of the singular set of the limit function was noticed by [Nagasaki and Suzuki (1990a)]. Several structures on the global bifurcation were clarified on (1.112), regarding λ as a bifurcation parameter. See [Suzuki (1992); Chang, Cheng,

and Lin (2003); Grossi and Takahashi (2010); Bartolucchi and Lin (2014)].
For example, each $0 < \lambda < 8\pi$ admits a unique classical solution. If Ω is
close to a ball there is no classical solution on $\lambda = 8\pi$. If Ω is convex, only
one singular limit lies on $\lambda = 8\pi$, and therefore, multiple blowup points do
not arise for any sequence (λ_k, v_k) of the solutions. See also [Senba and
Suzuki (2000)] for (1.111).

Blowup in infinite time was studied by [Senba and Suzuki (2002b);
Ohtsuka, Senba, and Suzuki (2007); Suzuki (2015b)]. Lemma 1.8 is shown
in [Suzuki (2015b)] for the case of $\mu_{ac}(dx, t) = 0$. Lemma 1.9 was shown
in [Ohtsuka, Senba, and Suzuki (2007)] for radially symmetric case, by
which the global-in-time behavior of the solution in this case is complete
classified. Thus, compact orbit global-in-time, blowup in infinite time, and
blowup in finite time arise accordingly as $0 < \lambda < 8\pi$, $\lambda = 8\pi$, and $\lambda > 8\pi$,
respectively. The Robin function $R = R(x)$ is real analytic in $x \in \Omega$ because
it satisfies

$$-\Delta R = 4e^{-2R} \text{ in } \Omega, \quad R|_{\partial\Omega} = -\infty.$$

See [Gustafsson (1979); Friedman (1982)] and the references therein.

The fact that bounded free energy (1.160) implies the independence of
$\mu = \mu(dx, t)$ in t, is done by [Senba and Suzuki (2002b)]. We refer to
[Brezis and Strauss (1973)] for elliptic L^1 regularity. Use of (1.164) was
adopted for the elliptic case by [Nagasaki and Suzuki (1990a)]. See the
next chapter. See also [Senba and Suzuki (2002b)] for this part of the
proof, in the case of (1.9)–(1.10).

Theory of gradient inequality is described in [Huang (2006)]. It arises in
accordance with several open questions on the system of chemotaxis, particularly, that in higher-space dimensions. In accordance with the possibility
of (1.160) we have the following result.

Theorem 1.8. *Let* $(u, v) = (u(\cdot, t), v(\cdot, t))$ *be the solution to (1.9) and
(1.12), where* $\Omega = B(0, 1) \subset \mathbf{R}^n$, $n \geq 3$, *is the unit ball. Then there is*
$0 < R < 1$ *independent of the solution such that (1.106) with* $\mathcal{S} \subset B_R$
implies (1.160), where

$$\mathcal{S} = \{x_0 \in \overline{B} \mid \exists x_k \to x_0, \exists t_k \uparrow +\infty \text{ such that } u(x_k, t_k) \to +\infty\}$$

denotes the blowup set.

For the proof we use the gradient inequality in the following form.

Lemma 1.13. *Let* $B = B_1 \subset \mathbf{R}^n$, $n \geq 3$, *be the unit ball, and assume*

$$-\Delta v = u > 0 \quad \text{in } B. \tag{1.170}$$

Then it holds that

$$\|\nabla v\|_2^2 \leq C_5 \left(\|\nabla v\|_{L^2(B\setminus B_{R_*})}^2 + \|u^{1/2}g\|_1 + \|u\|_1 \right) \tag{1.171}$$

for some $0 < R_* < 1$, where

$$g = u^{-1/2}\nabla u - u^{1/2}\nabla v.$$

Proof. Using polar coordinate, we have

$$(r^{n-1}v_r)_r + r^{n-3}\Lambda v = -r^{n-1}u$$

where Λ is the Laplace-Beltrami operator and $r = |x|$. With $\omega = x/r$ it holds that

$$\frac{1}{2}\frac{\partial}{\partial r}(r^{n-1}v_r)^2 + r^{2n-4}v_r\Lambda v = r^{2n-2}(u^{1/2}g \cdot \omega - u_r),$$

which implies

$$\frac{1}{2}\frac{d}{dr}\left(r^{2n-2}\int_{|\omega|=1}v_r^2 \right) - \frac{r^{2n-4}}{2}\frac{d}{dr}\int_{|\omega|=1}((-\Lambda)^{1/2}v)^2$$

$$= r^{2n-2}\int_{|\omega|=1}u^{1/2}g \cdot \omega - r^{2n-2}\frac{d}{dr}\int_{|\omega|=1}u$$

and hence

$$\frac{r^{2n-2}}{2}\int_{|\omega|=1}v_r^2 - \frac{r^{2n-4}}{2}\int_{|\omega|=1}((-\Lambda)^{1/2}v)^2$$

$$+(n-2)\int_0^r r^{2n-5}dr \cdot \int_{|\omega|=1}((-\Lambda)^{1/2}v)^2$$

$$= \int_0^r r^{2n-2}dr \cdot \int_{|\omega|=1}u^{1/2}g \cdot \omega - r^{2n-2}\int_{|\omega|=1}u$$

$$+(2n-2)\int_0^r r^{2n-3}dr \cdot \int_{|\omega|=1}u. \tag{1.172}$$

We write (1.172) as

$$\frac{r^{n-1}}{2}\int_{|\omega|=1}v_r^2 + r^{n-1}\int_{|\omega|=1}u + \frac{n-2}{r^{n-1}}\int_0^r r^{2n-5}dr \cdot \int_{|\omega|=1}((-\Lambda)^{1/2}v)^2$$

$$= \frac{1}{r^{n-1}}\int_0^r r^{2n-2}dr \cdot \int_{|\omega|=1}u^{1/2}g \cdot \omega + \frac{2(n-1)}{r^{n-1}}\int_0^r r^{2n-3}dr \cdot \int_{|\omega|=1}u$$

$$+\frac{r^{n-3}}{2}\int_{|\omega|=1}((-\Lambda)^{1/2}v)^2$$

to use Fubini's theorem. Since

$$\int_r^1 \frac{dr}{r^{n-1}} = \frac{1}{n-2}(r^{-n+2} - 1)$$

it holds that

$$\frac{1}{2}\|v_r\|_2^2 + \|u\|_1 + \int_0^1 r^{n-1} dr \cdot \int_{|\omega|=1} \frac{((-\Lambda)^{1/2}v)^2}{r^2}$$

$$\leq \int_0^1 \left(r^{n-2} + \frac{1}{2}\right) r^{n-1} dr \cdot \int_{|\omega|=1} \frac{((-\Lambda)^{1/2}v)^2}{r^2}$$

$$+ \frac{1}{n-2}\int_0^1 r \cdot r^{n-1} dr \cdot \int_{|\omega|=1} u^{1/2} g \cdot \omega + \frac{2(n-1)}{n-2}\int_0^1 r^{n-1} dr \cdot \int_{|\omega|=1} u$$

$$\leq \left(R_*^{n-2} + \frac{1}{2}\right) \int_0^1 r^{n-1} dr \cdot \int_{|\omega|=1} \frac{((-\Lambda)^{1/2}v)^2}{r^2}$$

$$+ \frac{3}{2}\|\nabla v\|_{L^2(B \setminus B_{R_*})}^2 + \frac{1}{n-2}\|u^{1/2}g\|_1 + \frac{2(n-1)}{n-2}\|u\|_1.$$

Then (1.171) follows if $R_*^{n-2} < 1/2$. □

Now we give the following proof.

Proof of Theorem 1.8. Equality (1.90) reads

$$\frac{d}{dt}\mathcal{F}(u) = -\mathcal{D}(u)$$

for

$$\mathcal{D}(u) = \int_\Omega u|\nabla(\log u - v)|^2 = \|g\|_2^2.$$

If $\mathcal{S} \subset B_R$, $0 < R < R^*$, it holds that

$$\sup_{t \geq 0} \|\nabla v(\cdot, t)\|_{L^2(B \setminus B_R)} \leq C_6$$

by Lemma 1.13 and the elliptic estimate for the Poisson part. Then we obtain

$$-\mathcal{F}(u) \leq C_7(\mathcal{D}^{1/2}(u) + 1) \tag{1.173}$$

by (1.171), using

$$\|u\|_1 = \lambda, \quad u(\log u - 1) \geq -1, \quad \|u^{1/2}g\|_1 \leq \|u\|_1^{1/2}\|g\|_2.$$

The function $y(t) = -\mathcal{F}(u)$ is monotone increasing. Assuming $y(0) \geq 2C_7$, we have

$$y(t) - C_7 \geq \frac{y(t)}{2}, \quad t \geq 0$$

and hence

$$\frac{dy}{dt} = \mathcal{D}(u) \geq \frac{1}{C_7^2}(y - C_7)^2 \geq \frac{y^2}{4C_7^2}.$$

Then $T < +\infty$ follows, a contradiction. We obtain $y(t) < 2C_7$ for $t \geq 0$, and hence (1.160). $\qquad\square$

To see the significance of $n \geq 3$ in Theorem 1.8, first, we note that Lemma 1.13 is invalid for $n = 2$. In fact, there is a family of (radially symmetric) solutions $\{(\lambda, v(x))\}$ to the Boltzmann-Poisson equation

$$-\Delta v = \frac{\lambda e^v}{\int_B e^v dx} \quad \text{in } B, \quad v|_{\partial B} = 0$$

where $B \subset \mathbf{R}^2$ is the unit disc. For this family we put

$$u = \frac{\lambda e^v}{\int_B e^v dx}$$

to obtain

$$u^{1/2}g = \nabla u - u\nabla v = 0, \quad \lim_{\lambda \uparrow 8\pi} \|\nabla v\|_2^2 \to +\infty,$$

and

$$\|\nabla v\|_{L^\infty(B\setminus B_r)} \leq C_8 = C_8(r)$$

for any $0 < r < 1$. Hence (1.171) does not occur.

The next observation is that inequality (1.173) for $v = v(r)$, $r = |x|$, is crucial. To see this, we follow the proof of Lemma 1.13 to

$$(rv_r)_r = -ru,$$

which implies

$$\frac{1}{2}\int_0^1 rv_r^2 dr + \int_0^1 ru\,dr = \int_0^1 \{(r\log\frac{1}{r})u^{1/2}g \cdot \omega + 2\log\frac{1}{r}\cdot u\}r\,dr.$$

Here, the last term on the right-hand side is not estimated above by

$$\int_0^1 (u\log u)\,rdr + C_9$$

because Young's inequality

$$uz \leq u\log u + e^{z-1}$$

is sharp.

Theory of gradient inequality roughly says that ω-limit set of compact orbit is a singleton in the gradient system with analytic nonlinearity. This principle was discovered by [Lojasiewicz (1963); Simon (1983)]. To conclude the present chapter, we describe the principle for the ordinary differential equations of single variable. Henceforth, $E = E(x)$ stands for a non-trivial real-analytic function of $x \in \mathbf{R}$ satisfying $E(0) = 0$. Assume that $x = x(t)$ is a global-in-time solution to

$$\dot{x} = -E'(x), \quad x(0) = x_0, \quad |x(t)| \leq C, \quad t \geq 0 \tag{1.174}$$

with the compact orbit $\mathcal{O} = \{x(t)\}_{t \geq 0}$, and let $\omega(x_0)$ be its ω-limit set. Now we show that $0 \in \omega(x_0)$ implies $\omega(x_0) = \{0\}$. To this end, it suffices to take the case $x_0 \neq 0$.

From the assumption, there is an integer $n \geq 0$ such that

$$E(x) = \sum_{k=n}^{\infty} a_k x^k, \quad a_n \neq 0,$$

which implies

$$|E(x)| = |a_n| \cdot |x|^n (1 + o(1)), \quad |E'(x)| = n|a_n||x|^{n-1}(1 + o(1))$$

as $x \to 0$. Then we obtain $c_0 > 0$ and $\delta > 0$ such that

$$|E'(x)| \geq c_0 |E(x)|^{1-\theta}, \quad |x| < \delta \tag{1.175}$$

for $\theta = \frac{1}{n} \in (0, \frac{1}{2}]$.

It holds that

$$\frac{d}{dt} E(x(t)) \leq 0 \tag{1.176}$$

by (1.174), while there is $t_k \uparrow +\infty$ such that

$$x(t_k) \to 0 \tag{1.177}$$

from the assumption. Since $x = 0$ is an isolated zero of $E = E(x)$, we obtain

$$E(x(t)) > 0, \quad \forall t \geq 0$$

by (1.176) and it follows also that

$$\lim_{t \uparrow +\infty} E(x(t)) = 0$$

from (1.177).

Putting $H(t) = E^\theta(x(t)) > 0$, we obtain

$$\dot{H} = \theta E^{\theta-1} E' \cdot \dot{x} = -\theta E^{\theta-1}|E'||\dot{x}| \tag{1.178}$$

and hence

$$|x(t)| < \delta \quad \Rightarrow \quad |\dot{x}(t)| \leq -\frac{1}{\theta c_0} \dot{H}(t)$$

by (1.175) and (1.178).

Here we have k_0 such that $|x(t_k)| < \delta$ for $k \geq k_0$, which implies

$$|x(t) - x(t_k)| \leq \frac{H(t_k)}{\theta c_0},$$

provided that $0 < t - t_k \ll 1$. Since $\lim_{k \to \infty} H(t_k) = 0$ there is $k_1 \geq k_0$ such that $|x(t_k)| < \delta/4$ and $\frac{H(t_k)}{\theta c_0} < \delta/4$ for $k = k_1$. Then, if there is $\tilde{t} > t_k$ satisfying

$$|x(\tilde{t})| = \delta/2, \quad |x(t)| < \delta/2, \ t_k < t < \tilde{t}$$

it follows that

$$|x(\tilde{t}) - x(t_k)| < \delta/4$$

and hence $|x(\tilde{t})| < \delta/2$, a contradiction. Hence it holds that $|x(t)| < \delta/2$ for $t \geq t_{k_1}$, which implies

$$\sup_{t \geq t_k} |x(t) - x(t_k)| \leq \frac{H(t_k)}{\theta c_0}. \tag{1.179}$$

Letting $t \uparrow +\infty$ and then $k \to \infty$ in (1.179) assures

$$\lim_{t \uparrow +\infty} |x(t)| = 0 \tag{1.180}$$

and hence $\omega(x_0) = \{0\}$.

Concerning the decay rate, we use (1.178), (1.175), and (1.180), to deduce

$$\dot{H} = -\theta E^{\theta-1} |E'|^2 \leq -c_0 \theta E^{1-\theta} = -c_0 \theta H^{\frac{1}{\theta}-1}$$

for $t \gg 1$, and then obtain

$$H(t) = \begin{cases} O(t^{-\frac{\theta}{1-2\theta}}), & \theta < 1/2 \\ O(\exp(-c_0\theta t)), & \theta = 1/2. \end{cases}$$

Chapter 2

2D Turbulence

In statistical mechanics micro-scopic states are classified by their state quantities, such as energies, temperatures, and so on. Then several equations are derived to describe the motion of their mean fields, regarding these quantities as parameters. In thermal equilibria, these parameters are equivalent through thermo-dynamical relations. The Boltzmann-Poisson equation arises in the stationary system of point vortices under the converting micro-canonical ensembles to canonical ensembles. We can observe also kinetic equations in semi-conductor physics and astrophysics which stand for the adiabatic limit of the master equation describing the mass conservation law. Derived from maximum entropy production principles, these models contain the Smoluchowski-Poisson equation of which stationary state is given by the Boltzmann-Poisson equation and their natural extensions to higher space dimensions. Elliptic eigenvalue problems with exponential nonlinearity in two-space dimensions, on the other hand, can cast several Euler-Lagrange equations concerning the kinetics of particles self-interacting with the electro-magnetic field created by themselves, particularly the ones provided with duality, which arises as a heritage of the gauge invariance of the Lagrangian. Then we can observe recursive hierarchy as well as quantized blowup mechanism for the blowup family of solutions.

2.1 Point Vortices

The stationary state of (1.9) with (1.12) is defined by the total mass conservation

$$\frac{d}{dt} \int_\Omega u \ dx = 0$$

and the free energy decreasing

$$\frac{d}{dt}\mathcal{F}(u) = -\int_\Omega u|\nabla(\log u - v)|^2 \, dx \leq 0$$

for $\mathcal{F} = \mathcal{F}(u)$ defined by (1.89). Thus it holds that

$$u > 0, \quad \|u\|_1 = \lambda, \quad \log u - v = \text{constant},$$

and hence

$$u = \frac{\lambda e^v}{\int_\Omega e^v \, dx},$$

which induces the Boltzmann-Poisson equation

$$-\Delta v = \frac{\lambda e^v}{\int_\Omega e^v \, dx}, \quad v|_{\partial\Omega} = 0. \tag{2.1}$$

Incompressible, non-viscous fluid is called the ideal fluid. In the space of three dimensions, the velocity

$$v = \begin{pmatrix} v_1(x,t) \\ v_2(x,t) \\ v_3(x,t) \end{pmatrix} \in \mathbf{R}^3, \qquad x = (x_1, x_2, x_3) \in \mathbf{R}^3$$

and the pressure $p = p(x,t) \in \mathbf{R}$ are subject to the Euler equation

$$v_t + (v \cdot \nabla)v = -\nabla p, \quad \nabla \cdot v = 0 \qquad \text{in } \mathbf{R}^3 \times (0,T) \tag{2.2}$$

without the outer force, where physical parameters are put to be one.

Under this flow, the particle $x_0 \in \mathbf{R}^3$ at $t = 0$ moves to $x = x(t) \in \mathbf{R}^3$ at $t = t$, satisfying

$$\frac{dx}{dt} = v(x,t), \quad x(0) = x_0. \tag{2.3}$$

Let $f = f(x,t) \in \mathbf{R}$ be a state quantity distributing in the space-time variables (x,t). Then the trajectory $h(t) = f(x(t),t)$ of the particle detected by this $f = f(x,t) \in \mathbf{R}$ satisfies

$$\frac{df}{dt} = f_t + v \cdot \nabla f = \frac{Df}{Dt} \tag{2.4}$$

by (2.3). From the right-hand side on (2.4) denoted by $\dfrac{Df}{Dt}$ is the material derivative, and the acceleration of this particle is defined by

$$\frac{dv}{dt}(x(t),t) = \frac{Dv}{Dt} = v_t + (v \cdot \nabla)v. \tag{2.5}$$

Since the relative force acting to this particle is equal to the minus of the pressure gradient, the first equation of (2.2) is due to Newton's equation of motion.

Let the solution to (2.3) and the volume of the domain $\omega \subset \mathbf{R}^3$ be $x(t) = T_t x_0$ and $|\omega|$, respectively. Then, Liouville's formula guarantees

$$\frac{d}{dt}|T_t(\omega)|_{t=0} = \int_\omega (\nabla \cdot v)(x, 0) \, dx. \tag{2.6}$$

Since the left-side on (2.6) indicates the rate of dilation of the fluid in ω at $t = 0$, the second equation of (2.2) describes that the fluid is incompressible. In the case of the rigid body, twice of the rotation

$$\nabla \times v = \begin{pmatrix} \frac{\partial v_3}{\partial x_2} - \frac{\partial v_2}{\partial x_3} \\ \frac{\partial v_1}{\partial x_3} - \frac{\partial v_3}{\partial x_1} \\ \frac{\partial v_2}{\partial x_1} - \frac{\partial v_1}{\partial x_2} \end{pmatrix}$$

of the velocity v indicates its angle velocity. We call this $\omega = \nabla \times v$ the vorticity for the fluid.

Euler's equation of motion, (2.2), implies

$$\omega_t + (v \cdot \nabla)\omega = (\omega \cdot \nabla)v, \quad \nabla \cdot v = 0 \qquad \text{in } \mathbf{R}^3 \times (0, T). \tag{2.7}$$

Under the two-dimensional flow described by

$$v_3 = 0, \quad v_1 = v_1(x_1, x_2, t), \quad v_2 = v_2(x_1, x_2, t), \tag{2.8}$$

it holds that

$$\omega = \nabla \times v = \begin{pmatrix} 0 \\ 0 \\ \frac{\partial v_2}{\partial x_1} - \frac{\partial v_1}{\partial x_2} \end{pmatrix},$$

and therefore, the first term on the right-hand side on (2.7) vanishes. For simplicity we use ω to indicate the two-dimensional scalar field $\omega = \frac{\partial v_2}{\partial x_1} - \frac{\partial v_1}{\partial x_2}$, which satisfies

$$\omega_t + (v \cdot \nabla)\omega = 0, \quad \nabla \cdot v = 0.$$

Then we obtain

$$\omega_t + \nabla \cdot (v\omega) = 0, \quad \nabla \cdot v = 0 \qquad \text{in } \mathbf{R}^2 \times (0, T). \tag{2.9}$$

The second equation of (2.9), on the other hand, is reduced to

$$\nabla \cdot v = \frac{\partial v_1}{\partial x_1} + \frac{\partial v_2}{\partial x_2} = 0. \tag{2.10}$$

In the whole space $\mathbf{R}^2$, this (2.10) guarantees the existence of the stream function $\psi = \psi(x_1, x_2, t) \in \mathbf{R}$ such that

$$v_1 = \frac{\partial \psi}{\partial x_2}, \quad dv_2 = -\frac{\partial \psi}{\partial x_1},$$

and hence

$$v = \nabla^{\perp}\psi, \quad \nabla^{\perp} = \begin{pmatrix} \frac{\partial}{\partial x_2} \\ -\frac{\partial}{\partial x_1} \end{pmatrix}. \tag{2.11}$$

Then we obtain

$$\omega = \frac{\partial v_2}{\partial x_1} - \frac{\partial v_1}{\partial x_2} = -\Delta\psi$$

by (2.11). Therefore, system (2.9) is reduced to the hyperbolic-elliptic system on (ω, ψ),

$$\omega_t + \nabla \cdot (\omega\nabla^{\perp}\psi) = 0, \quad -\Delta\psi = \omega \quad \text{in } \mathbf{R}^2 \times (0, T), \tag{2.12}$$

called the vortex equation. This (2.12) is provided with a formal similarity between the parabolic-elliptic system (1.9) with (1.12), that is, the Smolchowski-Poisson equation.

If $\Omega \subset \mathbf{R}^2$ is a bounded domain with smooth boundary $\partial\Omega$, we impose the null normal velocity on the boundary:

$$v_t + (v \cdot \nabla)v = -\nabla p, \quad \nabla \cdot v = 0 \quad \text{in } \Omega \times (0, T), \qquad \nu \cdot v|_{\partial\Omega} = 0.$$

If Ω is simply-connected, there is a stream function $\psi = \psi(x, t)$ such that $v = \nabla^{\perp}\psi$ and then the boundary condition $\nu \cdot v = 0$ is reduced to

$$\psi = \text{constant} \qquad \text{on } \partial\Omega \times (0, T). \tag{2.13}$$

We assume this constant to be zero, modifying ψ, and then it follows that

$$\omega_t + \nabla \cdot (\omega\nabla^{\perp}\psi) = 0, \quad -\Delta\psi = \omega \quad \text{in } \Omega \times (0, T), \quad \psi|_{\partial\Omega} = 0. \tag{2.14}$$

With the Green function $G = G(x, x')$ for the Poisson equation (1.12) we obtain

$$\omega_t + \nabla \cdot (\omega\nabla^{\perp}\psi) = 0, \quad \psi(\cdot, t) = \int_{\Omega} G(\cdot, x')\omega(x', t)dx' \quad \text{in } \Omega \times (0, T) \tag{2.15}$$

by (2.14).

System (2.15) takes a weak form due to the symmetry $G(x, x') = G(x', x)$. Thus, given $\varphi \in C^1(\overline{\Omega})$ with $\varphi|_{\partial\Omega} = 0$, we obtain

$$\frac{d}{dt}\int_{\Omega} \varphi\omega \, dx = \frac{1}{2}\iint_{\Omega \times \Omega} \rho_{\varphi} \cdot \omega \otimes \omega \, dxdx' \tag{2.16}$$

by (2.15), where $\omega \otimes \omega = (\omega \otimes \omega)(x, x', t) = \omega(x, t)\omega(x', t)$ and

$$\rho_{\varphi} = \rho_{\varphi}(x, x') = \nabla_x^{\perp} G(x, x') \cdot \nabla\varphi(x) + \nabla_{x'}^{\perp} G(x.x') \cdot \nabla\varphi(x').$$

Assuming $\omega(\cdot, t) \in L^1(\Omega)$, supp $\varphi \subset \Omega$ in (2.16), we use the interior regularity (1.84) to obtain

$$
\iint_{\Omega \times \Omega} \rho_\varphi \omega \otimes \omega \, dx dx'
$$
$$
= \iint_{\Omega \times \Omega} \nabla^\perp \Gamma(x - x') \cdot [\nabla \varphi(x) - \nabla \varphi(x')] \, \omega \otimes \omega \, dx dx'
$$
$$
+ \iint_{\Omega \times \Omega} \left[\nabla_x^\perp K(x, x') \cdot \nabla \varphi(x) + \nabla_{x'}^\perp K(x, x') \cdot \nabla \varphi(x') \right] \omega \otimes \omega \, dx dx'
$$
$$
= \lim_{\varepsilon \downarrow 0} \iint_{\Omega \times \Omega \setminus \{|x - x'| < \varepsilon\}} \nabla^\perp \Gamma(x - x') \cdot [\nabla \varphi(x) - \nabla \varphi(x')] \, \omega \otimes \omega \, dx dx'
$$
$$
+ \iint_{\Omega \times \Omega} \left[\nabla_x^\perp K(x, x') \cdot \nabla \varphi(x) + \nabla_{x'}^\perp K(x, x') \cdot \nabla \varphi(x') \right]
$$
$$
\omega \otimes \omega dx dx'. \tag{2.17}
$$

The right-hand side on (2.17) converges even to

$$
\omega(dx, t) = \sum_{i=1}^{\ell} \alpha_i \delta_{x_i(t)}(dx), \quad \alpha_i \in \mathbf{R}, \quad x_i(t) \in \Omega, \quad i = 1, \ldots, \ell, \tag{2.18}
$$

which results in

$$
\frac{dx_i}{dt} = \frac{1}{2} \alpha_i \nabla^\perp R(x_i) + \sum_{j \neq i} \alpha_i \alpha_j \nabla_x^\perp G(x_i, x_j), \quad 1 \leq i \leq \ell \tag{2.19}
$$

because φ is arbitrary, where $R(x) = K(x, x)$. The Hamiltonian $H = H_\ell(x_1, \cdots, x_\ell)$ in (1.116) is now modified to

$$
H(x_1, \ldots, x_\ell) = \frac{1}{2} \sum_{i=1}^{\ell} \alpha_i^2 R(x_i) + \sum_{1 \leq i < j \leq \ell} \alpha_i \alpha_j G(x_i, x_j)
$$
$$
= \frac{1}{2} \sum_{i=1}^{\ell} \sum_{j=1, j \neq i}^{\ell} \alpha_i \alpha_j G(x_i, x_j) + \frac{1}{2} \sum_{i=1}^{n} \alpha_i^2 R(x_i) \tag{2.20}
$$

and then (2.19) is written by the Hamilton system

$$
\alpha_i \frac{dx_{i1}}{dt} = \frac{\partial H}{\partial x_{i2}}, \quad \alpha_i \frac{dx_{i2}}{dt} = -\frac{\partial H}{\partial x_{i1}}, \quad 1 \leq i \leq \ell \tag{2.21}
$$

where $x_i = (x_{i1}, x_{i2})$.

2.2   Mean Field Limit

The Hamilton system in classical mechanics

$$\frac{dq_i}{dt} = \frac{\partial H}{\partial p_i}, \quad \frac{dp_i}{dt} = -\frac{\partial H}{\partial q_i}, \quad 1 \le i \le N \qquad (2.22)$$

is concerned on the position $(q_1, \ldots, q_N) \in \mathbf{R}^{3N}$ and the momentum $(p_1, \ldots, p_N) \in \mathbf{R}^{3N}$ in general coordinate, where $H = H(q_1, \ldots, q_N, p_1, \ldots, p_N)$ is called the Hamiltonian. This system (2.22) is proved with the conservation law

$$\frac{d}{dt} H(q(t), p(t)) = 0$$

and therefore, the set of orbits $\{\mathcal{O}\}$ formed by (2.22) is classified according to their energy $H = E$. Each $\mathcal{O}$ is a curve in the phase space $x = (p_1, \cdots, p_N, q_1, \cdots, q_N) \in \Gamma = \mathbf{R}^{6N}$, and the hypersurface $\Gamma_E = \{x \in \Gamma \mid H(x) = E\}$ in Γ is provided with the area element $d\Sigma(E)$. Then the micro-canonical measure $\mu^{E,N}(dx)$ on Γ_E is defined by the co-area formula

$$dx = dE \cdot \frac{d\Sigma(E)}{|\nabla H|}$$

where $dx = dq_1 \cdots dq_N dp_1 \ldots dp_N$ stands for the Lebesgue measure on $\mathbf{R}^{6N}$. Then we put

$$\mu^{E,N}(dx) = \frac{1}{W(E)} \cdot \frac{d\Sigma(E)}{|\nabla H|}, \quad W(E) = \int_{\{H=E\}} \frac{d\Sigma(E)}{|\nabla H|} \qquad (2.23)$$

for each $E \in \mathbf{R}$. This $\mu^{E,N}$ detects the events on Γ_E uniformly with respect to E, based on the principle of equal a priori probabilities, and $W(E)$ in (2.23) is called the weight factor.

Statistical ensembles are associated with thermo-dynamics if (2.22) describes the motion of gas molecules. In this case the particles are put on the space without any interactions outside through materials or thermodynamical energies. Thus it is concerned on the isolated system. The system of materially closed and thermodynamically open is called closed system. There, the second law of thermo-dynamics takes a different form from the entropy increasing. If the temperature T is constant, the free energy of Helmholtz decreases. The canonical ensemble is defined as an equivalent class of the same temperature, and the canonical measure is defined for each T by the principle of a priori equal probabilities.

Heat bath is used in a standard argument to derive the canonical measure, where Boltzmann's and thermo-dynamical relations

$$S = k \log W, \quad \frac{\partial S}{\partial E} = \frac{1}{T} \tag{2.24}$$

are applied. It holds that

$$\mu^{\beta,N}(dx) = \frac{e^{-\beta H} dx}{Z(\beta, N)}, \quad Z(\beta, N) = \int_\Gamma e^{-\beta H} dx,$$

where $\Gamma = \mathbf{R}^{6N}$ stands for the phase space, $E = H$ is the energy, and $\beta = T^{-1}$ denotes the inverse temperature.

Order structures in long term are observed for two-dimensional turbulence emerged from many point vortices. L. Onsager proposed a statistical mechanics based on the Hamilton structure of the system of point vortices. In this setting of (2.21), we have $\Gamma = \Omega^n$ where n denotes the number of point vortices. Let $X_n = (x_1, \cdots, x_n) \in \Omega^n$ be the element of the set of states and $\mu^n = \mu^n(dx_1 \cdots dx_n)$ be the statistical measure on point vortices. We take the case that $\omega(dx, t)$ is independent of t in (2.18) and the intensities α_i is a constant denoted by $\alpha > 0$:

$$\omega = \omega_n(dx), \quad \omega_n(x)dx = \sum_{i=1}^n \alpha \delta_{x_i}(dx).$$

From the principle of a priori equal probabilities, the measure

$$\rho_{1,i}^n(dx_i) = \int_{\Omega^{n-1}} \mu^n(dx_1 \ldots dx_{i-1} dx_{i+1} \cdots dx_n)$$

is independent of $1 \le i \le n$. Assuming $\rho_{1,i}^n(dx_i) = \rho_{1,i}^n(x_i)dx_i$, we call $\rho_1^n(x) = \rho_{1,i}^n(x_i)$ the one-point reduced probability density function for each $1 \le i \le n$. Similarly, we call

$$\rho_k^n(x_1, \ldots, x_k)dx_1 \cdots dx_k = \int_{\Omega^{n-k}} \mu^n(dx_{k+1} \cdots dx_n)$$

the k-point reduced probability density function.

Then the phase space mean is given by

$$\langle \omega_n(x) \rangle = \sum_{i=1}^n \int_{\Omega^n} \alpha \delta(x_i - x) \mu^n(dx_1 \ldots dx_n) = n\alpha \rho_1^n(x),$$

where $\rho_1^n = \rho_1^n(x)$ is the one point reduced probability density function. Let $\tilde{E}$ and $\tilde{\beta}$ be the energy and temperature of this system of n-point vortices, and take the limit $n \to \infty$ under the assumption of $\alpha n = 1$, $\alpha^2 n^2 \tilde{E} = E$, and $\alpha^2 n \tilde{\beta} = \beta$.

Assuming the limit

$$\lim_{n\to\infty} \langle \omega_n(x) \rangle = \rho(x) = \lim_{n\to\infty} \rho_1^n(x), \tag{2.25}$$

we can confirm

$$\rho = \frac{e^{-\beta\psi}}{\int_\Omega e^{-\beta\psi} dx}, \quad \psi = \int_\Omega G(\cdot, x')\rho(x') dx' \quad \text{in } \Omega. \tag{2.26}$$

This (2.26) is equivalent to

$$-\Delta v = \frac{\lambda e^v}{\int_\Omega e^v dx} \quad \text{in } \Omega, \qquad v = 0 \quad \text{on } \partial\Omega \tag{2.27}$$

where $v = \psi$ and $\lambda = -\beta$. Actually, ψ and ρ are due to the stream function and the point vortex density, constituting a duality in (2.26).

The convergence (2.25) is justified if $\{\rho_k^n\}$ has a uniform bound in k, and the solution to (2.27) is unique. These two requirements are valid, provided that $\beta = -\lambda > -8\pi$, which implies, simultaneously, propagation of chaos formulated by

$$\rho_k^n \quad \rightharpoonup \quad \rho^{\otimes k}, \quad \rho^{\otimes k}(x_1, \ldots, x_k) = \prod_{i=1}^{k} \rho(x_i)$$

in the sense of measures.

Above β is actually called the inverse temperature in the context of statistical mechanics, but is not associated with temperature in any physical sense. Actually, the Boltzmann and thermo-dynamical relations, there arises

$$\beta = \frac{\partial}{\partial E} \log W(E) \tag{2.28}$$

for the weight factor $W(E)$ of the micro-canonical measure defined by (2.23). From this relation and the co-area formula, it follows that

$$\Theta(E) \equiv \int_{\{H<E\}} dx_1 \ldots dx_n = \int_{-\infty}^{E} W(E') \, dE'$$

and hence (2.28) means

$$\beta = \frac{\Theta''(E)}{\Theta'(E)}. \tag{2.29}$$

Since $E \mapsto \Theta(E)$ is bounded and non-decreasing, this mapping has a point of inflection. In particular, negative inverse temperature $\beta < 0$ can happen in the range of $E \gg 1$.

In this way, L. Onsager suggested an ordered structure in negative inverse temperature. The quantized blowup mechanism of the Boltzmann-Poisson equation, on the contrary, shows the formation of singular limits of (2.27) at $\lambda = 8\pi\ell$ for $\ell \in \mathbf{N}$, of which singular points coincide with those of the Hamiltonian in (2.20). Thus there emerges a recursive hierarchy.

2.3 Patch Model

We continue to deal with non-viscous, non-compressible fluid with high Reynolds number occupied in a bounded, simply-connected domain denoted by $\Omega \subset \mathbf{R}^2$. Motion of this fluid is described by the Euler-Poisson equation

$$\omega_t + \nabla \cdot u\omega = 0, \quad \psi = -\omega, \quad u = \nabla^\perp \psi, \quad \psi|_{\partial\Omega} = 0 \qquad (2.30)$$

where

$$\nabla^\perp = \begin{pmatrix} \partial/\partial x_2 \\ -\partial/\partial x_1 \end{pmatrix}, \quad x = (x_1, x_2)$$

and u, ω, and ψ stand for the velocity, vorticity, and stream function, respectively. In the point vortex model

$$\omega(x, t) = \sum_{i=1}^{N} \alpha_i \delta_{x_i(t)}(dx) \qquad (2.31)$$

system (2.30) is reduced to

$$\alpha_i \frac{dx_i}{dt} = \nabla^\perp_{x_i} H_B, \quad i = 1, 2, \cdots, N \qquad (2.32)$$

associated with the Hamiltonian

$$H_N(x_1, \cdots, x_N) = \frac{1}{2} \sum_i \alpha_i^2 R(x_i) + \sum_{i<j} \alpha_i \alpha_j G(x_i, x_j), \qquad (2.33)$$

where $G = G(x, x')$ is the Green's function of $-\Delta$ provided with the Dirichlet boundary condition and

$$R(x) = \left[G(x, x') + \frac{1}{2\pi} \log |x - x'| \right]_{x'=x}.$$

In the previous section we take the case of single intensity, $\alpha_i = \alpha$. In the general case, the intensity of the i-th vortex is denoted by $\alpha_i = \alpha^i \alpha$ with $\alpha^i \in I = [-1, 1]$, and $\rho_\alpha(x)dx$ stands for the existence probability of the vortex at x with relative intensity α. It holds that

$$\int_\Omega \rho_\alpha(x)dx = 1, \quad \alpha \in I,$$

and in the limit $N \to \infty$ with $\alpha N = 1$, local mean of vortex distribution is given by

$$\omega(x) = \int_I \alpha \rho_\alpha(x) P(d\alpha), \quad x \in \Omega,$$

where $P(d\alpha)$ is the numerical density of the vortices with the relative intensity α. Under $H_N = E = $ constant, $\alpha^2 N \beta_N = \beta = $ constant, and

$N \to \infty$, mean field equation is derived by several arguments, see [Caglioti, Lions, Marchioro, and Pulvirenti (1992); Eyink, Spohn, and Chen (1993); Joyce and Montgomery (1973); Pointin and Lundgren (1976); Sawada and Suzuki (2008)], that is,

$$-\Delta\psi = \int_I \frac{\alpha e^{-\beta\alpha\psi}}{\int_\Omega e^{-\beta\alpha\psi)}\,dx} P(d\alpha), \quad \psi|_{\partial\Omega} = 0 \qquad (2.34)$$

with

$$\omega = -\psi, \quad \rho^\alpha = \frac{e^{-\beta\alpha\psi}}{\int_\Omega e^{-\beta\alpha\psi}}$$

where

$$\rho^\alpha(x) = \lim_{N\to\infty} \int_{\Omega^{N-1}} \mu_N^{\beta_N}(dx, dx_2, \cdots, dx_N)$$

$$\mu_N^{\beta_N}(dx_1, \cdots, dx_N) = \frac{1}{Z(N,\beta_N)} e^{-\beta_N H_N} dx_1 \cdots dx_N$$

$$Z(N,\beta_N) = \int_{\Omega^N} e^{-\beta_N H_N} dx_1 \cdots dx_N.$$

Since [Nagasaki and Suzuki (1990a)], structure of the set of solutions to (2.34) has been clarified in accordance with the Hamiltonian given by (2.33). Quasi-equilibria, on the other hand, are observed for several isolated systems with many components [10]. Thus we have a relatively stationary state, different from the equilibrium, which eventually approaches the latter. Relaxation indicates this time interval, from quasi-equilibrium to equilibrium. To approach relaxation dynamics of many point vortices, patch model

$$\omega(x,t) = \sum_{i=1}^{N_p} \sigma_i \chi_{\Omega_i(t)}(x) \qquad (2.35)$$

is used where $\chi_{\Omega_i(t)}$ denotes the indicator function of $\Omega_i(t)$. It describes detailed vortex distribution, where N_p, σ_i, and $\Omega_i(t)$ denote the number of patches, the vorticity of the i-th patch, and the domain of the i-th patch, respectively. Mean field equations for equilibrium and for relaxation time are derived by the principles of minimum entropy and maximum entropy production, respectively.

For the latter case, one obtains a system on $p = p(x, \sigma, t)$,

$$p_t + \nabla \cdot pu = \nabla \cdot D(\nabla p + \beta_p(\sigma - \omega)p\nabla\psi)$$

$$\beta_p = -\frac{\displaystyle\int_\Omega D\nabla\omega \cdot \nabla\psi \, dx}{\displaystyle\int_\Omega D\left(\int \sigma^2 p d\sigma - \omega^2\right)|\nabla\psi|^2 \, dx}$$

$$\omega = \int \sigma p \, d\sigma = -\Delta\psi, \quad \psi|_{\partial\Omega} = 0, \quad u = \nabla^\perp\psi \qquad (2.36)$$

with the diffusion coefficient $D = D(x, t) > 0$. Here we regard (2.31) as a limit of (2.35). First, point vortex model valid to the relaxation time is derived from (2.36), that is, a system on $\rho_\alpha = \rho_\alpha(x, t)$ in the form of

$$\rho_{\alpha t} + \nabla \cdot \rho_\alpha u = \nabla \cdot D(\nabla\rho_\alpha + \beta\alpha\rho_\alpha\nabla\psi)$$

$$\omega = \int_I \alpha\rho_\alpha P(d\alpha) = -\Delta\psi, \quad \psi|_{\partial\Omega} = 0, \quad u = \nabla^\perp\psi$$

$$\beta = -\frac{\displaystyle\int_\Omega D\nabla\omega \cdot \nabla\psi \, dx}{\displaystyle\int_\Omega D\int_I \alpha^2\rho_\alpha P(d\alpha)|\nabla\psi|^2 \, dx} \qquad (2.37)$$

Second, the stationary state of (2.37) is given by (2.34). Third, equation (2.37) coincides with the Brownian point vortex model. Finally, system (2.37) provided with the boundary condition

$$\frac{\partial\rho_\alpha}{\partial\nu} + \beta\alpha\rho_\alpha\frac{\partial\psi}{\partial\nu}\bigg|_{\partial\Omega} = 0 \qquad (2.38)$$

satisfies the requirements of isolated system in thermodynamics. In fact, averaging (2.37) implies

$$\omega_t + \nabla \cdot \omega u = \nabla \cdot D(\nabla\omega + \beta\omega_2\nabla\psi), \quad \frac{\partial\omega}{\partial\nu} + \beta\omega_2\frac{\partial\psi}{\partial\nu}\bigg|_{\partial\Omega} = 0$$

$$\omega = -\Delta\psi, \quad \psi|_{\partial\Omega} = 0, \quad u = \nabla^\perp\psi$$

$$\beta = -\frac{\displaystyle\int_\Omega D\nabla\omega \cdot \nabla\psi \, dx}{\displaystyle\int_\Omega D\omega_2|\nabla\psi|^2 \, dx} \qquad (2.39)$$

for

$$\omega = \int_I \alpha\rho_\alpha \, P(d\alpha), \quad \omega_2 = \int_I \alpha^2\rho_\alpha P(d\alpha). \qquad (2.40)$$

Then we obtain mass and energy conservations

$$\frac{d}{dt}\int_\Omega \omega \ dx = 0, \quad (\omega_t, \psi) = \frac{1}{2}\frac{d}{dt}(\omega, (-\Delta)^{-1}\omega) = 0 \qquad (2.41)$$

where (,) stands for the L^2-inner product. Assuming $\rho_\alpha = \rho_\alpha(x,t) > 0$, we write the first equation of (2.37) as

$$\rho_{\alpha t} + \nabla \cdot \rho_\alpha u = \nabla \cdot D\nabla\rho_\alpha \nabla(\log \rho_\alpha + \beta\alpha\psi).$$

Then it follows that

$$\frac{d}{dt}\int_\Omega \Phi(\rho_\alpha) \ dx + \beta\alpha(\rho_{\alpha t}, \psi) = -\int_\Omega D\rho_\alpha|\nabla(\log \rho_\alpha + \beta\alpha\psi)|^2 \ dx$$

from (2.38), where $\Phi(s) = s(\log s - 1) + 1$ for $s > 0$.

Hence it follows that

$$\frac{d}{dt}\int_\Omega \int_I \Phi(\rho_\alpha)P(d\alpha) \ dx$$
$$= -\int_\Omega \int_I D\rho_\alpha|\nabla(\log \rho_\alpha + \beta\alpha\psi|^2 P(d\alpha) \ dx \le 0 \qquad (2.42)$$

from (2.41), that is, entropy increasing.

In the vorticity patch model (2.35), the vorticity σ_i is uniform in the region $\Omega_i(t)$ with constant area, called vorticity patch. A patch takes a variety of forms as the time t varies. We collect all the vorticity patches in a small region, called cell. Cell area Δ thus takes the relation $|\Omega_i| \ll |\Omega|$. The probability that the average vorticity at x is σ is denoted by $p(x,\sigma,t)dx$, which satisfies

$$\int p(x,\sigma,t) \ d\sigma = 1. \qquad (2.43)$$

Let

$$\int_\Omega p(x,\sigma,t) \ dx = M(\sigma) \qquad (2.44)$$

be independent of t. Since

$$|\Omega| = \iint p(x,\sigma,t) \ dxd\sigma = \int M(\sigma) \ d\sigma, \qquad (2.45)$$

equality (2.44) means conservation of total area of patches of the vorticity σ. Then the macroscopic vorticity is defined by

$$\omega(x,t) = \int \sigma p(x,\sigma,t) \ d\sigma \qquad (2.46)$$

which is associated with the stream function $\psi = \psi(x,t)$ and the velocity $u = u(x,t)$ through

$$\omega = -\Delta\psi, \quad \psi|_{\partial\Omega} = 0, \quad u = \nabla^\perp\psi \qquad (2.47)$$

To formulate equilibrium, we apply the principle of minimum entropy, seeking minimal state of

$$S(p) = -\iint p(x,\sigma) \log p(x,\sigma) \, dx d\sigma$$

under the constraint (2.43), (2.44), and

$$E = \frac{1}{2} \int_\Omega \omega\psi \, dx.$$

With the Lagrange multipliers $(\beta_p, c(\sigma), \zeta(x))$, it follows that

$$\delta S - \beta_p \delta E - \int c(\sigma) \delta M(\sigma) d\sigma - \int_\Omega \zeta(x) \left[\delta \int p \, d\sigma \right] dx = 0,$$

which is reduced to

$$p(x,\sigma) = e^{-c(\sigma)-(\zeta(x)+1)-\beta_p \sigma\psi}. \tag{2.48}$$

Here, β_p and $c(\sigma)$ may be called inverse temperature and chemical potential, respectively. We put $c(0) = 0$ because of the degree of freedom of $c(\sigma)$ admitted by (2.45). Then it follows that

$$p(x,\sigma) = p(x,0)e^{-c(\sigma)-\beta_p \sigma\psi} \tag{2.49}$$

and hence (2.43) implies

$$p(x,\sigma) = \frac{e^{-c(\sigma)-\beta_p \sigma\psi}}{\int e^{-c(\sigma')-\beta_p \sigma'\psi} d\sigma'}. \tag{2.50}$$

From (2.44) and (2.49), similarly, it follows that

$$c(\sigma) = \log \left(\frac{\int_\Omega p(x,0)e^{-\beta_p \sigma\psi} dx}{\int_\Omega p(x,\sigma)dx} \right). \tag{2.51}$$

The equilibrium mean field equation of vorticity patch model is thus given by equations (2.46), (2.47), (2.50), and (2.51), which is reduced to

$$-\Delta\psi = \int \sigma M(\sigma) \frac{p(x,0)e^{-\beta_p \sigma\psi}}{\int_\Omega p(x,0)e^{-\beta_p \sigma\psi} dx} d\sigma, \quad \psi|_{\partial\Omega} = 0$$

$$\omega = \int_I \sigma p \, d\sigma = -\Delta\psi, \quad \int_\Omega p(x,\sigma) \, dx = M(\sigma). \tag{2.52}$$

One may use the principle of maximum entropy production to describe near from equilibrium dynamics. We apply the transport equation

$$p_t + \nabla \cdot pu = -\nabla \cdot J, \quad J \cdot \nu|_{\partial\Omega} = 0 \tag{2.53}$$

with the diffusion flux $J = J(x, \sigma, t)$ of $p = p(x, \sigma, t)$, where ν denotes the outer unit normal vector. We obtain the total patch area conservation for each σ

$$M_t = \frac{\partial}{\partial t} \int_\Omega p(x, \sigma, t) \, dx = 0$$

because $u \cdot \nu = 0$ on $\partial\Omega$ by (2.47). Equation (2.53) implies

$$\omega_t + \nabla \cdot (\omega u + J_\omega) = 0, \tag{2.54}$$

where $J_\omega = \int \sigma J(x, \sigma, t) d\sigma$ stands for the local mean vorticity flux. Since $J_\omega \cdot nu = 0$ on $\partial\Omega$, equation (2.54) implies conservation of circulation, $\Gamma = \int_\Omega \omega \, dx$. Furthermore, J_ω is associated with the detailed fluctuation of (ω, u) from (ω, u) by (2.12).

Here we ignore the diffusion energy $E_d = \frac{1}{2} \iint \frac{J^2}{p} \, d\sigma dx$ to take

$$E = \frac{1}{2} \int_\Omega \omega\psi \, dx$$

as the total energy of this system. Using maximum entropy production principle, we chose the flux J to maximize entropy production rate S under the constraint

$$\dot{E} = 0, \quad \int J \, d\sigma = 0, \quad \int \frac{J^2}{2p} \, d\sigma \le C(x, t) \tag{2.55}$$

where

$$S(p) = - \iint p(x, \sigma, t) \log p(x, \sigma, t) \, d\sigma dx.$$

Using Lagrange multipliers $(\beta_p, D, \zeta) = (\beta_p(t), D(x, t), \zeta(x, t))$, we obtain

$$\delta\dot{S} - \beta_p \delta\dot{E} - \int_\Omega D^{-1} \left[\delta \int \frac{J^2}{2p} \, d\sigma \right] dx - \int_\Omega \zeta \left[\delta \int J \, d\sigma \right] dx = 0. \tag{2.56}$$

Since

$$\dot{E} = \frac{dE}{dt} = \int_\Omega \psi\omega_t \, dx = \int_\Omega J_\omega \cdot \nabla\psi \, dx = \iint \sigma J \cdot \nabla\psi \, d\sigma dx$$

$$\dot{S} = \frac{dS}{dt} = - \iint p_t(\log p + 1) \, d\sigma dx = - \iint J \cdot \frac{\nabla p}{p} \, d\sigma dx$$

equation (2.56) is reduced to

$$J = -D(\nabla p + \beta_p \sigma p \nabla\psi + p\zeta). \tag{2.57}$$

From the constraint of (2.55), it follows that

$$0 = \int J \, d\sigma = - \int D(\nabla p + \beta_p \sigma p \nabla\psi + p\zeta) \, d\sigma = -D(\beta_p \omega \nabla\psi + \zeta)$$

and

$$0 = \iint \sigma J \cdot \nabla \psi \, d\sigma dx = \iint -\sigma D(\nabla p + \beta_p \sigma p \nabla \psi + p\zeta) \cdot \nabla \psi \, d\sigma dx$$

$$= \iint -\sigma D(\nabla p + \beta_p(\sigma p - p\omega)) \cdot \nabla \psi d\sigma dx$$

$$= -\int_\Omega D\nabla\omega \cdot \nabla \psi \, dx - \beta_p \int_\Omega D\left(\int \sigma^2 p \, d\sigma - \omega^2\right) |\nabla \psi|^2 \, dx$$

which implies

$$\zeta = -\beta_p \omega \nabla \psi \qquad (2.58)$$

and

$$\beta_p = -\frac{\int_\Omega D\nabla\omega \cdot \nabla \psi \, dx}{\int_\Omega D\left(\int \sigma^2 p \, d\sigma - \omega^2\right) |\nabla \psi|^2 \, dx}. \qquad (2.59)$$

Thus we end up with

$$p_t + \nabla \cdot \rho u = \nabla \cdot D(\nabla p + \beta_p(\sigma - \omega)p\nabla \psi)$$

$$\beta_p = -\frac{\int_\Omega D\nabla\omega \cdot \nabla \psi \, dx}{\int_\Omega D\left(\int \sigma^2 p \, d\sigma - \omega^2\right) |\nabla \psi|^2 \, dx}$$

$$D(\nabla p + \beta_p(\sigma - \omega)p\nabla \psi) \cdot \nu|_{\partial\Omega} = 0$$

$$\omega = \int_I \sigma p d\sigma = -\Delta\psi, \quad \psi|_{\partial\Omega} = 0, \quad u = \nabla^\perp \psi$$

by (2.53), (2.57), (2.58), and (2.59), where $D = D(x,t) > 0$.

2.4 Point Vortex Limit

Here we give a quantitative description of the limit process from vorticity patch model to point vortex model. First, we derive the equilibrium mean field equation of point vortices from that of vorticity patches. Then we derive relaxation equation for the point vortex model. Fundamental quantities of point vortex model are circulation $\alpha\tilde{\alpha}$, probability $\rho_{\tilde{\alpha}}(x,t)$, and number density $P(d\tilde{\alpha})$. Circulation of each vortex is set to be small to preserve total energy and total circulation in the mean field limit. In the vorticity patch model, on the other hand, vorticity σ and probability $p(x,\sigma,t)$ are the fundamental quantities.

Here we use the localization in order to transform vorticity patch to point vortex. Namely, we divide each patch into two patches with half area and the same vorticity. Then, we divide each patch into two patches with half area: one has doubled vorticity, and the other has null vorticity.

Under this procedure the number of nonzero patches is doubled, and their vorticities are also doubled. At the same time, the area of each patch becomes a quarter and the number of total patches is quadrupled, while the total circulation is preserved. First, we describe the detailed process for the stationary state of (2.35).

Let Ω be divided into many cells with uniform size Δ, and let each cell be composed of many patches. Let $N_k(x, \sigma)dxd\sigma$ be the number of patches in the cell after k-times of the above procedure, centered at x of which vorticity was originally σ, and let σ_k be the vorticity of these patches after k-times localization. We assume that the number of total vorticity patches in the cell,

$$N_k^c(\Delta) = \int N_k(x, \sigma) \, d\sigma,$$

is independent of x. Then the number of total patches in Ω, the total area of the patches, and the total circulation of the patches after k-times localization procedures, with original vorticity σ, are given by

$$N^k(\sigma)d\sigma = \int_\Omega N_k(x, \sigma) \, dx, \quad M_k(\sigma)d\sigma = |\Omega|\frac{N_k(\sigma)d\sigma}{\int N_k(\sigma')d\sigma'},$$

and

$$\gamma_k(\sigma)d\sigma = \sigma_k M_k(\sigma)d\sigma,$$

respectively.

We obtain

$$N_p = \iint N_0(x, \sigma) \, d\sigma dx,$$

recalling (2.35). Since

$$\sigma_k = 2^k\sigma, \tag{2.60}$$

it holds that

$$N_k(x, \sigma)dxd\sigma = (4^k - 2^k)N_0^c(\Delta)\delta_0(d\sigma) + 2^k N_0(x, \sigma) \, dxd\sigma. \tag{2.61}$$

From (2.61) the related probability

$$p_k(x, \sigma)dxd\sigma = \frac{N_k(x,)dxd\sigma}{N_k^c(\Delta)}$$

satisfies

$$p_k(x, \sigma)dxd\sigma = \frac{(4^k - 2^k)N_0^c(\Delta)\delta_0(d\sigma) + 2^k N_0(x, \sigma)dxd\sigma}{(4^k - 2^k)N_0^c(\Delta) + 2^k \int N_0(x, \sigma)d\sigma}$$

$$= \frac{(4^k - 2^k)N_0^c(\Delta)\delta_0(d\sigma) + 2^k N_0(x, \sigma)dxd\sigma}{4^k N_0^c(\Delta)}$$

and hence

$$\lim_{k \to \infty} p_k(x, \sigma) dx d\sigma = \delta_0(d\sigma). \tag{2.62}$$

We also have

$$M_k(\sigma) d\sigma = \int_\Omega p_k(x, \sigma) dx = \lim_{\Delta \to 0} \sum_{i=1}^{|\Omega|/\Delta} \frac{N_k(x_i, \sigma) dx d\sigma}{N_k^c(\Delta)} \cdot \Delta$$

which implies

$$M_k(\sigma) d\sigma = \frac{|\Omega|}{4^k N_p} \lim_{\Delta \to 0} \sum_{i=1}^{|\Omega|/\Delta} N_k(x_i, \sigma) d\sigma = \frac{|\Omega|}{4^k N_p} N_k(\sigma) d\sigma$$

$$= |\Omega| \left((1 - 2^{-k}) \delta_0(d\sigma) + 2^{-k} \frac{N_0(\sigma) d\sigma}{N_p} \right)$$

by $\dfrac{\Delta}{N_k^c(\Delta)} = \dfrac{|\Omega|}{4^k N_p}$ and (2.61). We have, therefore,

$$\lim_{k \to \infty} M_k(\sigma) d\sigma = |\Omega| \delta_0(d\sigma).$$

It holds also that

$$\gamma_k(\sigma) = \int_\Omega \sigma_k p_k(x, \sigma) dx = \int_\Omega \sigma p^0(x, \sigma) dx$$

$$= \sigma_k M_k(\sigma) d\sigma = \frac{\sigma |\Omega|}{N_p} N_0(\sigma) d\sigma$$

and

$$\omega_k(x) = \int \sigma_k p_k(x, \sigma) d\sigma = \int \sigma p_0(x, \sigma) d\sigma.$$

Fundamental quantities constituting of the mean field limit of point vortex model thus arise as $k \to \infty$.

To explore the relationship between the quantities in two models, we take regards to circulation of one patch, total circulation of patches with original vorticity σ, and local mean vorticity. Based on

$$\sigma_k \cdot \frac{|\Omega|}{4^k N_p} = \tilde{\alpha} \cdot \alpha, \quad k \gg 1$$

and (2.60), we reach the ansatz

$$\sigma |\Omega| = \tilde{\alpha}, \quad \frac{1}{2^k N_p} = \alpha, \quad 2^k N_p = N.$$

Similarly we use $\dfrac{\sigma|\Omega|}{N_p}N_0(\sigma)d\sigma = \tilde{\alpha}P(d\tilde{\alpha})$ to put

$$\frac{N_0(\sigma)d\sigma}{N_p} = \frac{M_0(\sigma)d\sigma}{|\Omega|} = P(d\tilde{\alpha})$$

by

$$\frac{\sigma|\Omega|}{N_p}N_0(\sigma)d\sigma = \sigma|\Omega| \cdot \frac{1}{2^k N_p} \cdot 2^k N_p \cdot \frac{N_0(\sigma)d\sigma}{N_p}$$

$$= \tilde{\alpha}\alpha N P(d\tilde{\alpha}) = \tilde{\alpha}P(d\tilde{\alpha}).$$

Finally, we use the identity on local mean vorticity

$$\int \sigma p_0(x,\sigma)d\sigma = \int \tilde{\alpha}\rho_{\tilde{\alpha}}(x)P(d\tilde{\alpha})$$

to assign

$$\frac{1}{|\Omega|}p_0(x,\sigma)d\sigma = \rho_{\tilde{\alpha}}(x)P(d\tilde{\alpha}),$$

regarding

$$\int \sigma p_0(x,\sigma)d\sigma = \int \sigma|\Omega| \cdot \frac{p_0(x,\sigma)}{|\Omega|}d\sigma = \int \tilde{\alpha}\rho_{\tilde{\alpha}}(x)P(d\tilde{\alpha}).$$

After k-times localization, the first equation in (2.52) takes the form

$$-\Delta\psi = \int \sigma_k M_k(\sigma)\frac{p_k(x,0)e^{-\beta_p\sigma_k\psi}}{\int_\Omega p_k(x,0)e^{-\beta_p\sigma_k\psi}}d\sigma$$

$$= \int \frac{\sigma|\Omega|}{N_p}N_0(\sigma)\frac{p_k(x,0)e^{-\beta_p 2^k\sigma\psi}}{\int_\Omega p_k(x,0)e^{-\beta_p 2^k\sigma\psi}dx}d\sigma$$

$$= \int \sigma|\Omega|\frac{p_k(x,0)e^{-\beta_p\frac{2^k}{|\Omega|}\sigma|\Omega|\psi}}{\int_\Omega p_k(x,0)e^{-\beta_p\frac{2^k}{|\Omega|}\sigma|\Omega|\psi}} \cdot \frac{N_0(\sigma)}{N_p}d\sigma. \qquad (2.63)$$

Then, the right-hand side on (2.63) is replaced by

$$\int \tilde{\alpha}\frac{p_k(x,0)e^{-\frac{\beta_N}{N}\tilde{\alpha}\psi}}{\int_\Omega p_k(x,0)e^{-\frac{\beta_N}{N}\tilde{\alpha}\psi}dx}P(d\tilde{\alpha})$$

for $\beta_N = 4^k\dfrac{N_p}{|\Omega|}\beta_p = N \cdot \dfrac{2^k\beta_p}{|\Omega|}$. Sending $k \to \infty$, we obtain the first equation of (2.34) with $\beta = \beta_N/N$ by (2.62). This means that the vorticity patch model is transformed by point vortex model applied to the mean field limit by taking the localization procedure.

2.5 Kinetic Model

Relaxation equation is derived to describe the kinetics of many self-interacting particles from quasi-equilibrium to equilibrium. We can derive also this equation for point vortex model from that for vorticity patch model. By (2.57), the value of the diffusion flux J for $\sigma = 0$ is

$$J(x, 0, t) = -D(x, t)(\nabla p(x, 0, t) + p(x, 0, t)\zeta(x, t)),$$

and hence

$$\zeta(x, t) = -\frac{D(x, t)^{-1} J(x, 0, t) + \nabla p(x, 0, t)}{p(x, 0, t)}.$$

Flux is thus given by

$$J(x, \sigma, t) = -D(x, t)\big(\nabla p(x, \sigma, t) + \beta_p(t)\sigma p(x, \sigma, t)\nabla\psi(x, t)$$

$$-p(x, \sigma, t)\frac{D(x, t)^{-1} J(x, 0, t) + \nabla p(x, 0, t)}{p(x, 0, t)}\big).$$

We reach

$$p_t + \nabla \cdot pu = \nabla \cdot D\left(\nabla p + \beta_p \sigma p \nabla\psi - p\left[\frac{D^{-1}J + \nabla p}{p}\right]_{\sigma=0}\right)$$

with

$$\beta_p = \beta_p(t) = -\frac{\displaystyle\int_\Omega D\nabla\omega \cdot \nabla\psi - D\omega\left[\frac{D^{-1}J + \nabla p}{p}\right]_{\sigma=0} \cdot \nabla\psi\, dx}{\displaystyle\iint D\sigma^2 p|\nabla\psi|^2\, d\sigma dx}.$$

Therefore, after k-times localization procedure, it holds that

$$(\sigma_k p_k)_t + \nabla \cdot \sigma_k p_k u = \nabla \cdot D\big(\nabla(\sigma_k p_k) + \beta_p(\sigma_k)^2 p_k \nabla\psi$$

$$-\sigma_k p_k\left[\frac{D^{-1}J_k + \nabla p_k}{p_k}\right]_{\sigma=0}\big).$$

Putting $\beta_N = 4^k \dfrac{N_p}{|\Omega|}\beta_p$, similarly, we obtain

$$(\tilde{\alpha}\rho_{\tilde{\alpha}}P(d\tilde{\alpha}))_t + \nabla \cdot (\tilde{\alpha}\rho_{\tilde{\alpha}}P(d\tilde{\alpha})u)$$

$$= \nabla \cdot \big(D\nabla(\tilde{\alpha}\rho_{\tilde{\alpha}}P(d\tilde{\alpha})) + \beta\tilde{\alpha}^2\rho_{\tilde{\alpha}}P(d\tilde{\alpha})\nabla\psi)\big), \tag{2.64}$$

from

$$\lim_{k\to\infty} p_k(x, \sigma, t) = \delta_0(d\sigma), \quad \lim_{k\to\infty} J_k(x, 0, t) = 0$$

$$\sigma_k p_k(x, \sigma, t) = \sigma p_0(x, \sigma, t) = \sigma|\Omega| \cdot \frac{p_0(x, \sigma, t)}{|\Omega|} \approx \tilde{\alpha}\rho_{\tilde{\alpha}}(x, t)P(d\tilde{\alpha})$$

$$(\sigma_k)^2 p_k(x, \sigma, t) = 2^k\sigma \cdot \sigma p_0(x, \sigma, t) = \frac{2^k}{|\Omega|} \cdot (\sigma|\Omega|)^2 \cdot \frac{p_0(x, \sigma, t)}{|\Omega|}$$

$$\approx \frac{2^k}{|\Omega|}\tilde{\alpha}^2\rho_{\tilde{\alpha}}(x, t)P(d\tilde{\alpha}).$$

Here we assume

$$\lim_{k \to \infty} J_k(x, 0, t) = 0,$$

because $\iint J_k(x, \sigma, t) d\sigma = 0$ and the zero-vorticity patch becomes domi-
nant in the system. Then we obtain (2.37) by (2.64).

If $P(d\tilde{\alpha}) = \delta_1(d\tilde{\alpha})$, it holds that $\omega = \omega_2$ in (2.39). Then we obtain

$$\omega_t + \nabla \cdot \omega \nabla^\perp \psi = \nabla \cdot (\nabla \omega + \beta \omega \nabla \psi)$$

$$\frac{\partial \omega}{\partial \nu} + \beta \omega \frac{\partial \psi}{\partial \nu} \bigg|_{\partial \Omega} = 0, \quad \omega|_{t=0} = \omega_0(x) \geq 0$$

$$-\Delta \psi = \omega, \quad \psi|_{\partial \Omega} = 0, \quad \beta = -\frac{\int_\Omega \nabla \omega \cdot \nabla \psi \, dx}{\int_\Omega \omega |\nabla \psi|^2 \, dx} \qquad (2.65)$$

assuming $D = 1$. Conservations of total mass and energy,

$$\|\omega(\cdot, t)\|_1 = \lambda, \quad (\psi(\cdot, t), \omega(\cdot, t)) = e, \qquad (2.66)$$

are derived from (2.41), while increase of entropy in (2.42) is reduced to

$$\frac{d}{dt} \int_\Omega \Psi(\omega) \, dx = -\int_\Omega \omega |\nabla(\log \omega - \beta \psi)|^2 \, dx \leq 0 \qquad (2.67)$$

with $\Psi(s) = s(\log s - 1) + 1$.

In the stationary state, we obtain $\log \omega + \beta \psi = $ constant by (2.67).
Hence it follows that

$$-\Delta \psi = \omega, \quad \psi|_{\partial \Omega} = 0, \quad \omega = \frac{\lambda e^{-\beta \psi}}{\displaystyle\int_\Omega e^{-\beta \psi} \, dx}$$

$$-\beta = \frac{\displaystyle\int_\Omega \nabla \omega \cdot \nabla \psi \, dx}{\displaystyle\int_\Omega \omega |\nabla \psi|^2 \, dx}, \quad e = \int_\Omega \omega \psi \, dx \qquad (2.68)$$

from (2.66). Here, the third equation implies the fourth equation as

$$(\nabla \omega, \nabla \psi) = -\beta \int_\Omega \omega |\nabla \psi|^2 \, dx.$$

Using $v = \beta \psi$ and $\mu = \dfrac{\beta \lambda}{\int_\Omega e^{-\beta \psi} dx}$, therefore, system (2.68) is reduced to

$$-\Delta v = \mu e^{-v}, \quad v|_{\partial \Omega} = 0, \quad \frac{e}{\lambda^2} = \frac{\displaystyle\int_\Omega |\nabla v|^2 \, dx}{\left(\displaystyle\int_{\partial \Omega} -\frac{\partial v}{\partial \nu} \, dS\right)^2}. \qquad (2.69)$$

In fact, to see the third equality of (40), we note

$$e = (\omega, \psi) = \beta^{-1} \cdot \frac{\lambda \int_\Omega e^{-v} v \, dx}{\int_\Omega e^{-v} \, dx}$$

which implies

$$\mu = \frac{\lambda}{\int_\Omega e^{-v} \, dx} \cdot \frac{\lambda}{e} \cdot \frac{\int_\Omega e^{-v} v \, dx}{\int_\Omega e^{-v} \, dx}$$

$$= \frac{\lambda^2}{e} \cdot \frac{\int_\Omega e^{-v} v \, dx}{\left(\int_\Omega e^{-v}\right)^2}$$

and hence

$$\frac{e}{\lambda^2} = \frac{1}{\mu} \cdot \frac{\int_\Omega e^{-v} v \, dx}{\left(\int_\Omega e^{-v} \, dx\right)^2} = \frac{\|\nabla v\|_2^2}{\left(\int_{\partial\Omega} -\frac{\partial v}{\partial \nu} \, dS\right)^2}.$$

If $\mu < 0$, system (2.69), except for the third equation, is equivalent to the Gel'fand equation

$$-\Delta w = \sigma e^w, \qquad w|_{\partial\Omega} = 0$$

with $\sigma = -\mu$. If Ω is simply-connected, there is a non-compact family of solutions with $\mu \uparrow 0$ which are uniformly bounded near the boundary, and therefore,

$$\lim_{\mu \uparrow 0} \frac{e}{\lambda^2} = +\infty.$$

If $\mu \leq 0$, on the other hand, system (2.69) except for the third equation admits a unique solution, denoted by $v = v_\mu(x)$. By (2.66), and the standard theory of dynamical systems, it must hold that

$$\lim_{\mu \uparrow \infty} \frac{\|\nabla v_\mu\|_2^2}{\left(\int_{\partial\Omega} -\frac{\partial v_\mu}{\partial \nu} \, dS\right)^2} = 0 \qquad (2.70)$$

for the existence of global-in-time, compact orbit to (2.65), for any $\lambda > 0$ and $e > 0$ in (2.66). If $\Omega = B \equiv \{x \in \mathbf{R}^2 \mid |x| < 1\}$, we have $v = v(r)$, $r = |x|$, and the solution has an explicit form. Then the above property actually is examined.

2.6   Boltzmann-Poisson Equation — Classical Theory

Semilinear elliptic equations with exponential nonlinearity, particularly in
two space dimension are provided with a geometric structure, which is revealed by complex analysis. The Hamiltonian $H_\ell = H_\ell(x)$, $x = (x_1, \cdots, x_\ell)$
is observed also in this context. Through this recursive hierarchy, we see
how the shape of the domain Ω controls the global bifurcation of the solution set.

A typical example of such an equation is the Boltzmann-Poisson equation (2.27) where $\Omega \subset \mathbf{R}^2$ is a bounded domain with smooth boundary
$\partial\Omega$. The unknown $v = v(x)$ and the parameter λ are associated with the
mean field of stream function and negative inverse temperature. Below we
shall also use the other form, equivalent to (2.27) under the changing of the
parameter,

$$-\Delta v = \lambda e^v \quad \text{in } \Omega, \qquad v = 0 \quad \text{on } \partial\Omega \qquad (2.71)$$

called the Gel'fand equation.

Putting $u = v + \log\lambda$ in (2.71) yields

$$-\Delta u = e^u \quad \text{in } \Omega. \qquad (2.72)$$

We identify $x = (x_1, x_2) \in \Omega \subset \mathbf{R}^2$ with $z = x_1 + \imath x_2 \in \mathbf{C}$ to use the
complex conjugate $\bar{z} = x_1 - \imath x_2$ in (2.72). Then it follows that

$$u_{z\bar{z}} = -\frac{1}{4}e^u$$

and hence the complex-valued function

$$s = u_{zz} - \frac{1}{2}u_z^2 \qquad (2.73)$$

satisfies

$$s_{\bar{z}} = u_{zz\bar{z}} - u_z u_{z\bar{z}} = -\frac{1}{4}e^u u_z + \frac{1}{4}u_z e^u = 0. \qquad (2.74)$$

This (2.74) means that $s = s(z, \bar{z})$ define by (2.73) is a holomorphic function
of $z \in \Omega \subset \mathbf{C}$: $s = s(z)$.

Equation (2.73), on the other hand, is a Riccati equation concerning u_z,
and therefore, $\varphi = e^{-u/2}$ for fixed $\bar{z}$ is a solution to the linear equation

$$\varphi_{zz} + \frac{1}{2}s\varphi = 0. \qquad (2.75)$$

Given $x^* = (x_1^*, x_2^*) \in \Omega$, let $z^* = x_1^* + \imath x_2^*$, and $\{\varphi_1(z), \varphi_2(z)\}$ be a system
of fundamental solutions to (2.75) satisfying

$$\left(\varphi_1, \frac{\partial\varphi_1}{\partial z}\right)\bigg|_{z=z^*} = (1, 0), \qquad \left(\varphi_2, \frac{\partial\varphi_2}{\partial z}\right)\bigg|_{z=z^*} = (0, 1). \qquad (2.76)$$

These $\varphi_1(z)$ and $\varphi_2(z)$ are holomorphic functions of z if Ω is simply-connected, and are analytic functions for the other case. In both cases, the function $\varphi = e^{-u/2}$ has a local representation of $\overline{z}$ in the form of

$$\varphi = e^{-u/2} = \overline{f_1}(\overline{z})\varphi_1(z) + \overline{f_2}(\overline{z})\varphi_2(z) \qquad (2.77)$$

through the functions $\overline{f_1}$ and $\overline{f_2}$ in $\overline{z}$.

These $\overline{f_1}(\overline{z})$ and $\overline{f_2}(\overline{z})$ are independent of z, and determined through the Wronskian

$$W(g, h) = gh_z - g_z h$$

concerning z. In fact, we have

$$W(\varphi_1, \varphi_2) = \varphi_1\varphi_{2z} - \varphi_{1z}\varphi_2 \equiv 1$$

by (2.76), and therefore,

$$\overline{f_1}(\overline{z}) = W(\varphi, \varphi_2) = \varphi\varphi_{2z} - \varphi_z\varphi_2$$
$$\overline{f_2}(\overline{z}) = W(\varphi_1, \varphi) = \varphi_1\varphi_z - \varphi_{1z}\varphi \qquad (2.78)$$

by (2.77) and (2.75), regarding $\varphi = \varphi(z, \overline{z})$. Since the left-hand side on (2.78) is independent of z, we obtain

$$\overline{f_1}(\overline{z}) = \varphi(z^*, \overline{z}), \quad \overline{f_2}(\overline{z}) = \varphi_z(z^*, \overline{z}), \qquad (2.79)$$

putting $z = z^*$.

Now we put the complex conjugate of (2.75). Since φ is real-valued, the functions $\overline{s}(\overline{z}) = \overline{s(z)}$ of $\overline{z}$ satisfies

$$\varphi_{\overline{z}\overline{z}} + \frac{1}{2}\overline{s}\varphi = 0. \qquad (2.80)$$

This time we fix z and regard the right-hand sides on (2.79) as the functions in $\overline{z}$. By (2.78), these functions are linear combinations of φ and φ_z, both of which satisfy (2.80) as the functions of $\overline{z}$. Consequently, these $\overline{f_1}(\overline{z})$ and $\overline{f_2}(\overline{z})$ are the solutions to (2.80). The same argument guarantees that $\{\overline{\varphi}_1, \overline{\varphi}_2\}$, made by $\overline{\varphi}_1(\overline{z}) = \overline{\varphi_1(z)}$ and $\overline{\varphi}_2(\overline{z}) = \overline{\varphi_2(z)}$, forms a fundamental system of the solutions to (2.80) satisfying

$$\left(\overline{\varphi}_1, \frac{\partial\overline{\varphi}_1}{\partial\overline{z}}\right)\Big|_{\overline{z}=\overline{z}^*} = (1, 0), \quad \left(\overline{\varphi}_2, \frac{\partial\overline{\varphi}_2}{\partial\overline{z}}\right)\Big|_{\overline{z}=\overline{z}^*} = (0, 1).$$

Therefore, $\overline{f_1}(\overline{z})$ and $\overline{f_2}(\overline{z})$ are some linear combinations of $\overline{\varphi}_1(\overline{z})$ and $\overline{\varphi}_2(\overline{z})$ as the functions of $\overline{z}$.

To provide their simple explicit forms, let $x^* = (x_1^*, x_2^*) \in \Omega$ be a critical, say a maximum, point of $u = u(x)$. Then, from $\nabla u(x^*) = 0$ it follows that

$$\overline{f_1}(\overline{z}^*) = \varphi(z^*, \overline{z}^*) = e^{-u/2}\Big|_{x=x^*}$$

$$\frac{\partial \overline{f_1}}{\partial \overline{z}}(\overline{z}^*) = \varphi_{\overline{z}}(z^*, \overline{z}^*) = \frac{\partial}{\partial \overline{z}} e^{-u/2}\Big|_{x=x^*} = 0$$

$$\overline{f_2}(\overline{z}^*) = \varphi_z(z^*, \overline{z}^*) = \frac{\partial}{\partial z} e^{-u/2}\Big|_{x=x^*} = 0$$

$$\frac{\partial \overline{f_1}}{\partial \overline{z}}(\overline{z}^*) = \varphi_{z\overline{z}}(z^*, \overline{z}^*) = -\frac{1}{4}\Delta e^{-u/2}\Big|_{x=x^*} = -\frac{1}{8} e^{-u/2}\Delta u\Big|_{x=x^*}$$

$$= \frac{1}{8} e^{u/2}\Big|_{x=x^*},$$

which implies

$$\overline{f_1}(\overline{z}) = c\overline{\varphi_1}(\overline{z}), \quad \overline{f_2}(\overline{z}) = \frac{c^{-1}}{8}\overline{\varphi_2}(\overline{z})$$

for $c = e^{-u/2}\big|_{x=x^*}$.

Taking complex conjugate, we obtain

$$f_1 = c\varphi_1, \quad f_2 = \frac{c^{-1}}{8}\varphi_2 \tag{2.81}$$

and then (2.77) is reduced to

$$e^{-u/2} = c|\varphi_1|^2 + \frac{c^{-1}}{8}|\varphi_2|^2 \tag{2.82}$$

by (2.81). Letting $\psi_1 = c^{1/2}8^{1/4}\varphi_1$ and $\psi_2 = c^{-1/2}8^{-1/4}\varphi_2$ in (2.82), we obtain

$$\left(\frac{1}{8}\right)^{1/2} e^{u/2} = \left\{ c\left(\frac{1}{8}\right)^{-1/2}|\varphi_1|^2 + c^{-1}\left(\frac{1}{8}\right)^{1/2}|\varphi_2|^2 \right\}^{-1}$$

$$= \frac{1}{|\psi_1|^2 + |\psi_2|^2}. \tag{2.83}$$

Let $F = \psi_2/\psi_1$. Then, since

$$W(\psi_1, \psi_2) = W(\varphi_1, \varphi_2) = 1$$

it holds that

$$\frac{|F'|}{1 + |F|^2} = \frac{W(\psi_1, \psi_2)}{|\psi_1|^2 + |\psi_2|^2} = \left(\frac{1}{8}\right)^{1/2} e^{u/2} \tag{2.84}$$

by (2.83). This $F(z)$ is also an analytic function in Ω and is called the Liouville integral of $u(x)$ in (2.72).

We have $u = v + \log \lambda$ with $v|_{\partial\Omega} = 0$ in (2.84), and therefore, (2.71) is reduced to

$$\rho(F)|_{\partial\Omega} = \left(\frac{\lambda}{8}\right)^{1/2}$$

where

$$\rho(F) = \frac{|F'|}{1 + |F|^2}. \tag{2.85}$$

2.7 Geometric Structure

The left-hand side on (2.85) denoted by $\rho(F)$ stands for the spherical derivative of the analytic function $F = F(z)$. Let $S^2 \subset \mathbf{R}^3$ be the round sphere with south and north poles at $(0,0,0)$ and $(0,0,1)$, respectively, and let $\tau : S^2 \to \mathbf{C} \bigcup \{\infty\}$ be the stereographic projection. Then, under the conformal mapping $\overline{F} = \tau^{-1} \circ F : \Omega \to S^2$, the standard metrices on $\Omega \subset \mathbf{C} \cong \mathbf{R}^2$ and S^2 denoted by $ds^2 = dx_1^2 + dx_2^2$ and $d\Sigma^2$, respectively, are so related as $d\Sigma^2 = \rho(F)^2 ds^2$, or,

$$\frac{d\Sigma}{ds} = \rho(F).$$

The Gel'fand equation (2.71) is thus equivalent to finding a conformal mapping $\overline{F} : \Omega \to S^2$ satisfying

$$\frac{d\Sigma}{ds}\bigg|_{\partial\Omega} = \left(\frac{\lambda}{8}\right)^{1/2}. \tag{2.86}$$

This $\overline{F}$ is not unique, as $\rho(F)$ is invariant under the orthogonal transformation T on S^2. Thus it holds that $\rho(G) = \rho(F)$ for the analytic function $G = G(z)$ on Ω defined by $T \circ \overline{F} = \overline{G}$.

Let $\overline{F} : \Omega \to S^2$ be free from bifurcation points and $\omega \subset\subset \Omega$ be a subdomain. Then the immersed length of the image of $\partial\omega$ and the immersed area of the image of ω under $\overline{F}$, are given by

$$\ell_1 = \int_{\partial\omega} \rho(F) ds \tag{2.87}$$

and

$$m_1 = \int_\omega \rho(F)^2 dx, \tag{2.88}$$

respectively. If $\overline{F} : \Omega \to S^2$ is univalent on ω, it maps ω into a subdomain in S^2 denoted by $\tilde{\omega}$. If ω is simply-connected and $\tilde{\omega}$ is homeomorphic to a disc furthermore, there arises the isoperimetric inequality on S^2 between ℓ_1 and m_1, that is,

$$\ell_1^2 \geq 4m_1 \left(\pi - m_1\right). \tag{2.89}$$

Let $p = e^u$ and put

$$\ell(\partial\omega) = \int_{\partial\omega} p^{1/2} \, ds = \sqrt{8}\ell_1, \quad m(\omega) = \int_\omega p \, dx = 8m_1. \tag{2.90}$$

Then inequality (2.89) means

$$\ell(\partial\omega)^2 \geq \frac{1}{2} m(\omega) \left(8\pi - m(\omega)\right) \tag{2.91}$$

by (2.84)–(2.88), which is called Bol's inequality.

Inequality (2.91) is valid even if ω is not simply-connected nor $\overline{F} : \Omega \to S^2$ is univalent on ω. We recall that u is a solution to (2.72) and hence C^2 function $p = p(x) > 0$ satisfies

$$-\Delta \log p \leq p \quad \text{in } \Omega. \tag{2.92}$$

Theorem 2.1. *Let $p = p(x) > 0$ be a C^2 function in $\Omega \subset \mathbf{R}^2$ satisfying (2.92), and let $\omega \subset\subset \Omega$ be an open set. Define $\ell(\partial\omega)$ and $m(\omega)$ by (2.90). Then there holds that (2.91) under one of the following conditions:*

(1) Ω is simply-connected.
(2) Ω is C^1, p is C^1 on $\overline{\Omega}$, and is constant on $\partial\Omega$.

For the proof of the first part, we use Nehari's isoperimetric inequality indicated as follows.

Theorem 2.2. *Let $\Omega \subset \mathbf{R}^2$ be a simply-connected domain and $h = h(x)$ be a harmonic function on Ω. Then it holds that*

$$\left\{\int_{\partial\omega} e^{h/2} ds\right\}^2 \geq 4\pi \int_\omega e^h dx \tag{2.93}$$

where ω is a subdomain in Ω.

Proof. Since Ω is simply-connected, there is a conjugate harmonic function of $h = h(x)$ denoted by $g = g(x)$, $x = (x_1, x_2)$, such that $f = h + \imath g$ is a holomorphic function of $z = x_1 + \imath x_2 \in \Omega$. Then, taking a holomorphic function $\zeta = \zeta(z)$ on Ω such that $\zeta' = e^{f/2}$, we obtain $|\zeta'|^2 = e^h$.

This $\zeta(z)$ maps $\omega \subset\subset \Omega$ conformally to a domain $\hat{\omega}$ in a plane, possibly with self-crossing, and then the immersed length $\partial\hat{\omega}$ and the immersed area $\hat{\omega}$ are given by

$$\int_{\partial\omega} e^{h/2}ds = \int_{\partial\omega} |g'|\, ds \quad \text{and} \quad \int_{\omega} e^h dx = \int_{\omega} |g'|^2\, dx,$$

respectively. Dividing several subsets, we see that this $\hat{\omega}$ satisfies the standard isoperimetric inequality for the set in the plane, which means (2.93). □

Proof of Theorem 2.1. Deforming ω continuously, this theorem is reduced to the case that $\partial\omega$ is C^1. First, we assume the simply-connectedness of Ω, to define $h = h(x)$ by

$$\Delta h = 0 \quad \text{in } \omega, \qquad h = \log p \quad \text{on } \partial\omega. \tag{2.94}$$

Then $q = pe^{-h}$ satisfies

$$-\Delta \log q \le qe^h \quad \text{in } \omega, \qquad q = 1 \quad \text{on } \partial\omega \tag{2.95}$$

by (2.92).

Here we take the open set $\{q > t\} = \{x \in \omega \mid q(x) > t\}$ and define the non-increasing, right-continuous function

$$K(t) = \int_{\{q>t\}} qe^h dx, \quad \mu(t) = \int_{\{q>t\}} e^h dx \tag{2.96}$$

of $t \ge 1$. Then, the coarea formula implies

$$-K'(t) = \int_{\{q=t\}} \frac{qe^h}{|\nabla q|} ds = t\int_{\{q=t\}} \frac{e^h}{|\nabla q|} ds = -t\mu'(t) \quad \text{a.e. } t. \tag{2.97}$$

We have, on the other hand, by Sard's lemma and Green's formula, that

$$\int_{\{q>t\}} (-\Delta \log q)dx = \int_{\{q=t\}} \frac{|\nabla q|}{q} ds = \frac{1}{t} \int_{\{q=t\}} |\nabla q|\, ds \quad \text{a.e. } t > 1.$$

Then (2.95) implies

$$\frac{1}{t} \int_{\{q=t\}} |\nabla q|\, ds \le \int_{\{q>t\}} qe^h dx = K(t) \quad \text{a.e. } t > 1.$$

By these inequalities, together with Schwarz's inequality and Nehari's isoperimetric inequality we obtain

$$-K'(t)K(t) \ge \frac{1}{t} \int_{\{q=t\}} |\nabla q|\, ds \cdot t \int_{\{q=t\}} \frac{e^h}{|\nabla q|} ds$$

$$\ge \left\{ \int_{\{q=t\}} e^{h/2} ds \right\}^2$$

$$\ge 4\pi \int_{\{q>t\}} e^h dx = 4\pi\mu(t) \quad \text{a.e. } t > 1. \tag{2.98}$$

From (2.97) and (2.98) it follows that

$$\frac{d}{dt}\left\{\mu(t)t - K(t) + \frac{K(t)^2}{8\pi}\right\} = \mu(t) + \frac{1}{4\pi}K(t)K'(t) \le 0 \qquad (2.99)$$

for a.e. $t > 1$.

Here we have $K(t+0) = K(t) \le K(t-0)$ by the definition, and it holds also that

$$j(t) \equiv K(t) - \mu(t)t = \int_{\{q>t\}}(q-t)e^h dx \qquad (2.100)$$

is continuous as $j(t+0) = j(t) = j(t-0)$. Hence we obtain

$$\left[-j(t) + \frac{K(t)^2}{8\pi}\right]_{t=1}^{\infty} = j(1) - \frac{K(1)^2}{8\pi} \le 0 \qquad (2.101)$$

by (2.99). Here we have $K(1)^2 \le m(\omega)^2$ and

$$j(1) = \int_{\{q>1\}}(q-1)e^h dx \ge \int_{\omega}(q-1)e^h dx = m(\omega) - \int_{\omega}e^h dx$$

in (2.101), and therefore,

$$m(\omega) - \frac{m(\omega)^2}{8\pi} \le \int_{\omega}e^h dx \le \frac{1}{4\pi}\left(\int_{\partial\omega}e^{h/2}ds\right)^2 = \frac{\ell(\partial\omega)^2}{4\pi}$$

which means (2.91).

In the second case of Ω, we take a simply-connected domain $\hat{\Omega}$ including Ω. Next, we extend $p = p(x)$ outside Ω using the constant on $\partial\Omega$. Then, it holds that

$$-\Delta \log p \le p \quad \text{in } \hat{\Omega} \qquad (2.102)$$

in the sense of distributions. Then the above argument is valid in $\hat{\Omega}$, and we obtain (2.91) for any open set $\omega \subset\subset \Omega$. $\qquad\square$

Isoperimetric inequality for the first eigenvalue $\nu_1(p,\Omega)$ of

$$-\Delta\varphi = \nu p\varphi \quad \text{in } \Omega, \qquad \varphi = 0 \quad \text{on } \partial\Omega \qquad (2.103)$$

emerges from Theorem 2.1. First, the mini-max principle guarantees

$$\nu_1(p,\Omega) = \inf\left\{\int_{\Omega}|\nabla v|^2 dx \mid v \in H_0^1(\Omega), \int_{\Omega}v^2 p\, dx = 1\right\}. \qquad (2.104)$$

Here, the total area of the surface (Ω, pds^2) is given by

$$\sigma = \int_{\Omega}p.$$

We compare $\nu_1(p,\Omega)$ in (2.104) to the case of equality in (2.92) with radially symmetric and the same total area of p. Thus we take $p* = p^*(|x|)$ satisfying

$$-\Delta \log p^* = p^* \quad \text{in } \Omega^*, \qquad \int_{\Omega^*} p^* dx = \sigma \qquad (2.105)$$

for $\Omega^* = B(0,1) \equiv \{x \in \mathbf{R}^2 \mid |x| < 1\}$, and set the eigenvalue problem

$$-\Delta \varphi = \nu p^* \varphi \quad \text{in } \Omega^*, \qquad \varphi = 0 \quad \text{on } \partial\Omega^*. \qquad (2.106)$$

The first eigenvalue of (2.106) is given by

$$\nu_1(p^*, \Omega^*) = \inf \left\{ \int_{\Omega^*} |\nabla v|^2 \, dx \mid v \in H_0^1(\Omega^*), \int_\Omega v^2 p^* \, dx = 1 \right\}. \qquad (2.107)$$

As we shall show in §2.9 below, each $\sigma \in (0, 8\pi)$ admits a unique $p^* = p^*(|x|)$ satisfying (2.105).

The reference domain Ω^* need not to be the unit disc. If $\Omega^* = B(0, R)$ for $R > 0$, then any $\sigma \in (0, 8\pi)$ admits a unique $p^* = p^*(|x|)$ satisfying (2.106). Then $\nu_1(p^*, \Omega^*)$ defined by (2.107) does not depend on $R > 0$.

Theorem 2.3. *For $p \in C(\overline{\Omega})$ in Theorem 2.1 it holds that*

$$\sigma = \int_\Omega p \, dx < 8\pi \quad \Rightarrow \quad \nu_1(p, \Omega) \geq \nu_1(p^*, \Omega^*) \qquad (2.108)$$

The equality in (2.108) holds if and only if Ω is a disc and $p = p(|x|)$.

The proof of Theorem 2.3 relies on the theory of rearrangement. In fact, since the first eigenfunction $\varphi = \varphi(x)$ of (2.104) may be positive definite in Ω, we take a Schwarz-like symmetrization of φ, which is used for the proof of the Faber-Krahn inequality concerning the case of $p \equiv 1$. In this monograph, we call this process the Bandle symmetrization.

Here, we apply Bol's inequality to each level set of $\varphi(x)$ to compare the Rayleigh quotients for $\nu_1(p, \Omega)$ and $\nu_1(p^*, \Omega^*)$. At this process we note that if $\omega \subset\subset \Omega$ is a concentric ball of Ω for $(\Omega^*, p^*) = (\Omega, p)$ in (2.105), the equality arises in (2.90)–(2.91), which is the only case for this equality.

More precisely, since $\varphi > 0$ in Ω and $\varphi = 0$ on $\partial\Omega$, for each $t > 0$ the boundary of the open set $\Omega_t = \{x \in \Omega \mid \varphi(x) > t\}$ is included inside Ω as

$$\partial\Omega_t \subset\subset \Omega, \quad t > 0. \qquad (2.109)$$

Then we take a concentric disc $\Omega_t^* \subset\subset \Omega^*$ of Ω^* such that

$$\int_{\Omega_t^*} p^* dx = \int_{\Omega_t} p \, dx$$

and define the Bandle symmetrization of $\varphi(x)$ by

$$\varphi^*(x) = \sup\{t \mid x \in \Omega_t^*\}, \quad x \in \Omega^*.$$

This φ^* satisfies

$$\varphi^* = \varphi^*(|x|), \ \varphi^* > 0 \quad \text{in } \Omega^*, \qquad \varphi^* = 0 \text{ on } \partial\Omega^*.$$

The coarea formula ensures that the transformation $\varphi \mapsto \varphi^*$ is equi-measurable from $(\Omega, p\, ds^2)$ to $(\Omega^*, p^* ds^2)$. In particular, it holds that

$$\int_\Omega \varphi^2 p \, dx = \int_{\Omega^*} \varphi^{*2} p^* dx. \tag{2.110}$$

For the proof of the decreasing of the Dirichlet norm,

$$\int_\Omega |\nabla\varphi|^2 \, dx \geq \int_{\Omega^*} |\nabla\varphi^*|^2 \, dx, \tag{2.111}$$

we decompose the integral on the left-hand side into the line integral on $\partial\Omega_t$ and $\int_0^\infty \cdot dt$ by the coarea formula. Then we apply Bol's inequality, regarding (2.109). Then we apply the above argument using Sard's lemma and the differentiation and integration of the monotone function in t.

2.8   sup + inf Inequality

Sharp forms of geometric isoperimetric inequalities depend on the topology of the domain. Here we use the simplest form of Alexandroff's inequality, a generalization of Bol's inequality, for which a geometric proof is known.

Lemma 2.1. *Let $B = B(0,1) \subset \mathbf{R}^2$ be the unit disc and $V(x) > 0$ be a continuous function in Ω. Assume that $v = v(x)$ satisfies*

$$-\Delta v = V(x)e^v \quad \text{in } B$$

and put

$$\ell = \int_{\partial B} e^{v/2} \, ds, \quad m = \int_B e^v \, dx.$$

Then each $K_0 \in \mathbf{R}$ admits

$$\ell^2 \geq (2\alpha - K_0 m)m, \tag{2.112}$$

provided that $\alpha = 2\pi - \omega_{K_0}^+(B) > 0$, where

$$\omega_{K_0}^+(B) = \int_{K > K_0} (K(x) - K_0)e^v dx, \quad K(x) = V(x)/2.$$

Let $p = e^v$, $0 < p = p(x) \in C^2(B) \cap C(\overline{B})$, in Lemma 2.1. We introduce the metric $d\sigma^2 = p(x)ds^2$ to $B = B(0,1) \subset \mathbf{R}^2$. Then the Gauss curvature K, the total area m, and the length ℓ of the boundary are given by

$$K = -\frac{\Delta \log p}{2p}, \quad m = \int_B p \, dx, \quad \ell = \int_{\partial B} p^{1/2} \, ds. \tag{2.113}$$

Then this lemma guarantees for

$$m_\mu^+(B) = \int_{\{K > \mu\}} (K(x) - \mu) \, p \, dx, \quad \mu \in \mathbf{R}$$

that

$$\alpha = 2\pi - m_\mu^+(B) > 0 \quad \Rightarrow \quad \ell^2 \geq (2\alpha - \mu m)m. \tag{2.114}$$

If $0 < p = p(x) \in C^2(B) \cap C(\overline{B})$ satisfies (2.92), it holds that $K = K(x) \leq 1/2$, and therefore, $m_\mu^+(B) = 0$ for $\mu = 1/2$. Thus, (2.114) implies Bol's inequality,

$$\ell^2 \geq \frac{1}{2}m(8\pi - m). \tag{2.115}$$

Using (2.114), we can make a sharper form of inequality (2.115) under the condition (2.92). For this purpose, we take μ_* in

$$\int_{\{K > \mu_*\}} p \, dx \leq m/2, \quad \int_{\{K < \mu_*\}} p \, dx \leq m/2.$$

Since $\{K > \mu_*\} \neq B$ and $\{K < \mu_*\} \neq B$ we have

$$\{K \leq \mu_*\} \neq \emptyset, \quad \{K \geq \mu_*\} \neq \emptyset.$$

It holds also that

$$\frac{a}{2} \leq K(x) \leq \frac{b}{2} \quad \Rightarrow \quad \frac{a}{2} \leq \mu_* \leq \frac{b}{2} \tag{2.116}$$

where $a, b > 0$ are constants.

We define $C, D \geq 0$ by

$$\int_{\{K > \mu_*\}} p \, dx = \frac{m}{2} - C, \quad \int_{\{K < \mu_*\}} p \, dx = \frac{m}{2} - D.$$

Then

$$\alpha \equiv 2\pi - m_{\mu_*}^+(B) = 2\pi - \int_{\{K > \mu_*\}} (K - \mu_*) p \, dx$$

satisfies

$$2\alpha - \mu_* m = 4\pi - 2\int_{\{K > \mu_*\}} (K - \mu_*)p \, dx - \mu_* m$$

$$= 4\pi - 2\int_{\{K \geq \mu_*\}} (K - \mu_*)p \, dx - \mu_* m$$

$$= 4\pi - 2\int_{\{K \geq \mu_*\}} Kp \, dx + 2\mu_* D. \tag{2.117}$$

Since

$$\int_{\{K < \mu_*\}} Kp \, dx \geq \frac{a}{2}\int_{\{K < \mu_*\}} p \, dx = \frac{a}{2}\left(\frac{m}{2} - D\right)$$

$$\geq \frac{a}{2b}\int_B Kp \, dx - \frac{a}{2}D,$$

on the other hand, it holds that

$$\int_{\{K \geq \mu_*\}} Kp \, dx \leq \left(1 - \frac{a}{2b}\right)\int_B Kp \, dx + \frac{a}{2}D. \tag{2.118}$$

By (2.117) and (2.118) we obtain

$$2\alpha - \mu_* m \geq 4\pi - 2\left(1 - \frac{a}{2b}\right)\int_B Kp \, dx + (2\mu_* - a)D$$

$$\geq 4\pi - 2\left(1 - \frac{a}{2b}\right)\int_B Kp \, dx. \tag{2.119}$$

Here we put

$$4\pi < \alpha_0 < \frac{4\pi}{1 - \frac{a}{2b}}, \quad \gamma_0 = 4\pi - \left(1 - \frac{a}{2b}\right)\alpha_0 > 0.$$

Then inequality (2.119) guarantees

$$\int_B Kp \, dx \leq \alpha_0/2 \quad \Rightarrow \quad 2\alpha - \mu_* m \geq \gamma_0. \tag{2.120}$$

We obtain the following lemma by this inequality (2.120).

Lemma 2.2. *Let $B = B(0,1) \subset \mathbf{R}^2$ be the unit disc, and $a, b > 0$ be constants. Then there are $\alpha_0 > 4\pi$ and $C_1 > 0$ such that*

$$-\Delta v = V(x)e^v, \ a \leq V(x) \leq b \quad in \ B, \quad \int_B V(x)e^v \leq \alpha_0 \tag{2.121}$$

implies $v(0) \leq C_1$.

Proof. We assume (2.121), and apply (2.120) for $p = e^v$ and $K = V/2$. In fact, the function

$$f(r) = 4\pi - 2 \int_{\{K > \mu_*\} \cap B_r} (K - \mu_*) p \, dx - \mu_* \int_{B_r} p \, dx$$

is non-increasing in $r \in [0, 1)$, and it holds that

$$f(r) \geq f(1) = 4\pi - 2 \int_{\{K > \mu_*\}} (K - \mu_*) p \, dx - \mu_* \int_B p \, dx$$

$$= 2\alpha - \mu_* m \geq \gamma_0. \tag{2.122}$$

Let

$$A(r) = \int_0^r dr' \int_{\partial B(0, r')} p \, ds = \int_{B(0, r)} p \, dx.$$

From (2.114) for each $0 \leq r < 1$ it follows that

$$A'(r) = \int_{\partial B(0, r)} p \, ds \geq \frac{1}{2\pi r} \left\{ \int_{\partial B(0, r)} p^{1/2} ds \right\}^2 \geq \frac{1}{2\pi r} f(r) A(r). \tag{2.123}$$

Then (2.123) implies, for each $0 < r_0 < 1$, that

$$-\frac{1}{2\pi} \log r_0 \leq \int_{r_0}^1 \frac{A'(r)}{f(r) A(r)} \, dr = \int_{r_0}^1 \frac{1}{f(r)} \{\log A(r)\}' \, dr$$

$$= \frac{\log A(1)}{f(1)} - \frac{\log A(r_0)}{f(r_0)} + \int_{r_0}^1 \frac{\log A(r)}{f(r)^2} f'(r) \, dr. \tag{2.124}$$

Here we have

$$A(r_0) = \pi r_0^2 p(0)(1 + o(1)), \quad f(r_0) = 4\pi - O\left(r_0^2\right), \qquad r_0 \downarrow 0$$

and hence

$$\lim_{r_0 \downarrow 0} \left\{ \frac{\log A(r_0)}{f(r_0)} - \frac{1}{2\pi} \log r_0 \right\} = \lim_{r_0 \downarrow 0} \frac{1}{4\pi} \log \frac{A(r_0)}{r_0^2}$$

$$= \frac{1}{4\pi} \left(\log p(0) + \log \pi \right). \tag{2.125}$$

By (2.124)–(2.125) it holds that

$$v(0) = \log p(0) \leq \frac{4\pi \log A(1)}{f(1)} - \log \pi$$

$$+ 4\pi \int_0^1 \frac{\log A(r)}{f(r)^2} f'(r) \, dr. \tag{2.126}$$

For the third term of the right-hand side on (2.126) we use

$$A(r) \leq A(1) = \int_B p \, dx \leq \frac{1}{a} \int_B V p \, dx \leq \frac{\alpha_0}{a}$$

derived from (2.121). Then (2.122) implies

$$\frac{\log A(r)}{f(r)^2} f'(r) = \frac{\log \frac{\alpha_0}{A(r)} + \log \frac{1}{\alpha_0}}{f(r)^2}(-f'(r))$$

$$\leq \gamma_0^{-2} \log \frac{\alpha_0}{A(r)} \cdot (-f'(r)). \qquad (2.127)$$

Here we have

$$0 \leq -f'(r) = 2 \int_{\{K > \mu_*\} \cap \partial B_r} (K - \mu_*) p \, ds + \mu_* \int_{\partial B_r} p \, ds$$

$$\leq \left(\frac{3}{2}b - a\right) \int_{\partial B_r} p \, ds = \left(\frac{3}{2}b - a\right) A'(r) \qquad (2.128)$$

by (2.116). Writing $t = A(r)/\alpha_0$ in (2.127)–(2.128), we obtain

$$\int_0^1 \frac{\log A(r)}{f(r)^2} \cdot f'(r) \, dr \leq \gamma_0^{-2} \left(\frac{3}{2}b - a\right) \cdot \alpha_0 \cdot \int_0^{1/a} \log \frac{1}{t} \, dt < +\infty$$

and the result follows.   □

We emphasize that $\alpha_0 > 4\pi$ in (2.120) is assured by the estimate of $V(x) > 0$ below. Without this condition, we have a similar conclusion but with $\alpha_0 < 4\pi$, which is not sufficient to guarantee residual vanishing for blowup family of the solutions. This result is concerned on the case that $V(x)$ is restricted to a compact set in $C(\overline{\Omega})$, where α_0 can be arbitrary close to 8π. For the proof of this property of quantization we apply the following theorem for the rescaled solution, which actually follows from Lemma 2.2.

Theorem 2.4. *Let $\Omega \subset \mathbf{R}^2$ be a bounded domain, $V = V(x) \in C(\overline{\Omega})$, and $K \subset \Omega$ be a compact set. Then, given $a, b > 0$, we have $c_1 = c_1(a, b) \geq 1$ and $c_2 = c_2(a, b, \mathrm{dist}\,(K, \partial\Omega)) > 0$ such that*

$$-\Delta v = V(x)e^v, \ a \leq V(x) \leq b \quad in \ \Omega \qquad (2.129)$$

implies

$$\sup_K v + c_1 \inf_\Omega v \leq c_2. \qquad (2.130)$$

Proof. By the covering argument, this theorem is reduced to the case of $\Omega = B(x_0, r)$, $K = \{x_0\}$, and $c_2 = c_2(a, b, r)$. We assign $x_0 = 0$ and take

$$\tilde{v}(x) = v(rx) + 2\log r,$$

which satisfies (2.129) for $\Omega = B(0,1)$ and $\tilde{V}(x) = V(rx)$. Thus it suffices to take the case that $\Omega = B = B(0,1)$ and $K = \{0\}$. We shall show

$$v(0) + \frac{c_1}{2\pi} \int_{\partial B} v \, ds \leq c_2 \qquad (2.131)$$

which implies (2.130) in this case. Namely, below we show (2.131) if (2.129) holds for $\Omega = B$.

Noting $\alpha_0 > 4\pi$ in Lemma 2.2, we take $c_1 \geq 1$ in $4\pi(c_1 + 1)/c_1 = \alpha_0$. In the case of

$$\int_B V e^v \, dx > 4\pi(c_1 + 1)/c_1 \qquad (2.132)$$

we define $r_0 \in (0,1)$ by $\int_{B(0,r_0)} V e^v dx = 4\pi(c_1 + 1)/c_1$, and $r_0 = 1$ if (2.132) does not hold. Thus we have $0 < r_0 \leq 1$ satisfying

$$\int_{B(0,r_0)} V e^v dx \leq 4\pi(c_1 + 1)/c_1. \qquad (2.133)$$

Then, we have (2.129) for $\tilde{u}(x) = u(r_0 x) + 2\log r_0$, $\tilde{V}(x) = V(r_0 x)$, and $\Omega = B = B(0,1)$, together with

$$\int_B \tilde{V} e^{\tilde{v}} dx \leq 4\pi(c_1 + 1)/c_1 = \alpha_0.$$

Hence Lemma 2.2 implies

$$\tilde{v}(0) = v(0) + 2\log r_0 \leq c_0. \qquad (2.134)$$

Let

$$G(r) = v(0) + \frac{c_1}{2\pi r} \int_{\partial B(0,r)} v \, ds + 2(c_1 + 1)\log r$$

$$= v(0) + c_1 \int_0^{2\pi} v(re^{i\theta}) \, d\theta + 2(c_1 + 1)\log r.$$

Since v is super-harmonic, we have

$$\frac{1}{2\pi r} \int_{\partial B(0,r)} v \, ds \leq v(0),$$

which implies

$$G(r_0) \leq (c_1 + 1)v(0) + 2(c_1 + 1)\log r_0 \leq (c_1 + 1)c_0 \qquad (2.135)$$

by (2.134). If $r_0 = 1$ in (2.135) then it follows that (2.131) from

$$G(1) = v(0) + \frac{c_1}{2\pi} \int_{\partial B} v \, ds.$$

If $r_0 < 1$ we have (2.133), and hence

$$\int_{B(0,r)} V e^v dx \geq 4\pi (c_1 + 1)/c_1, \quad r_0 \leq r \leq 1. \qquad (2.136)$$

Here we obtain

$$G'(r) = c_1 \int_0^{2\pi} v_r(r, \theta) \, d\theta + \frac{2(c_1 + 1)}{r}$$

$$= \frac{1}{r} \left\{ \frac{c_1}{2\pi} \int_{\partial B(0,r)} v_r \, ds + 2(c_1 + 1) \right\}$$

and

$$\int_{\partial B(0,r)} v_r \, ds = \int_{B(0,r)} \Delta v \, dx = - \int_{B(0,r)} V e^v \, dx,$$

and therefore,

$$G'(r) \leq 0, \quad r_0 \leq r \leq 1$$

by (2.136). Now, (2.135) implies

$$v(0) + \frac{1}{2\pi} \int_{\partial B} v \, ds = G(1) \leq G(r_0) \leq (c_1 + 1)c_0$$

and hence (2.131). $\qquad \qquad \qquad \square$

2.9   Radial Solutions

If $f = f(v)$ is locally Lipschitz continuous and $\Omega = B(0, R) \subset \mathbf{R}^n$ is a ball, any positive solution $v = v(x)$ to

$$-\Delta v = f(v), \ v > 0 \quad \text{in } \Omega, \qquad v = 0 \quad \text{on } \partial\Omega \qquad (2.137)$$

is radially symmetric. Since $\lambda e^v > 0$ in the Gel'fand problem (2.71), we obtain $v > 0$ in Ω by the maximum principle. In particular, any solution is radially symmetric if Ω is a ball. Here we continue to take the case $n = 2$, and classify the solution to (2.71) for $\Omega = B(0, 1)$:

$$-\Delta v = \lambda e^v \quad \text{in } B = B(0, 1), \qquad v = 0 \quad \text{on } \partial B.$$

Since $v = v(r)$ for $r = |x|$, the equation in (2.71) is reduced to

$$v'' + \frac{1}{r} v' + \lambda e^v = 0, \quad r > 0, \qquad v'(0) = 0. \qquad (2.138)$$

Following (2.71), equation (2.138) is provided with the scaling invariance. Thus if $v_0 = v_0(r)$, then

$$v_\alpha(r) = v_0(e^{\alpha/2} r) + \alpha \qquad (2.139)$$

defined for $\alpha \in \mathbf{R}$ is also a solution. We shall find a solution to (2.138) denoted by $v_0(r)$, and adjust $\alpha \in \mathbf{R}$ such that $v = v_\alpha(r)$ in (2.139) satisfies the boundary condition in (2.71), $v_\alpha(1) = 0$, that is,

$$v_0(e^{\alpha/2}) + \alpha = 0. \tag{2.140}$$

For this purpose we use $s = \log r$ to write (2.138) as

$$\frac{d^2}{ds^2}(v + 2s) + \lambda e^{v+2s} = 0. \tag{2.141}$$

Thus, $u = v + 2s + \log \lambda$ in (2.141) satisfies (2.72) in one space dimension,

$$u'' + e^u = 0, \quad -\infty < s < \infty. \tag{2.142}$$

Since (2.142) implies

$$\left\{ u'' - \frac{1}{2}(u')^2 \right\}' = 0,$$

we seek $u = u(s)$ satisfying

$$u'' - \frac{1}{2}(u')^2 = -2. \tag{2.143}$$

We note that this (2.143) casts the one-dimensional Liouville's integral.

Here, $\ell = \dfrac{2 - u'(s/2)}{4}$ satisfies the logistic equation

$$\ell' = (1 - \ell)\ell \tag{2.144}$$

and we take a solution of this (2.144) in the form of

$$\ell(s) = \frac{1}{2}\left(1 + \tanh \frac{s}{2}\right).$$

Then a solution to (2.138) is obtained as

$$v_0 = -2\log \cosh s + \log \frac{2}{\lambda} - 2s = \log\left\{ \frac{8/\lambda}{(r^2 + 1)^2} \right\}.$$

Now, equality (2.140) means

$$\frac{8}{\lambda} = \frac{(e^\alpha + 1)^2}{e^\alpha},$$

and therefore, the solution to (2.71) on $\Omega = \Omega^* \equiv \left\{ x \in \mathbf{R}^2 \mid |x| < 1 \right\}$ is given by

$$v = v^*_{\lambda\pm}(x) = \log\left\{ \frac{8\beta_\pm/\lambda}{\left(1 + \beta_\pm |x|^2\right)^2} \right\}$$

$$\beta_\pm = \frac{4}{\lambda}\left\{ 1 - \frac{\lambda}{4} \pm \left(1 - \frac{\lambda}{2}\right)^{1/2} \right\}. \tag{2.145}$$

Regarding (2.145), we see that the solution is unique for $\lambda = 2$ and is given by

$$v_{\lambda+}^*(x) = v_{\lambda-}^*(x) = 2 \log \frac{2}{1 + |x|^2}.$$

Also, there are two solutions for $0 < \lambda < 2$, and no solution for $\lambda > 2$. The total set of solutions, denoted by $\mathcal{C}^* = \{(\lambda, v)\}$, is homeomorphic to $\mathbf{R}$ in the function space $\mathbf{R}_+ \times C(\overline{\Omega}^*)$, and in the limit $\lambda \downarrow 0$ there arises

$$\lim_{\lambda \downarrow 0} v_{\lambda-}^*(x) = 0 \qquad \text{uniformly on } \overline{\Omega}^*$$

$$\lim_{\lambda \downarrow 0} v_{\lambda+}^*(x) = 4 \log \frac{1}{|x|} \quad \text{locally uniformly in } x \in \overline{\Omega}^* \setminus \{0\}.$$

In other words, the endpoints of $\mathcal{C}^*$ are $(0,0)$ and $(0, v_*)$ with the latter outside the above function space $\mathbf{R}_+ \times C(\overline{\Omega}^*)$, where

$$v_*(x) = 4 \log \frac{1}{|x|}. \tag{2.146}$$

This $v_* = v_*(x)$ in (2.146) is called the singular limit of the solution, which is not a solution to (2.71) anymore.

Radial solutions in (2.145) take the form

$$\left(\frac{\lambda}{8}\right)^{1/2} e^{v/2} = \left(\frac{e^u}{8}\right)^{1/2} = \frac{\mu^{1/2}}{|x|^2 + \mu}$$

$$\mu = \mu_\pm = \frac{8}{\lambda}\left\{1 - \frac{\lambda}{4} \mp \sqrt{1 - \frac{\lambda}{2}}\right\} = \beta_\pm^{-1}, \tag{2.147}$$

and therefore, their Liouville integrals are given by $F(z) = C_\pm z$ for

$$C_\pm = \mu_\pm^{-1/2} = \left\{\frac{1}{\lambda}\left\{4 - \lambda \pm 2\sqrt{4 - 2\lambda}\right\}\right\}^{1/2}.$$

The immersed length of $\overline{F}(\partial\Omega^*)$ and the immersed area of $\overline{F}(\Omega^*)$ given by (2.87) and (2.88), respectively, are equal to

$$\ell_1(\partial\Omega^*) = \int_{\partial\Omega^*} \left(\frac{\lambda e^v}{8}\right)^{1/2} ds = 2\pi \left(\frac{\lambda}{8}\right)^{1/2}$$

$$m_1(\Omega^*) = \int_{\Omega^*} \frac{\lambda e^v}{8} \, dx = \frac{\pi}{1 + \mu_\pm} = \frac{\sigma}{8}$$

for $\sigma = \int_\Omega \lambda e^v dx$. Therefore, as the solution varies from $(0,0)$ to $(0, v_*)$ along $\mathcal{C}^*$, the value σ increases monotonically from 0 to 8π. The value $\lambda = 2$ on $\mathcal{C}^*$ corresponds to $\sigma = 4\pi$, while λ increases from 0 to 2 and then decreases from 2 to 0, monotonically.

2.10   Laplace-Beltrami Operators

For a solution to (2.71) denoted by $v = v(x)$, its linearized operator is a self-adjoint operator in $L^2(\Omega)$ given by

$$L_v = -\Delta - \lambda e^v, \quad D(L) = H_0^1(\Omega) \cap H^2(\Omega).$$

The form $\mathcal{C}^*$ of the total set of the solution on $\Omega = \Omega^*$ is associated with the linearized stability of $v_{-\lambda}^*$ for $0 < \lambda \leq 2$. Thus, the first eigenvalue of $L_{v_{\lambda-}^*}$ is positive for $0 < \lambda < 2$ and 0 at $\lambda = 2$. This property is equivalently stated using $\nu_1(p^*, \Omega^*)$ in Theorem 2.3 for $p^* = \lambda e^{v_{\lambda-}^*}$, that is, $\nu_1(p^*, \Omega^*) > 1$ for $0 < \lambda < 2$ and $\nu_1(p^*, \Omega^*) = 1$ for $\lambda = 2$.

This statement is again translated via (2.147) as

$$0 < \sigma < 4\pi \quad \Rightarrow \quad \nu_1(p^*, \Omega^*) > 1, \tag{2.148}$$

where

$$\sigma = \int_{\Omega^*} p^* dx, \quad p^*(x) = \frac{8\mu}{(|x|^2 + \mu)^2}, \quad \mu > 0. \tag{2.149}$$

Theorem 2.3, therefore, implies the following result.

Theorem 2.5. *Under the assumption of Theorem 2.1, it holds that* $\nu_1(p, \Omega) > 1$ *if* $\sigma = \displaystyle\int_\Omega p\, dx < 4\pi$.

It is known that the eigenvalue problem (2.106) for (2.149) is reduced to the study on the associated Legendre equation through separation of variables, and therefore, all eigenvalues and eigenfunctions are described by the associated Legendre functions. This expression again assures (2.148).

In fact, using

$$\varphi(x) = \Phi(\xi)e^{\imath m\theta}, \quad x = re^{\imath\theta}, \quad \xi = \frac{\mu - r^2}{\mu + r^2}, \quad \Lambda = 1/\nu,$$

we obtain

$$\left[(1 - \xi^2)\Phi_\xi\right]_\xi + \left[2/\Lambda - m^2/(1 - \xi^2)\right]\Phi = 0, \quad \xi_\mu < \xi < 1$$
$$\Phi(1) = 1, \quad \Phi(\xi_\mu) = 0 \tag{2.150}$$

in (2.106) and (2.149), where $\xi_\mu = (\mu - 1)/(\mu + 1)$. The first equation of (2.150) is the associated Legendre equation. Then we obtain the associated Legendre function of the first order denoted by $\Phi = \Phi(\xi)$, requiring $\Lambda = 1$, $m = 0$, and $\Phi(1) = 1$. Then from the positivity of the first eigenfunction we have $\nu_1(p^*, \Omega^*) > 1$ if and only if

$$\Phi(\xi) > 0, \quad \xi_\mu < \xi < 1. \tag{2.151}$$

Since this Φ is given by $P_0(\xi) = \xi$, and therefore, (2.151) means $\xi_\mu > 0$. Hence (2.148) is verified again by

$$\sigma < 4\pi \quad \Leftrightarrow \quad \mu > 1 \quad \Leftrightarrow \quad \xi_\mu > 0.$$

The associated Legendre equation arises under the separation of variables using polar coordinate for the eigenvalue problem on the Laplacian in three space dimension, $\Delta = \dfrac{\partial^2}{\partial x_1^2} + \dfrac{\partial^2}{\partial x_2^2} + \dfrac{\partial^2}{\partial x_3^2}$. The reason why we find this equation in the study of (2.106) for (2.149) lies in the fact that this $p^* = p^*(x)$ is associated with the Liouville integral $F(z) = \mu^{-1/2}z$ through

$$\left(\frac{p^*}{8}\right)^{1/2} = \rho(F).$$

Thus, the surface $(\Omega^*, p^* ds^2)$ with (2.149) is realized in the round surface $(S^2, d\Sigma^2)$ through the stereographic projection $\tau : S^2 \to \mathbf{C} \cup \{\infty\}$, and then (2.106) is transformed into

$$-\Delta_{S^2}\overline{\varphi} = \frac{\nu}{8}\overline{\varphi} \text{ in } \omega, \quad \overline{\varphi} = 0 \text{ on } \partial\omega \qquad (2.152)$$

by $\overline{\varphi} = \varphi \circ \tau$, where Δ_{S^2} is the Laplace-Beltrami operator on $(S^2, d\Sigma)$ and $\omega \subset S^2$ is a disc realized in S^2 with the center at south pole $(0, 0, 0)$ and area σ. Then we obtain the associated Legendre equation (2.150) under the separation of variables, using the polar coordinate in $\mathbf{R}^3$.

Thus, the Bandle symmetrization in §2.7 is nothing but the Schwarz symmetrization executed on the surface of the Gauss curvature $K \leq 1/2$ with the reference surface $(S^2, d\Sigma^2)$. More precisely, under the assumption of Theorem 2.3 we take the above described disc $\omega \subset S^2$. Given $\varphi = \varphi(x)$ on Ω we define a function $\varphi^* = \varphi^*(x)$ on ω by

$$\varphi^*(x) = \sup \{t \mid x \in \omega_t\},$$

where ω_t is the concentric disc of ω in $(S^2, d\Sigma^2)$ such that

$$\int_{\omega_t} dv = \int_{\{\varphi > t\}} p \, dx,$$

where dv stands for the area element of $(S^2, d\Sigma^2)$. Then, properties (2.110) and (2.111) mean that if $\varphi = \varphi(x) > 0$ is C^1 on $\overline{\Omega}$ and takes 0 on $\partial\Omega$ then it follows that

$$\int_\Omega \varphi^2 p \, dx = \int_\omega \varphi^{*2} \, dv, \quad \int_\Omega |\nabla\varphi|^2 \, dx \geq \int_\omega |\nabla\varphi^*|^2 \, dv.$$

2.11 Mean Value Theorems

The Liouville integral (2.84) assures that if (2.72) holds in the domain $\Omega \subset \mathbf{R}^2$ then there is an analytic function $F(z)$ such that $\rho(F) = \left(\dfrac{e^u}{8} \right)^{1/2}$. This property is to be compared with the classical fact that if u is harmonic in Ω, $\Delta u = 0$, then there is an analytic function $F(z)$ such that $u = \operatorname{Re} F$. Actually, we can show the mean value theorems for spherically sub-harmonic and super-harmonic functions by

$$-\Delta u \leq e^u \quad \text{and} \quad \Delta u \leq e^u,$$

respectively, which implies a Harnack inequality.

Theorem 2.6. *Let $\Omega \subset \mathbf{R}^2$ be an open set. Then the C^2 function $u = u(x)$ in Ω satisfies*

$$-\Delta u \leq e^u \quad in \ \Omega \tag{2.153}$$

if and only if

$$u(x_0) \leq \frac{1}{|\partial B(x_0, r)|} \int_{\partial B(x_0, r)} u \, ds - 2 \log \left\{ 1 - \frac{1}{8\pi} \int_{B(x_0, r)} e^u \, dx \right\}_+ \tag{2.154}$$

holds for any $B(x_0, r) \subset\subset \Omega$. Similarly,

$$\Delta u \leq e^u \quad in \ \Omega$$

if and only if

$$u(x_0) \geq \frac{1}{|\partial B(x_0, r)|} \int_{\partial B(x_0, r)} u \, ds - 2 \log \left\{ 1 + \frac{1}{8\pi} \int_{B(x_0, r)} e^u \, dx \right\}$$

for any $B(x_0, r) \subset\subset \Omega$.

Proof. Below we only show that (2.153) implies (2.154). It suffices to derive

$$\log p(x_0) \leq \frac{1}{|\partial B|} \int_{\partial B} \log p \, ds - 2 \log \left(1 - \frac{1}{8\pi} \int_B p \, dx \right)_+ \tag{2.155}$$

for $p = e^u \in C^2(\Omega)$ satisfying (2.92) and $B = B(x_0, r) \subset\subset \Omega$.

For this purpose we take $\omega = B$ in the proof of Theorem 2.1, and define the harmonic function $h(x)$ by (2.94) to put $q = pe^{-h}$. Then, for $K(t)$, $\mu(t)$, and $j(t) \geq 0$ defined by (2.96) and (2.100) it holds that (2.101). Similarly to (2.101), furthermore, we obtain

$$j(t) \leq \frac{K(t)^2}{8\pi}, \quad t > 1, \tag{2.156}$$

which means

$$\mu(t) \geq \frac{K(t)^2}{t}\left(\frac{1}{K(t)} - \frac{1}{8\pi}\right), \quad 1 < t < t_0 \qquad (2.157)$$

for $t_0 = \max_{\overline{B}} q$.

The function

$$J(t) = \frac{\mu(t)}{K(t)} - \frac{\mu(t)}{8\pi} = \frac{1}{t} - \frac{j(t)}{tK(t)} - \frac{\mu(t)}{8\pi}$$

defined for $1 < t < t_0$, on the other hand, is right-continuous:

$$J(t+0) = J(t). \qquad (2.158)$$

It holds also that

$$J(t-0) - J(t) = \frac{j(t)}{t}\left\{\frac{1}{K(t)} - \frac{1}{K(t-0)}\right\} + \frac{1}{8\pi}\left(\mu(t) - \mu(t-0)\right)$$

$$= \frac{j(t)}{t}\left\{\frac{1}{j(t) + \mu(t)t} - \frac{1}{j(t) + \mu(t-0)t}\right\} + \frac{1}{8\pi}\left(\mu(t) - \mu(t-0)\right)$$

$$= -\left(\mu(t) - \mu(t-0)\right)\cdot\left\{\frac{j(t)}{K(t)K(t-0)} - \frac{1}{8\pi}\right\}$$

together with

$$K(t-0) - K(t) = \int_{\{q=t\}} qe^h dx = t\left(\mu(t-0) - \mu(t)\right) \geq 0.$$

Therefore, we obtain

$$J(t-0) - J(t) \leq -\left(\mu(t-0) - \mu(t)\right)\cdot\left(\frac{1}{8\pi} - \frac{j(t)}{K(t)^2}\right) \leq 0 \qquad (2.159)$$

by (2.156). Equality (2.97), furthermore, implies

$$J'(t) = \mu'(t)\left(\frac{1}{K(t)} - \frac{1}{8\pi}\right) - \mu(t)\frac{K'(t)}{K(t)^2}$$

$$\geq \mu'(t)\cdot\frac{t\mu(t)}{K(t)^2} - \mu(t)\cdot\frac{K'(t)}{K(t)^2}$$

$$= 0, \quad \text{a.e. } t \in (1, t_0) \qquad (2.160)$$

by (2.156).

Combining (2.158), (2.159), (2.160), and (2.97), we obtain

$$\lim_{t\uparrow t_0} J(t) = \lim_{t\downarrow 0}\frac{\mu(t)\left(1 - \frac{K(t)}{8\pi}\right)}{K(t)} = \lim_{t\uparrow t_0}\frac{\mu'(t)}{K'(t)} = \frac{1}{t_0}$$

$$\geq J(t) = \mu(t)\left(\frac{1}{K(t)} - \frac{1}{8\pi}\right) \qquad (2.161)$$

for $1 \le t \le t_0$ and then

$$\frac{1}{t_0} \ge \frac{K(t)^2}{t} \left(\frac{1}{K(t)} - \frac{1}{8\pi} \right)_+^2 \tag{2.162}$$

follows from (2.157) and (2.161). Inequality (2.162) for $t = 1$ means

$$\frac{1}{t_0} \ge \left(1 - \frac{K(1)}{8\pi} \right)_+^2 \ge \left(1 - \frac{m}{8\pi} \right)_+^2, \quad m = \int_B p \, dx \tag{2.163}$$

and hence

$$\left(1 - \frac{m}{8\pi} \right)_+^{-2} \ge t_0 = \max_{\overline{B}} p e^{-h} \ge p(x_0) e^{-h(x_0)},$$

or,

$$\log p(x_0) \le h(x_0) - 2\log \left(1 - \frac{1}{8\pi} \int_B p dx \right)_+. \tag{2.164}$$

Now we apply the mean value theorem for harmonic functions

$$h(x_0) = \frac{1}{|\partial B|} \int_{\partial B} h \, ds = \frac{1}{|\partial B|} \int_{\partial B} \log p \, ds$$

in (2.164), to infer (2.155). $\qquad \square$

Mean value theorem implies the Harnack inequality similarly to the theory of harmonic functions.

Theorem 2.7. *Let* $B = B(0, R) \subset \mathbf{R}^2$ *and* $v = v(x) \in C^2(B) \cap C(\overline{B})$ *satisfy*

$$0 \le -\Delta v \le \lambda e^v, \quad v \ge 0 \quad in \ B. \tag{2.165}$$

Then it holds that

$$v(0) \le \frac{R + |x|}{R - |x|} v(x) - 2\log \left(1 - \frac{1}{8\pi} \int_B \lambda e^v \, dx \right)_+, \quad x \in B. \tag{2.166}$$

Proof. The second inequality of (2.165) implies

$$-\Delta \log p \le p \ \text{in} \ B$$

for $p = \lambda e^v$ and hence

$$\log p(0) \le \frac{1}{|\partial B|} \int_{\partial B} \log p \, ds - 2\log \left(1 - \frac{\lambda}{8\pi} \right)_+$$

by (2.155), which means

$$v(0) \le \frac{1}{|\partial B|} \int_{\partial B} v ds - 2\log \left(1 - \frac{\lambda}{8\pi} \right)_+. \tag{2.167}$$

The function $v(x) \geq 0$, on the other hand, is super-harmonic, and therefore, it holds that

$$v(re^{i\theta}) \geq \frac{1}{2\pi} \int_0^{2\pi} \frac{R^2 - r^2}{R^2 - 2Rr\cos(\theta - \varphi) + r^2} v(Re^{i\varphi})\, d\varphi$$

$$\geq \frac{R - r}{R + r} \frac{1}{|\partial B|} \int_{\partial B} v\, ds, \quad 0 \leq r < R. \tag{2.168}$$

Inequality (2.166) holds by (2.167)–(2.168).   □

The standard argument guarantee the following Harnack principle from the Harnack inequality.

Theorem 2.8. *Assume that $\Omega \subset \mathbf{R}^2$ is an open set, $v_k = v_k(x)$, $k = 1, 2, \cdots$, are C^2 functions, and*

$$0 \leq -\Delta v_k \leq \lambda_k e^{v_k}, \quad v_k \geq 0, \quad in\ \Omega.$$

Then one of the following alternatives holds, where

$$\mathcal{S} = \{x_0 \in \Omega \mid \exists x_k \to x_0\ s.t.\ v_k(x_k) \to +\infty\}$$

denotes the interior blowup set.

(1) $\{v_k\}$ is locally uniformly bounded in Ω.
(2) $v_k \to +\infty$ holds locally uniformly in Ω.
(3) $\mathcal{S} \neq \emptyset$ and $\sharp\mathcal{S} \leq \liminf_k \left[\dfrac{1}{8\pi} \int_\Omega \lambda_k e^{v_k} dx\right]$.

The following theorem is derived from (2.154).

Theorem 2.9. *Let $B = B(0, R) \subset \mathbf{R}^2$ with $R > 0$, and $p = p(x) \in C(\overline{B}) \cap C^2(B)$ satisfy*

$$-\Delta \log p \leq p\ in\ B, \quad \int_B p \leq 4\pi.$$

Then it follows that

$$\frac{p(0)}{1 + r^2 p(0)/8} \leq \frac{1}{|\partial B_r|} \int_{\partial B_r} p^{1/2} ds, \quad 0 < r < R. \tag{2.169}$$

Proof. It suffices to prove (2.169) for $r = R$. In fact, we can apply Theorem 2.6 for $u = \log p$ and $m(r) = \int_{B(0,r)} e^u dx \le 4\pi,\ 0 < r < R$. Inequality (2.154) means

$$u(0) \le \frac{1}{|\partial B_r|} \int_{\partial B_r} u\, ds - 2\log\left(1 - \frac{m(r)}{8\pi}\right), \quad 0 < r < R,$$

and therefore, Jensen's inequality implies

$$p(0) \le \left(1 - \frac{m(r)}{8\pi}\right)^{-2} \exp\left(\frac{1}{|\partial B_r|} \int_{\partial B_r} u\, ds\right)$$

$$\le \left(1 - \frac{m(r)}{8\pi}\right)^{-2} \frac{1}{|\partial B_r|} \int_{\partial B_r} p\, ds = \frac{1}{2\pi r}\left(1 - \frac{m(r)}{8\pi}\right)^{-2} m'(r).$$

Thus, for $m \equiv m(R) = \int_{B(0,R)} p\, dx$ it holds that

$$p(0)R^2 = 2\int_0^R p(0)r\, dr \le \frac{1}{\pi}\int_0^R \frac{m'(r)}{(1 - m(r)/(8\pi))^2}\, dr$$

$$= 8m(8\pi - m)^{-1}. \tag{2.170}$$

Here, Bol's inequality reads

$$\ell^2 \ge \frac{1}{2}m(8\pi - m), \quad \ell = \int_{\partial B} p^{1/2}\, ds. \tag{2.171}$$

Since $m \le 4\pi$, we obtain $m \le m_-$, where

$$m_- = 4\pi\left(1 - \sqrt{1 - j^2}\right), \quad j = \ell/(2\sqrt{2}\pi)$$

stands for the smaller solution to $M^2 - 8\pi M + 2\ell^2 = 0$. Then

$$p(0)R^2 \le 8m_-(8\pi - m_-)^{-1} \tag{2.172}$$

holds by (2.170), inequality (2.169) for $r = R$. □

2.12   Recursive Control of the Hamiltonian

Complex structure of the Boltzmann-Poisson equation induces mass quantization and Hamiltonian control of the family of blowup solutions. We recall that $G = G(x, x')$ be the Green function to the Poisson equation (1.12)

$$-\Delta G(\cdot, x') = \delta_{x'} \quad \text{in } \Omega, \qquad G(\cdot, x') = 0 \quad \text{on } \partial\Omega \tag{2.173}$$

where $\Omega \subset \mathbf{R}^2$ is a bounded domain with smooth boundary $\partial\Omega$ and $\delta_{x'} = \delta_{x'}(dx)$ is the delta function with the support at $x = x' \in \Omega$. The Robin function is given by

$$R(x) = \left[G(x, x') + \frac{1}{2\pi} \log |x - x'| \right]_{x'=x} \tag{2.174}$$

and

$$H = H_\ell(x_1, \cdots, x_\ell) = \frac{1}{2} \sum_{j=1}^{\ell} R(x_j) + \sum_{1 \le i < j \le \ell} G(x_i, x_j) \tag{2.175}$$

denotes the ℓ-th Hamiltonian of point vortices, where $\ell \in \mathbf{N}$. The fundamental solution to $-\Delta$ in two-space dimension, $-\Delta\Gamma = \delta$, is given by

$$\Gamma(x) = \frac{1}{2\pi} \log \frac{1}{|x|}, \tag{2.176}$$

and we obtain

$$-\Delta w = 0 \quad \text{in } \Omega, \qquad w = -\Gamma(\cdot - x') \quad \text{on } \partial\Omega$$

for $x' \in \Omega$ and $w = G(\cdot, x') - \Gamma$. Hence this $w = w(x)$ is smooth on $\overline{\Omega}$ and so is $R = R(x)$ in Ω. It holds also that

$$R|_{\partial\Omega} = -\infty. \tag{2.177}$$

Theorem 2.10. *Let $\Omega \subset \mathbf{R}^2$ be a bounded domain with smooth boundary $\partial\Omega$, and (λ_k, v_k), $k = 1, 2, \cdots$, be the classical solution to (2.71) for $\lambda = \lambda_k$, $v = v_k$:*

$$-\Delta v_k = \lambda_k e^{v_k} \ in \ \Omega, \quad v_k = 0 \ on \ \partial\Omega \tag{2.178}$$

satisfying

$$\lambda_k \downarrow 0. \tag{2.179}$$

Then, we have a subsequence and $\ell = 0, 1, \cdots, +\infty$ such that

$$\sigma_k \equiv \int_\Omega \lambda_k e^{v_k} dx \to 8\pi\ell. \tag{2.180}$$

According to the value ℓ the solution v_k behaves as follows:

(1) $\ell = 0$. [uniform convergence to 0]: It holds that $\lim\limits_{k\to\infty} \|v_k\|_\infty = 0$.

(2) $0 < \ell < +\infty$. [ℓ-point blowup]: There is a set of ℓ-points denoted by $x_j^* \in \Omega$, $1 \le j \le \ell$, such that

$$v_k \to v_0 \quad loc. \ unif. \ in \ \overline{\Omega} \setminus \mathcal{S}, \qquad (2.181)$$

where $\mathcal{S} = \{x_1^*, \ldots, x_\ell^*\}$ and

$$v_0(x) = 8\pi \sum_{j=1}^{\ell} G(x, x_j^*). \qquad (2.182)$$

This $\mathcal{S}$ is the blowup set of $\{v_k\}$:

$$\mathcal{S} = \{x_0 \in \overline{\Omega} \mid \exists x_k \to x_0 \ s.t. \ v_k(x_k) \to +\infty\}$$

and there arises

$$\nabla_{x_j} H_\ell(x_1^*, \cdots, x_\ell^*) = 0, \quad 1 \le j \le \ell. \qquad (2.183)$$

(3) $\ell = +\infty$. [entire blowup]: It holds that $v_k \to +\infty$ locally uniformly in Ω.

Via the Liouville integral, equation (2.71) is transformed into the problem to find a conformal mapping $\overline{F} = \tau^{-1} \circ F : \Omega \to S^2$ satisfying (2.86). Then $\dfrac{\sigma}{8} = \displaystyle\int_\Omega \rho(F)^2 dx$ and $s = \left(\dfrac{\lambda}{8}\right)^{1/2} |\partial\Omega|$ stand for the immersed area of $\overline{F}(\Omega)$ and the immersed length of $\overline{F}(\partial\Omega)$, respectively. Since the total area of this S^2 is π, convergence (2.180) illustrates the ℓ-covering of S^2 by $\overline{F}(\Omega)$ as $s \downarrow 0$.

In the Gel'fand equation (2.178), we have $\|v_k\|_\infty \to +\infty$ only if (2.179). Concerning the Boltzmann-Poisson equation (2.27), on the other hand, the second case of Theorem 2.10 is important, stated as follows. We shall provide a proof using complex function theory below.

Theorem 2.11. Let $\Omega \subset \mathbf{R}^2$ be a bounded domain with smooth boundary $\partial\Omega$, and (λ_k, v_k), $k = 1, 2, \cdots$, $\lambda = \lambda_k$, be a sequence of the classical solutions to (2.27):

$$-\Delta v_k = \frac{\lambda_k e^{v_k}}{\int_\Omega e^{v_k} dx} \ in \ \Omega, \quad v_k = 0 \ on \ \partial\Omega, \qquad (2.184)$$

satisfying

$$\lim_{k \to \infty} \lambda_k = \lambda_0 \in (0, +\infty), \quad \lim_{k \to \infty} \|v_k\|_\infty = +\infty. \qquad (2.185)$$

Then, it holds that $\lambda_0 = 8\pi\ell$, $\ell \in \mathbf{N}$. Passing to a subsequence, there is a set of ℓ-interior points $x_1^*, \cdots, x_\ell^*$ satisfying (2.183). Here, $\mathcal{S} = \{x_1^*, \cdots, x_\ell^*\}$ is the blowup set of $\{v_k\}$, and it holds that (2.181) for $v_0(x)$ defined by (2.182).

Proof. Since the L^1 norm of the right-hand side on (2.184) is bounded, the elliptic estimate guarantees

$$\|v_k\|_{W^{1,q}} = O(1), \quad 1 \le q < 2 = \frac{n}{n-1} \qquad (2.186)$$

by $n = 2$. Now we apply the method of moving plane to (2.184).

First, we take the case that Ω is convex. Since $v_k > 0$ in Ω, there is a family $A = \{T\}$ of simplices with uniform shape such that $v_k(x)$ takes the maximum in $T \in A$ at the vertex P closest to $\partial\Omega$, and furthermore,

$$\omega \equiv \Omega \cap \hat{\omega} \subset \bigcup_{T \in A} T,$$

where $\hat{\omega}$ is an open set containing $\partial\Omega$. This family $A = \{T\}$ is determined by Ω, independent of the nonlinearity. We call this property the monotone decreasing of v_k near $\partial\Omega$. Since $\|v_k\|_1 \le C_1$ follows from (2.186) this property implies

$$\|v_k\|_{L^\infty(\omega)} = O(1). \qquad (2.187)$$

For general Ω, we take a disc outscribing $\partial\Omega$. Then, regarding its center the origin, we apply the Kelvin transformation. Due to $n = 2$, we have the monotonicity of v_k near $\partial\Omega$.

Using (2.187) and local elliptic regularity near the boundary to (2.184), we obtain the uniform boundedness of any partial derivatives of $\{v_k\}$ near $\partial\Omega$. In particular, the holomorphic functions

$$s_k = v_{kzz} - \frac{1}{2}v_{kz}^2$$

of $z = x_1 + ix_2$ are uniformly bounded near $\partial\Omega$, and then the classical maximum principle and Montel's theorem guarantee that the family $\{s_k(z)\}$ takes a subsequence, denoted by the same symbol, such that $s_k(z) \to s_0(z)$ locally uniformly in Ω as $k \to \infty$. Using

$$u_k = v_k + \log \lambda_k - \log \int_\Omega e^{v_k} dx, \qquad (2.188)$$

we have

$$-\Delta u_k = e^{u_k} \quad \text{in } \Omega \qquad (2.189)$$

and

$$s_k = u_{kzz} - \frac{1}{2}u_{kz}^2, \qquad (2.190)$$

where the argument in §2.6 is applicable.

Let $x_k = (x_{1k}, x_{2k}) \in \Omega$ be the maximum point of $v_k(x)$, and let $\{\varphi_{1k}(z), \varphi_{2k}(z)\}$ be the fundamental system of solutions to

$$\varphi_{zz} + \frac{1}{2} s_k(z)\varphi = 0$$

satisfying

$$\left(\varphi_{1k}, \frac{\partial \varphi_{1k}}{\partial z}\right)\bigg|_{z=z_k^*} = (1,0), \qquad \left(\varphi_{2k}, \frac{\partial \varphi_{2k}}{\partial z}\right)\bigg|_{z=z_k^*} = (0,1)$$

for $z_k^* = x_{1k} + \imath x_{2k}$. Then, we obtain

$$e^{-u_k/2} = \tilde{c}_k |\varphi_{1k}|^2 + \frac{\tilde{c}_k^{-1}}{8} |\varphi_{2k}|^2 \tag{2.191}$$

for $\tilde{c}_k = e^{-u_k(x_k)/2}$. Since (2.188) equality (2.191) means

$$e^{-v_k/2} = c_k |\varphi_{1k}|^2 + \frac{\sigma_k c_k^{-1}}{8} |\varphi_{2k}|^2 \tag{2.192}$$

for $c_k = e^{-v_k(x_k)/2}$ and $\sigma_k = \frac{\lambda_k}{\int_\Omega e^{v_k}}$.

We may assume $\lim_{k\to\infty} x_k = x_0 \equiv (x_{10}, x_{20}) \in \Omega$, regarding (2.187). Then we define the fundamental system of solutions $\{\varphi_{10}(z), \varphi_{20}(z)\}$ to

$$\varphi_{zz} + \frac{1}{2} s_0(z)\varphi = 0$$

by

$$\left(\varphi_{10}, \frac{\partial \varphi_{10}}{\partial z}\right)\bigg|_{z=z_0^*} = (1,0), \qquad \left(\varphi_{20}, \frac{\partial \varphi_{20}}{\partial z}\right)\bigg|_{z=z_0^*} = (0,1),$$

where $z_0^* = x_{10}^* + \imath x_{20}^*$. It thus holds that

$$\varphi_{1k} \to \varphi_{10}, \qquad \varphi_{2k} \to \varphi_{20}$$

locally uniformly in Ω. Here, we have $v_k(x_k) = \|v_k\|_\infty \to +\infty$ by (2.185), which implies

$$\lim_{k\to\infty} c_k = 0 \tag{2.193}$$

in (2.192). Then we may assume

$$\lim_{k\to\infty} \sigma_k c_k^{-1} = \gamma > 0 \tag{2.194}$$

by (2.187). The blowup set $\mathcal{S}$ of $\{v_k\}$ thus coincides with the zero set of φ_{20} in Ω. Since zeros of the analytic function $\varphi_{20}(z)$ does not take accumulation points in Ω, the set $\mathcal{S}$ is finite. There arises also that (2.181) for $v_0 = v_0(x)$ defined by

$$e^{-v_0/2} = \gamma |\varphi_{20}|^2.$$

By the elliptic regularity, this convergence is valid up to their derivatives of any order.

We have also $\sigma_k \to 0$ by (2.193) and (2.194), which implies

$$\lim_{k \to \infty} \int_\Omega e^{v_k} = +\infty. \tag{2.195}$$

Taking the limit in (2.184), we obtain

$$-\Delta v_0 = 0 \text{ in } \Omega \setminus \mathcal{S}, \quad v_0 = 0 \text{ on } \partial\Omega.$$

This $v_0(x)$ takes each element in $\mathcal{S} = \{x_1^*, \cdots, x_\ell^*\}$ as an isolated singular point. Since $v_0 \geq 0$ we have $a_j > 0$, $1 \leq j \leq \ell$, for which

$$u_0(x) = v_0(x) + \sum_{j=1}^{\ell} a_j \log|x - x_j^*| \tag{2.196}$$

is harmonic in Ω.

Now we make $k \to \infty$ in (2.190), to obtain

$$s_0 = v_{0zz} - \frac{1}{2} v_{0z}^2.$$

This $s_0(z)$ is thus a holomorphic function in z, and the singularities on the right-hand side at $z = x_{1j}^* + \iota x_{2j}^*$ for $x_j^* = (x_{1j}^*, x_{2j}^*)$, is removable. By (2.196), we take an expansion of the right-hand side. Vanishings of the pole $z = z_j^*$ of the second and the first orders imply $a_j = 4$ equivalent to $m_j = 8\pi$ and (2.183), respectively. $\square$

By (2.193)–(2.194) it holds that

$$\|v_k\|_\infty = -2\log\sigma_k + 2\log\gamma + o(1), \quad \sigma_k = \frac{\lambda_k}{\int_\Omega e^{v_k}} \tag{2.197}$$

in (2.184)–(2.185). We can show, on the other hand, the existence of a local maximizer $x = x_k^j$ of $v_k(x)$ such that $\lim_{k \to \infty} x_k^j = x_j^*$. By the above argument using this $x = x_k^j$ in (2.185), we obtain an analogous result to (2.197), that is,

$$v_k(x_k^j) = -2\log\sigma_k + 2\log\gamma_j + o(1), \quad 1 \leq j \leq \ell \tag{2.198}$$

for some $\gamma_j > 0$. Equality (2.198) means the concentration at each blowup point of $\{v_k\}$ with the same magnitude.

2.13 Convex Domains

The Hamiltonian for the system of point vortices and its relatives control the blowup mechanism of solutions in several elliptic problems including those in higher space dimensions other than the Boltzmann-Poisson equation. Critical points of the Hamiltonian are determined by the shape of the domain, regardless of the form of nonlinearities. Hence the profile of the solution to an elliptic equation affects that of the other equation through the properties of the Hamiltonian.

Here, $\Omega \subset \mathbf{R}^n$ is a bounded domain with smooth boundary $\partial\Omega$ and $G = G(x, y)$ is the Green's function:

$$-\Delta_x G(\cdot, y) = \delta_y \quad \text{in } \Omega, \qquad G(\cdot, y) = 0 \quad \text{on } \partial\Omega, \qquad y \in \Omega. \qquad (2.199)$$

The fundamental function of $-\Delta$ is taken as

$$\Gamma(x) = \begin{cases} \frac{1}{2\pi} \log \frac{1}{|x|}, & n = 2 \\ \frac{1}{(n-2)\omega_n} |x|^{2-n}, & n \geq 3, \end{cases} \qquad (2.200)$$

where $\omega_n = \displaystyle\int_{|x|=1} d\sigma_x$ is the surface volume of the n-dimensional unit ball. Then

$$R(x) = [G(x, y) - \Gamma(x - y)]_{y=x} \qquad (2.201)$$

denotes the Robin function. Let $V = V(x)$ be a C^1 function in $x \in \overline{\Omega}$, and

$$H_\ell(x_1, \cdots, x_\ell) = A \sum_{i=1}^{\ell} (R(x_i) + V(x_i)) \Lambda_i^2$$

$$+ B \sum_{1 \leq i < j \leq \ell} G(x_i, x_j) \Lambda_i \Lambda_j \qquad (2.202)$$

is a generalized Hamiltonian defined for $\ell \in \mathbf{N}$ and $x = (x_1, \cdots, x_\ell) \in \Omega^\ell \setminus D$, where $D = \{(x_1, \cdots, x_\ell) \mid \exists i \neq j, \ x_i = x_j\}$ and $A, B, \Lambda_i > 0$ are constants.

Theorem 2.12. *If Ω is convex and $R(x) + V(x)$ is concave, then $H_\ell = H_\ell(x)$ defined by (2.202) for $x = (x_1, \cdots, x_\ell)$ does not have a critical point in Ω if $\ell \geq 2$.*

It is known that if Ω is convex, then $R(x)$ is concave and takes a unique critical point in Ω as a maximizer. Therefore, if $\Omega \subset \mathbf{R}^2$ is convex, only one point blowup of the family of solutions to the Boltzmann-Poisson equation or the Gel'fand equation is admitted, and furthermore, the singular limit as $\lambda \downarrow 0$ is unique in (2.71).

For the proof of Theorem 2.12 we use the following identity.

Lemma 2.3. *The Green function* $G = G(x, y)$ *defined by (2.199) satisfies*

$$\int_{\partial\Omega} (x - p) \cdot \nu_x \frac{\partial G}{\partial \nu_x}(x, a) \frac{\partial G}{\partial \nu_x}(x, b) \, d\sigma_x$$
$$= (2 - n)G(a, b) + (p - a) \cdot \nabla_x G(a, b) + (p - b) \cdot \nabla_x G(b, a) \quad (2.203)$$

for $p \in \mathbf{R}^n$, $a, b \in \Omega$, *and* $a \neq b$.

Putting $p = a$ in (2.203), we obtain

$$\int_{\partial\Omega} (x - a) \cdot \nu_x \frac{\partial G}{\partial \nu_x}(x, a) \frac{\partial G}{\partial \nu_x}(x, b) \, d\sigma_x$$
$$= (2 - n)G(a, b) + (a - b) \cdot \nabla_x G(b, a). \quad (2.204)$$

Since (2.201) implies

$$G(x, y) = \Gamma(x - y) + R(x) + o(1), \quad y \to x$$

we obtain

$$(a - b) \cdot \nabla_x G(b, a) = -(b - a) \cdot \nabla \Gamma(b - a) + o(1)$$
$$(2 - n)G(a, b) = (2 - n)\Gamma(b - a) + (2 - n)R(a) + o(1), \quad b \to a.$$

Since

$$x \cdot \nabla \Gamma(x) = \begin{cases} (2 - n)\Gamma(x), & n \geq 3 \\ -\frac{1}{2\pi}, & n = 2 \end{cases}$$

equality (2.204) is reduced to the Brezis-Peletier identity,

$$\int_{\partial\Omega} (x - a) \cdot \nu_x \left\{ \frac{\partial G}{\partial \nu_x}(x, a) \right\}^2 d\sigma_x = \begin{cases} (2 - n)R(a), & n \geq 3 \\ \frac{1}{2\pi}, & n = 2. \end{cases} \quad (2.205)$$

Lemma 2.3 is proven by the following lemma. Here, inequality (2.206) is a special case of (2.203) for $p = b$. Then (2.203) is obtained by adding (2.206) and the inner product of (2.207) with $b - p$.

Lemma 2.4. *For* $a, b \in \Omega$ *in* $a \neq b$, *it holds that*

$$\int_{\partial\Omega} [(x - b) \cdot \nu_x] \frac{\partial G}{\partial \nu_x}(x, a) \frac{\partial G}{\partial \nu_x}(a, b) \, d\sigma_x$$
$$= (2 - n)G(a, b) + (b - a) \cdot \nabla_x G(a, b) \quad (2.206)$$

and

$$\int_{\partial\Omega} \nu_x \frac{\partial G}{\partial \nu_x}(x, a) \frac{\partial G}{\partial \nu_x}(x, b) \, d\sigma_x = -[\nabla_x G(a, b) + \nabla_x G(b, a)]. \quad (2.207)$$

Proof. First, we show (2.207). Let $\Omega_\varepsilon = \Omega \setminus \left(B(a,\varepsilon) \bigcup B(b,\varepsilon) \right), 0 < \varepsilon \ll 1$, and

$$v = \nabla_x G(\cdot, b). \tag{2.208}$$

Since $v = v(x)$ is harmonic in $x \in \Omega \setminus \{b\}$, it holds that

$$\int_{\Omega_\varepsilon} v \Delta G(\cdot, a) - G(\cdot, a) \Delta v \, dx = \int_{\partial \Omega_\varepsilon} v \frac{\partial G}{\partial \nu}(\cdot, a) - G(\cdot, a) \frac{\partial v}{\partial \nu} \, d\sigma$$

and hence

$$\int_{\partial \Omega} v \frac{\partial G}{\partial \nu}(\cdot, a) - G(\cdot, a) \frac{\partial v}{\partial \nu} \, d\sigma$$

$$= \int_{\partial B(a,\varepsilon)} v \frac{\partial G}{\partial \nu}(\cdot, a) - G(\cdot, a) \frac{\partial v}{\partial \nu} \, d\sigma$$

$$+ \int_{\partial B(b,\varepsilon)} v \frac{\partial G}{\partial \nu}(\cdot, a) - G(\cdot, a) \frac{\partial v}{\partial \nu} \, d\sigma. \tag{2.209}$$

In this (2.209) we use $G(\cdot, a)|_{\partial \Omega} = 0$ and

$$G(x, y) = \Gamma(x - y) + K(x, y), \quad K \in C^\infty(\Omega \times \Omega)$$

$$\nabla \Gamma(x) = -\frac{1}{\omega_n |x|^{n-1}} \cdot \frac{x}{|x|}, \quad \frac{\partial}{\partial r} \nabla \Gamma(x) = \frac{n-1}{\omega_n |x|^n} \cdot \frac{x}{|x|} \tag{2.210}$$

for $r = |x|$. By (2.208) we obtain

$$\int_{\partial \Omega} \nu_x \frac{\partial G}{\partial \nu_x}(x, a) \frac{\partial G}{\partial \nu_x}(x, b) \, d\sigma_x = -v(a) + I + o(1)$$

$$I = \int_{\partial B(b,\varepsilon)} v \frac{\partial G}{\partial \nu}(\cdot, a) - G(\cdot, a) \frac{\partial v}{\partial \nu} \, d\sigma \tag{2.211}$$

as $\varepsilon \downarrow 0$. It holds also that

$$I = -\frac{1}{\omega_n \varepsilon^{n-1}} \int_{\partial B(b,\varepsilon)} v \frac{\partial G}{\partial \nu}(\cdot, a) + \frac{n-1}{\varepsilon} G(x, a) v \, d\sigma + o(1) \tag{2.212}$$

by (2.208) and (2.210). In this (2.212) we apply $\nu = (x - b)/\varepsilon$ and

$$G(x, a) = G(b, a) + (x - b) \cdot \nabla_x G(b, a) + o(\varepsilon), \quad x \in \partial B(b, \varepsilon)$$

$$\int_{\partial B(b,\varepsilon)} \nu \, d\sigma = 0, \quad \frac{1}{\omega_n \varepsilon^{n-1}} \int_{\partial B(b,\varepsilon)} \nu_i \nu_j \, d\sigma = \frac{\delta_{ij}}{n},$$

to obtain

$$I = -\frac{1}{\omega_n \varepsilon^{n-1}} \int_{\partial B(b,\varepsilon)} n \nu \frac{\partial G}{\partial \nu}(x, a) \, d\sigma_x + o(1)$$

$$= -\nabla_x G(b, a) + o(1). \tag{2.213}$$

Then, equality (2.207) follows from (2.208), (2.211), and (2.213).

Next, we show (2.206). Let

$$w(x) = (x - b) \cdot \nabla_x G(x, b). \tag{2.214}$$

Since $w = w(x)$ is harmonic in $x \in \Omega \setminus \{b\}$ it holds that

$$\int_{\partial \Omega_\varepsilon} w \frac{\partial G}{\partial \nu}(\cdot, a) - G(\cdot, a) \frac{\partial w}{\partial \nu} \, d\sigma = 0.$$

Then, similarly to (2.211) we obtain

$$\int_{\partial \Omega} [(x - b) \cdot \nu_x] \frac{\partial G}{\partial \nu_x}(x, a) \frac{\partial G}{\partial \nu_x}(x, b) \, d\sigma_x = -w(a) + II$$

$$II = \int_{\partial B(b,\varepsilon)} w \frac{\partial G}{\partial \nu}(\cdot, a) - G(\cdot, a) \frac{\partial w}{\partial \nu} \, d\sigma. \tag{2.215}$$

Here we have $w(x) = O(\varepsilon^{2-n})$ uniformly in $x \in \partial B(b, \varepsilon)$ as $\varepsilon \downarrow 0$, and therefore,

$$II = -\frac{1}{\varepsilon} \int_{\partial B(b,\varepsilon)} G(x, a)[(x - b) \cdot \nabla w(x)] \, d\sigma_x + o(1).$$

It holds also that

$$(x - b) \cdot \nabla w(x) = (x - b) \cdot \nabla \Gamma(x - b)$$
$$+ \sum_{i,j=1}^{n} \frac{\partial^2 \Gamma(x - b)}{\partial x_i \partial x_j}(x_i - b_i)(x_j - b_j) + o(\varepsilon) \tag{2.216}$$

by (2.210), and (2.200) implies

$$x \cdot \nabla \Gamma(x) + \sum_{i,j=1}^{n} \frac{\partial^2 \Gamma(x)}{\partial x_i \partial x_j} x_i x_j = (n - 2)^2 \Gamma(x). \tag{2.217}$$

By (2.216), (2.217), and $G(b, a) = G(a, b)$, we obtain

$$II = -\frac{(n-2)^2}{\varepsilon} \int_{\partial B(b,\varepsilon)} \Gamma(x - b) G(x, a) \, d\sigma_x + o(1)$$
$$= -(n - 2)G(a, b) + o(1). \tag{2.218}$$

Equality (2.206) now follows from (2.214), (2.215), and (2.218).   □

We are ready to give the following proof.

Proof of Theorem 2.12. Let $\ell \geq 2$ and assume $\nabla_{x_i} H_\ell(x_1^*, \ldots, x_\ell^*) = 0$, $1 \leq i \leq \ell$, for $(x_1^*, \cdots, x_\ell^*) \in \Omega^\ell \setminus D$, that is,

$$A(\nabla R(x_i^*) + \nabla V(x_i^*)\Lambda_i^2) + B \sum_{j \neq i} \nabla_x G(x_i^*, x_j^*)\Lambda_i\Lambda_j = 0, \quad 1 \leq i \leq \ell.$$

We take the inner product of the right-hand side and $p - x_i^*$, which results in, by $G(x, y) = G(y, x)$,

$$\sum_{i=1}^\ell \sum_{j \neq i} (p - x_i^*) \cdot \nabla_x G(x_i^*, x_j^*)\Lambda_i\Lambda_j = \sum_{1 \leq i < j \leq \ell} \big\{ (p - x_i^*) \cdot \nabla_x G(x_i^*, x_j^*)$$

$$+ (p - x_j^*) \cdot \nabla_y G(x_i^*, x_j^*) \big\} \Lambda_i\Lambda_j.$$

By Lemma 2.3, on the other hand, it holds that

$$(p - x_i^*) \cdot \nabla_x G(x_i^*, x_j^*) + (p - x_j^*) \cdot \nabla_y G(x_i^*, x_j^*)$$

$$= \int_{\partial\Omega} (x - p) \cdot \nu_x \frac{\partial G}{\partial \nu_x}(x, x_i^*) \frac{\partial G}{\partial \nu_x}(x, x_j^*) \, d\sigma_x$$

$$+ (n - 2)G(x_i^*, x_j^*) \tag{2.219}$$

for $j \neq i$.

In the right-hand side on (2.219) there arises $G(x_i^*, x_j^*) > 0$ and

$$\left. \frac{\partial G}{\partial \nu_x}(\cdot, y) \right|_{\partial\Omega} < 0, \quad y \in \Omega.$$

Since Ω is convex, on the other hand, it holds that

$$(x - p) \cdot \nu_x \geq 0, \quad (x - p) \cdot \nu_x \not\equiv 0 \quad \text{for any } (p, x) \in \Omega \times \partial\Omega.$$

Thus we obtain

$$\sum_{i=1}^\ell (p - x_i^*) \cdot (\nabla R(x_i^*) + \nabla V(x_i^*))\Lambda_i^2 < 0. \tag{2.220}$$

From the assumption, on the other hand, the function $R(x) + V(x)$ is concave in Ω. Since $R|_{\partial\Omega} = -\infty$, there is a maximizer of $R(x) + V(x)$ in Ω denoted by $\hat{x} \in \Omega$, and it holds that

$$(x - \hat{x}) \cdot (\nabla R(x) + \nabla V(x)) \leq 0, \quad x \in \Omega.$$

In particular, we obtain

$$\sum_{i=1}^\ell (x_i^* - p) \cdot (\nabla R(x_i^*) + \nabla V(x_i^*))\Lambda_i^2 \leq 0 \tag{2.221}$$

for $p = \hat{x}$, which contradicts to (2.220). $\qquad\square$

2.14   Notes

Foundations of thermodynamics and statistical mechanics are described in
Chapter 2.2 of [Suzuki (2015)]. Onsager's point of view on the negative inverse temperature is seen in [Onsager (1949)]. Several methods of deriving
(2.26) are known. See Chapter 2.2 of [Suzuki (2015)] and the references
therein. Uniform estimate of $\{\rho_k^n\}$ and uniqueness of the solution to (2.26)
for both $0 < \lambda < 8\pi$ are proven by [Caglioti, Lions, Marchioro, and Pulvirenti (1992)] and [Suzuki (1992)], respectively.

Relaxation theory described in Sections 2.3, 2.4, and 2.5 is due to
[Sawada and Suzuki (2017)]. We study relaxation dynamics of the mean
field of many point vortices from quasi-equilibrium to equilibrium. Maximum entropy production principle implies four consistent equations concerning relaxation-equilibrium states and patch-point vortex models. Point
vortex relaxation equation coincides with Brownian point vortex equation in
micro-canonical setting. Mathematical analysis to point vortex relaxation
equation is done in accordance with the Smoluchowski-Poisson equation.
Study on the Boltzmann-Poisson equation (2.33), especially, on the single
intensity case, is described in [Suzuki (2015)]. Equation (2.37) is derived
from the Brownian point vortex model in [Chavanis (2008)]. Principle
of maximum entropy and principle of maximum entropy production were
used by [Robert (1991); Robert and Sommeria (1991)] and [Robert and
Sommeria (1992); Robert and Rosier (1997)] to formulate equilibrium and
near from equilibrium dynamics, respectively.

Point vortex model is regarded as a special case of vorticity patch model,
where the patch size shrinks to zero in [Chavanis, Sommeria, and Robert
(1996)]. In the canonical setting there arises (2.65) with constant β. This
model is derived in [Chavanis (2008)] using Brownian point vortices and
mathematical analysis is done in [Suzuki (2014)]. See [Sawada and Suzuki
(2017)] for equality (2.70) in the case of $\Omega = B$.

Writing

$$\sigma = \int_\Omega \lambda e^v \, dx$$

in (2.71), we obtain

$$-\Delta v = \frac{\sigma e^v}{\int_\Omega e^v dx} \quad \text{in } \Omega, \qquad v = 0 \quad \text{on } \partial\Omega. \qquad (2.222)$$

Rewriting σ to λ in (2.222) yields the Boltzmann-Poisson equation (2.27),

$$-\Delta v = \frac{\lambda e^v}{\int_\Omega e^v dx} \quad \text{in } \Omega, \qquad v = 0 \quad \text{on } \partial\Omega. \qquad (2.223)$$

In (2.223) the parameter λ casts the geometric quantity in Section 2.7 and the inverse temperature in the statistical mechanics of point vortices in Section 2.2. This parameter also stands for a coupling constant in the self-dual gauge theory. See Chapter 4 of [Lions (1997)], Chapter 7 of [Marchioro and Pulvirenti (1994)], Chapter 1 of [Tarantello (2008)], and Chapter 5 of [Yang (2001)].

The first and the second cases of Theorem 1.1 are due to [Bandle (1976)] and [Bartolucchi and Lin (2014)], respectively. Unless Ω is simply-connected, inequality (2.91) is strict. Theorem 2.2 is shown in [Nehari (1958)].

Theorem 2.3 is proven by [Bandle (1971)]. We follow the argument of [Talenti (1976)] using Sard's lemma and differentiation and integration of monotone functions to avoid the exceptional case of $\partial \Omega_t$ with more than one Hausdorff dimension. A variant of the Bandle rearrangement was adopted in [Suzuki (1992)] to show the uniqueness of the solution to (2.27) to $0 < \lambda < 8\pi$, particularly for simply-connected Ω.

The proof of Lemma 2.1 is given in Section 1.2 of [Burago and Zalgaller (1988)] and Section I.3 of [Bandle (1980)]. Lemma 2.2 is a basis of Theorem 2.4, the sup + inf inequality, and is shown in [Shafrir (1992)]. Residual vanishing and collapse mass quantization for the family of solutions to (2.129) with $V(x)$ restricted to a compact set in $C(\overline{\Omega})$ is proven by [Li and Shafrir (1994)]. In the alternative proof of Theorem 2.4, we assume $\|\nabla V\|_\infty \leq C_2$ additionally [Brezis, Li, and Shafrir (1993)]. In this case inequality (2.130) is valid to $c_1 = 1$. The proof is based on the blowup analysis, applicable to higher dimensional case [Suzuki and Takahashi (2017)].

Radial symmetry of the solution $v = v(x)$ to (2.137) described in Section 2.9 is due to [Gidas, Ni, and Nirenberg (1979)]. The proof relies on the method of moving planes. Radially symmetric solution on the annulus $\Omega = A \equiv \{x \in \mathbf{R}^2 \mid a < |x| < 1\}$, $0 < a < 1$ takes the Liouville integral in the form of $F(z) = Cz^\alpha$. By this analytic function, radially symmetric solutions are classified and are given explicitly by (2.84). Based on these solutions, the bifurcation theory and variational analysis exhibit the structure of non-radial solutions. Actually, any mode solutions emerge from radially symmetric solutions, which eventually blowup with the number of blowup points equal to their nodes [Lin (1989); Nagasaki and Suzuki (1990b); Kan (2013)]. This phenomenon has a generalization that if $\Omega \subset \mathbf{R}^2$ is multiply-connected and $k \in \mathbf{N}$ is given, there is a family of classical solutions which exhibits k blowup points as $\lambda \downarrow 0$ ([del Pino, Kowalczyk, and Musso (2005)]).

The description of the beginning of Section 2.10 concerning the shape of C^* and the linearized stability of v_-^* in $0 < \lambda \leq 2$ is justified by the bifurcation theory developed by [Crandall and Rabinowitz (1975)]. For a direct proof of (2.148) using associated Legendre functions, see Chapter 3.1.1 of [Bandle (1980)]. Viewing the Bandle symmetrization as the Schwarz symmetrization on round sphere is effectively used in [Suzuki (1992)] for the proof of the uniqueness of the solution to (2.27) for $0 < \lambda < 8\pi$.

Mean value theorems for spherically sub-harmonic and super-harmonic functions are given by [Suzuki (1990)]. Theorem 2.9 is due to [Bandle (1976)]. Some applications to Theorems 2.7 and 2.9 to $-\Delta v = V(x)e^v$ and $u_t - \Delta u = e^u$ are found in [Itoh (1989); Cheng and Lin (1997); Chen and Lin (1998)].

Theorem 2.10 is due to [Nagasaki and Suzuki (1990a)]. More general nonlinearity is treated by [Nagasaki and Suzuki (1990a); Ye (1997)]. The fact that $\|v_k\|_\infty \to +\infty$ arises with (2.179) in (2.178) can be proven by an identity due to Obata [Obata (1971)]. See [Spruck (1988); Nagasaki and Suzuki (1990a)] for this proof. The elliptic L^1 estimate of (2.186) follows from truncation method and duality argument. See [Stampacchia (1965); Brezis and Strauss (1973)]. The decreasing property of the positive solution to the semilinear elliptic equation used in the proof of Theorem 2.11 is shown in [Gidas, Ni, and Nirenberg (1979)]. The general property (2.187) under the uniform L^1 estimate is due to [de Figueiredo, Lions, and Nussbaum (1982)]. See Section 7.3.6 of [Suzuki and Senba (2011)] for the Kelvin transformation and also Chapter 9 of [Gilbarg and Trudinger (1983)] for interior and boundary elliptic estimates. Equality (2.196) follows from the classical result on non-negative harmonic functions by [Serrin and Weinberger (1960)]. One may use the L^1 elliptic estimate instead as in ns90. In the other argument, one notices that the right-hand side on (2.184) converges $*$-weakly in $\mathcal{M}(\overline{\Omega}) = C(\overline{\Omega})'$, passing to a subsequence. From (2.195), furthermore, the limit measure $\mu(dx)$ as the support contained in $\mathcal{S}$, and therefore, a finite sum of delta functions:

$$\mu(dx) = \sum_{j=1}^{\ell} m_j \delta_{x_j^*}(dx).$$

Here, by a theorem of [Brezis and Merle (1991)] it holds that $m_j > 0$ for $1 \leq j \leq \ell$, or, more precisely, $m_j \geq 4\pi$. Since (2.184) means

$$v_k(x) = \int_\Omega G(x,x')\mu_k(x)\,dx, \quad \mu_k(x) = \frac{\lambda_k e^{v_k}\,dx}{\int_\Omega e^{v_k}}$$

we obtain

$$v_0(x) = \sum_{j=1}^{\ell} m_j G(x, x_j^*) \quad \text{in } \overline{\Omega} \setminus \mathcal{S}.$$

With $\Gamma(x)$ in (2.176), the function $u_0(x)$ defined by (2.196) for $a_j = m_j/(2\pi)$ is harmonic in Ω. The last part of the proof of Theorem 2.11, $m_j = 8\pi$, and (2.183), is derived also by a Pohozaev identity [Ma and Wei (2001)]. A similar result to Theorem 2.10 is also derived there, for the Gel'fand equation with inhomogeneous coefficient,

$$-\Delta v = \lambda V(x) e^v \text{ in } \Omega, \quad v = 0 \text{ on } \partial\Omega$$

with the Hamiltonian replaced by

$$H_\ell(x_1, \cdots, x_\ell) = \frac{1}{2} \sum_{j=1}^{\ell} R(x_j) + \sum_{1 \le i < j \le \ell} G(x_i, x_j) + \frac{1}{8\pi} \sum_{j=1}^{\ell} \log V(x_j).$$

Here, we note that the method of Liouville integral is invalid to this problem. The existence of a local maximizer x_j^k of $v_k = v_k(x)$ such that $\lim_{k \to \infty} x_j^k = x_j^*$ follows from the result in [Li and Shafrir (1994)].

Theorem 2.12 is due to [Grossi and Takahashi (2010)]. The concavity of $R(x)$ and its uniqueness of the critical point in Ω, realized as a non-degenerate maximizer, is shown in [Caffarelli and Friedman (1989)] for $n = 2$. The other proof in complex function theory is also known [Haegi (1951); Gustafsson (1979)]. Concerning this concavity property for $n \ge 3$ we refer to [Cardaliaguet and Tahraoui (2002)].

In the Gel'fand problem (2.71) there is an upper bound of λ for the existence of the solution [Crandall and Rabinowitz (1975)]. One dimensional manifold is generated in $\lambda - v$ space from the trivial solution $(\lambda, v) = (0, 0)$, which is composed of minimal solutions. Each solution v on this manifold is linearized stable. Furthermore, for each $\varepsilon > 0$ there is a constant which bounds uniformly to any solution v for $\lambda \ge \varepsilon$ (see [Nagasaki and Suzuki (1990a)]). From these estimates and a topological argument based on the degree theory, we see that the trivial solution $(0, 0$ and the singular limit $(0, 8\pi G(\cdot, x_0)$ with $\nabla R(x_0) = 0$ lies on the same connected component of the set of solutions in $\lambda - v$ space, if Ω is convex.

If $\Omega \subset \mathbf{R}^2$ is simply-connected and $x_0 \in \Omega$ is a non-degenerate critical point of $R(x)$, there is a unique one-dimensional manifold of the solution in $\lambda - v$ space with the end point $(0, 8\pi G(\cdot, x_0))$ ([Suzuki (1993)]). Under this condition, the non-degeneracy and the Morse index of the solution near this singular limit are equal to those of $R(x)$ ([Gladiali, Grossi, Ohtsuka, and Suzuki (2014)]). Therefore, if Ω is convex, the unique non-minimal solution to (2.71) for $0 < \lambda \ll 1$ is non-degenerate and its Morse index is one.

If the blowup point is single, $\ell = 1$ in (2.71), the value

$$\sigma_k = \lambda_k \int_\Omega e^{v_k}\, dx$$

converges to 8π below or above, based on the property of the critical point x_0 of $R(x)$. These alternatives are studied in details by [Chang, Cheng, and Lin (2003)] in accordance with the shape of Ω. In the former case, there is a one-dimensional manifold in $\lambda - v$ space with the singular limit and the trivial solution as the end points [Suzuki (1992)]. Even if Ω is convex, the latter case arises if it is thin. Inequality (2.205) is used in [Brezis and Peletier (1989); Han (1991)] in the study of the Yamabe equation

$$-\Delta u = u^{\frac{n+2}{n-2}},\ u > 0 \quad \text{in } \Omega, \qquad u = 0 \quad \text{on } \partial\Omega$$

for $n \geq 3$.

Sections 12.6, 12.7, and 12.8 of [Suzuki (2015)] are devoted to more detailed structures of the Gel'fand equation (2.71), or the Boltzmann-Poisson equation (2.27), derived from real analytic methods, that is, ε-regularity [Brezis and Merle (1991)], classification of the entire solution [Chen and Li (1991)], and quantized blowup mechanism [Li and Shafrir (1994)]. These theorems are stated as follows.

Theorem 2.13. *Let $\Omega \subset \mathbf{R}^2$ be a bounded domain and $V_k = V_k(x)$, $k = 1, 2, \cdots$, be a sequence of measurable functions satisfying $0 \leq V_k(x) \leq C$ in Ω. Let $v_k = v_k(x)$ satisfies*

$$-\Delta v_k = V_k(x)e^{v_k} \quad \text{in } \Omega, \qquad \int_\Omega e^{v_k}\, dx \leq C.$$

Then, there is a subsequence denoted by the same symbol satisfying one of the following alternatives:

(1) $\{v_k\}$ is locally uniformly bounded in Ω.

(2) $v_k \to -\infty$ locally uniformly in Ω.

(3) There is a finite set $\mathcal{S} = \{x_j^*\} \subset \Omega$, $m_j \geq 4\pi$ such that $v_k \to -\infty$ locally uniformly in $\Omega \setminus \mathcal{S}$ and

$$V_k(x)e^{v_k}\, dx \rightharpoonup \sum_j m_j \delta_{x_j^*}(dx) \quad \text{in } \mathcal{M}(\Omega), \qquad (2.224)$$

where $\mathcal{S} = \{x_0 \in \Omega \mid \exists x_k \to x_0 \text{ s.t. } v_k(x_k) \to +\infty\}$ is the blowup set of $\{v_{k+}\}$ in Ω.

Theorem 2.14. *If*

$$-\Delta v = e^v \quad in \ \mathbf{R}^2, \qquad \int_{\mathbf{R}^2} e^v dx < +\infty, \qquad (2.225)$$

then it holds that

$$v(x) = \log\left\{ \frac{8\mu^2}{\left(1 + \mu^2 \left|x - x_0\right|^2\right)^2} \right\}, \qquad x_0 \in \mathbf{R}^2, \ \mu > 0. \qquad (2.226)$$

Theorem 2.15. *Let* $v_k = v_k(x)$, $k = 1, 2, \ldots$ *satisfy*

$$-\Delta v_k = V_k(x)e^{v_k}, \ 0 \le V_k(x) \le C_1 \quad in \ \Omega, \qquad \int_\Omega e^{v_k} dx \le C_2,$$

where $C_1, C_2 > 0$ *are constants,* $V_k = V_k(x)$ *is continuous, and* $V_k \to V$
uniformly on $\overline{\Omega}$. *Then, passing to a sub-sequence, we have the following alternatives:*

(1) $\{v_k\}$ *is locally uniformly bounded in* Ω.

(2) $v_k \to -\infty$ *locally uniformly in* Ω.

(3) *We have a finite set* $\mathcal{S} = \{a_i\} \subset \Omega$ *and* $m_i \in \mathbf{N}$ *such that* $v_k \to -\infty$
locally uniformly in $\Omega \setminus \mathcal{S}$ *and*

$$V_k(x)e^{v_k} dx \rightharpoonup \sum_i 8\pi m_i \delta_{a_i}(dx)$$

in $\mathcal{M}(\Omega)$. *Here,* $\mathcal{S}$ *is the blowup set of* $\{v_k\}$ *in* Ω.

Chapter 3

Fluid Motion

Macroscopic state of particles that constitute self-interacting fluid is formulated by a system of equations provided with self-duality between the particle density and the field distribution. Here, formation of the field is subject to the action in distance and is associated with the Poisson equation. The stationary state is then described by a nonlinear eigenvalue problem with non-local term because of mass conservation, and this problem is provided with the variational structure from the energy conservation law. These structures are quite similar to the 2D Smoluchowski-Poisson equation and 2D Boltmann-Poisson equation. Method of scaling thus detects the critical exponents with critical mass for mass quantization in high space dimensions. Hence we will meet Euler-Poisson equation, plasma confinement problem, and degenerate parabolic equation derived from the Tsallis entropy.

3.1 Self-gravitating Fluids

A solid can be put on a desk on which its some part does not lie. This phenomenon is due to the *tangential force*. Fluid is a continuum, of which tangential force is absent at the equilibrium. The ideal fluid, on the other hand, is a continuum of which tangential force is absent even when it is moving.

As is confirmed in §2.1, if $v = v(x, t)$ is the velocity of a fluid, the position $x = x(t) \in \mathbf{R}^3$ of a particle in this fluid is subject to (2.3):

$$\frac{dx}{dt} = v(x, t), \quad x(s) = \xi. \tag{3.1}$$

Then we define the propagator $\{T(t, s)\}$ by

$$x(t) = T(t, s)\xi.$$

The acceleration vector of this particle is given by (2.5), and hence the Euler equation arises as the equation of motion,

$$\rho \frac{Dv}{Dt} = \rho F - \nabla p,$$ (3.2)

where ρ, p, and F denote the density, the pressure, and the outer force, respectively. In fact, the left-hand side of (3.2) indicates the mass times acceleration vector, while ρF on the right-hand side is compensated by the gradient of the pressure, $-\nabla p$.

The other derivation of (3.2) uses the equation of continuity

$$\rho_t + \nabla \cdot \rho v = 0.$$ (3.3)

This equation describes the mass conservation,

$$\frac{d}{dt} \int_\omega \rho \, dx = - \int_{\partial \omega} \nu \cdot j \, ds,$$ (3.4)

because $j = \rho v$ is the flux of ρ, where ω is an arbitrary subdomain with smooth boundary.

To confirm (3.2) from this point of view, we assume that the rate of change of momentum of a moving piece of ideal fluid is equal to the total force acting on it. This total force is composed of the surface and the volume forces, and, hence it follows that

$$\frac{d}{dt} \int_{T(t,s)\omega} \rho v \, dx = - \int_{T(t,s)\omega} p\nu \, dS + \int_{T(t,s)\omega} \rho F \, dx$$ (3.5)

where ν denotes the outer unit normal vector. Regarding $x = T(t,s)\xi$ as a transformation of variables

$$\xi \in \mathbf{R}^3 \quad \mapsto \quad x \in \mathbf{R}^3,$$ (3.6)

we have

$$J_t(\xi) = 1 + (t-s)\nabla \cdot v(\xi, s) + o(t-s), \quad t \to s$$ (3.7)

by Liouville's formula, where

$$J_t = \det\left(\frac{\partial x_i}{\partial \xi_j}\right)$$

denotes the Jacobian.

At $t = s$, therefore, the left-hand side of (3.5) is equal to

$$\frac{d}{dt} \int_\omega (\rho v)(x(\xi, t), t)|J_t(\xi)| \, d\xi \bigg|_{t=s} = \int_\omega \frac{D}{Dt}(\rho v) + \rho v(\nabla \cdot v) \, dx$$

$$= \int_\omega \left(\frac{D\rho}{Dt} + \rho \nabla \cdot v\right) v + \rho \frac{Dv}{Dt} \, dx.$$ (3.8)

Using the equation of continuity (3.3), we obtain

$$\int_\omega \rho \frac{Dv}{Dt}\, dx = \int_\omega -\nabla p + \rho F\, dx$$

by the divergence formula. Then (3.2) follows because the sub-domain ω is arbitrary.

The equation of continuity (3.4) can be derived from

$$0 = \frac{d}{dt} \int_{T(t,s)\omega} \rho(x,t)\, dx \bigg|_{t=s} \tag{3.9}$$

which indicates the mass conservation in the Lagrange coordinate. In fact, under the transformation (3.6), equality (3.9) reads as

$$0 = \frac{d}{dt} \int_\omega \rho(x(\xi,t),t)|J_t(\xi)|\, d\xi \bigg|_{t=s} = \int_\omega \frac{D\rho}{Dt} + \rho \nabla \cdot v\, dx$$

$$= \int_\omega \rho_t + \nabla \cdot \rho v\, dx \tag{3.10}$$

similarly to (3.8). Then (3.3) follows because the subdomain ω is arbitrary.

In the barotropic fluid the density is a function of the pressure, represented by the state equation $\rho = \rho(p)$. In this case, using

$$P(p) = \int^p \frac{d\xi}{\rho(\xi)}$$

in (3.2), we obtain

$$v_t + (v \cdot \nabla)v = -\nabla P + F. \tag{3.11}$$

From the thermodynamical law, it follows that

$$T d_e S = dU + p dV. \tag{3.12}$$

The fluid is said to be isentropic if $d_e S = 0$. Since $V = 1/\rho$ we have $dU = \dfrac{p}{\rho^2} d\rho$ by (3.12) in this case, which means

$$dw = \frac{1}{\rho} dp \tag{3.13}$$

for $w = U + \dfrac{p}{\rho}$. This w, called the specific enthalpy, is a thermodynamical quantity. Hence it holds that $w = w(p)$ by (3.13), and also a functional relation between ρ and p. This relation is indicated as the state equation $p = p(\rho)$. Hence an isentropic fluid is barotropic.

If the outer force is a potential, we have $F = \nabla\varphi$ with a scalar field φ in (3.11). Hence it holds that

$$\frac{Dv}{Dt} = -\nabla Q, \quad Q = P - \varphi \qquad (3.14)$$

which implies

$$\omega_t + \nabla \times (v \cdot \nabla)v = 0, \quad \omega = \nabla \times v.$$

Since

$$(v \cdot \nabla)v = \frac{1}{2}\nabla|v|^2 - v \times \omega \qquad (3.15)$$

and

$$\begin{aligned}\nabla \times (v \times \omega) &= (\omega \cdot \nabla)v - (v \cdot \nabla)\omega + v\nabla \cdot \omega - \omega\nabla \cdot v \\ &= -(v \cdot \nabla)\omega + (\omega \cdot \nabla)v - \omega\nabla \cdot v\end{aligned}$$

it holds that

$$\frac{D\omega}{Dt} = (\omega \cdot \nabla)v - (\nabla \cdot v)\omega. \qquad (3.16)$$

Equation (3.3) means

$$\frac{D\rho}{Dt} = -(\nabla \cdot v)\rho,$$

and, therefore, we obtain

$$\frac{D}{Dt}\left(\frac{\omega}{\rho}\right) = \left(\frac{\omega}{\rho} \cdot \nabla\right)v \qquad (3.17)$$

by (3.16).

For the moment we assume that the fluid is occupied in the whole space $\mathbf{R}^n$ with rapid decays at ∞ and also sufficient regularities of all physical quantities. First, (3.3) implies $\rho \geq 0$ from $\rho|_{t=0} \geq 0$. The total mass conservation,

$$M = \int_{\mathbf{R}^n} \rho \, dx,$$

also follows from this equation. Assuming (3.11) with $F = 0$, we now derive the total energy conservation

$$E = \int_{\mathbf{R}^n} \frac{\rho}{2}|v|^2 + Q(\rho) \, dx, \quad Q''(\rho) = \frac{p'(\rho)}{\rho}. \qquad (3.18)$$

In fact, in this case, equation (3.11) is reduced to

$$(\rho v)_t + \nabla \cdot \rho v \otimes v + \nabla p = 0 \qquad (3.19)$$

by (3.3) where $v \otimes v = (v^i v^j)$ for $v = (v^j)_{1 \leq j \leq n}$. Here we have

$$\int_{\mathbf{R}^n} [\nabla \cdot \rho v \otimes v] \cdot v \, dx = \int_{\mathbf{R}^n} [\partial_j (\rho v^i v^j)] \, v^i \, dx = -\int_{\mathbf{R}^n} \rho v^j v^i \partial_j v^i \, dx$$

$$= -\frac{1}{2} \int_{\mathbf{R}^n} \rho v^j \partial_j |v|^2 \, dx = \frac{1}{2} \int_{\mathbf{R}^n} |v|^2 \, \nabla \cdot (\rho v) \, dx = -\frac{1}{2} \int_{\mathbf{R}^n} |v|^2 \, \rho_t \, dx$$

and

$$\int_{\mathbf{R}^n} (\rho v)_t \cdot v \, dx = \int_{\mathbf{R}^n} \rho_t |v|^2 + \rho v_t \cdot v \, dx = \int_{\mathbf{R}^n} \rho_t |v|^2 + \frac{1}{2} \rho \partial_t |v|^2 \, dx,$$

which results in

$$\int_{\mathbf{R}^n} [(\rho v)_t + \nabla \cdot \rho v \otimes v] \cdot v \, dx = \frac{1}{2} \int_{\mathbf{R}^n} |v|^2 \, \rho_t + \rho \partial_t |v|^2 \, dx$$

$$= \frac{1}{2} \frac{d}{dt} \int_{\mathbf{R}^n} |v|^2 \, \rho \, dx. \qquad (3.20)$$

Next, we have

$$\int_{\mathbf{R}^n} \nabla p \cdot v \, dx = \int_{\mathbf{R}^n} \nabla Q(\rho) \cdot \rho v \, dx = \int_{\mathbf{R}^n} Q'(\rho) \rho_t \, dx = \frac{d}{dt} \int_{\mathbf{R}^n} Q(\rho) \, dx. \qquad (3.21)$$

Then (3.18) follows from (3.19)–(3.21).

If the state equation is given by

$$p = A\rho^\gamma, \quad A > 0, \ 1 < \gamma < 2 \qquad (3.22)$$

it follows that

$$Q(\rho) = \frac{A\rho^\gamma}{\gamma - 1}, \quad E = \int_{\mathbf{R}^3} \frac{\rho}{2} |v|^2 + \frac{A\rho^\gamma}{\gamma - 1} \, dx.$$

In (3.22), $A > 0$ is a constant determined by the total entropy of the system, and $1 < \gamma < 2$ is the adiabatic constant. The isentropic fluid is assumed here, while $\gamma = 5/3, 7/5, \ldots$ when the gas is mono-atomic, bi-atomic, and so on, and $\gamma = 4/3$ in the case of the excellent radiational pressure.

The self-gravitating fluid is associated with

$$F = \nabla \Gamma * \rho$$

in (3.2)–(3.3) where

$$\Gamma(x) = \frac{1}{(n-2)\omega_n |x|^{n-2}}, \quad n \geq 3 \qquad (3.23)$$

with ω_n standing for the $(n-1)$-dimensional volume of the unit ball in $\mathbf{R}^n$ and

$$(\Gamma * \rho)(x) = \int_{\mathbf{R}^n} \Gamma(x - y)\rho(y) \, dy.$$

These relations are combined with the state equation (3.22), and thus there arises the Euler-Poisson equation

$$\rho_t + \nabla \cdot (\rho v) = 0, \quad \rho\,(v_t + (v \cdot \nabla)v) + \nabla p + \rho \nabla \Phi = 0$$
$$\Delta \Phi = \rho, \quad p = A\rho^\gamma \quad \text{in } \mathbf{R}^n \times (0, T) \tag{3.24}$$

with $\Phi = -F$.

The first equation of (3.24) guarantees the non-negativity $\rho \geq 0$ from that of the initial value, $\rho|_{t=0} \geq 0$. The total mass conservation,

$$M = \int_{\mathbf{R}^n} \rho \, dx,$$

follows also from this equation, while the total energy conservation

$$E = \int_{\mathbf{R}^n} \frac{\rho}{2}\,|v|^2 + \frac{p}{\gamma - 1}\,dx - \frac{1}{2}\langle \Gamma * \rho, \rho \rangle \tag{3.25}$$

is derived using (3.19), similarly to (3.18). In fact, we have, besides (3.20)–(3.21),

$$\int_{\mathbf{R}^n} \rho \nabla \Phi \cdot v \, dx = - \int_{\mathbf{R}^n} \Phi \nabla \cdot (v\rho) \, dx = \int_{\mathbf{R}^n} \Phi \rho_t \, dx = -\frac{1}{2}\frac{d}{dt}\langle \Gamma * \rho, \rho \rangle$$

by $\Phi = -\Gamma * \rho$. Hence $\dfrac{dE}{dt} = 0$ follows for E defined by (3.25). This variational structure is associated with higher dimensional Trudinger-Moser inequality

$$\inf \{\mathcal{F}(\rho) \mid \rho \geq 0, \ \|\rho\|_1 = \lambda_* \} = 0, \quad \gamma = 2 - \frac{2}{n}, \tag{3.26}$$

where $\lambda_* = \lambda_*(n) > 0$ is the critical mass determined by the next section and

$$\mathcal{F}(\rho) = \int_\Omega \frac{A\rho^\gamma}{\gamma - 1}\,dx - \frac{1}{2}\langle \Gamma * \rho, \rho \rangle. \tag{3.27}$$

Inequality (3.26) then implies

$$\sup_{t \geq 0} \left\{ \int_{\mathbf{R}^n} \rho^\gamma + \rho|v|^2 \, dx + \langle \Gamma * \rho, \rho \rangle \right\} \leq C$$

in (3.24), provided that $\gamma > 2 - \frac{2}{n}$ or $\gamma = 2 - \frac{2}{n}$ and $\lambda < \lambda_*$. Here, the critical exponent and the critical mass are derived from the scaling invariance of (3.24),

$$\rho_\mu(x, t) = \mu^{\frac{2}{2-\gamma}} \rho(\mu x, \mu^{\frac{1}{2-\gamma}} t), \quad v_\mu(x, t) = \mu^{\frac{\gamma-1}{2-\gamma}} v(\mu x, \mu^{\frac{1}{2-\gamma}} t)$$

which results in

$$\|\rho_\mu\|_1 = \mu^{\frac{2}{2-\gamma} - n}\|\rho\|_1, \quad E(\rho_\mu, v_\mu) = \mu^{\frac{2\gamma}{2-\gamma} - n} E(\rho, v).$$

The classical equilibrium of (3.24) is defined by putting $v = 0$ and $\partial_t \cdot = 0$. This formulation means

$$\frac{A\gamma}{\gamma - 1}\nabla\rho^{\gamma-1} + \nabla\Phi = 0 \qquad \text{in } \{\rho > 0\},$$

and, therefore,

$$\frac{A\gamma}{\gamma - 1}\rho^{\gamma-1} - \Gamma * \rho = \text{constant in each component of } \{\rho > 0\}$$

$$\rho \geq 0 \quad \text{in } \mathbf{R}^n, \qquad \int_{\mathbf{R}^n} \rho = M. \tag{3.28}$$

Putting

$$q = \frac{1}{\gamma - 1}, \quad u = \left(\frac{\gamma - 1}{A\gamma}\right)^{(\gamma-1)/(\gamma-2)} \rho^{\gamma-1}, \quad \Omega = \{u > 0\},$$

thus we obtain

$$-\Delta u = u^q, \ u > 0 \ \text{in } \Omega, \quad u = 0, \ \Gamma * u^q = \text{constant on } \partial\Omega$$

$$\int_\Omega u^q \, dx = \left(\frac{A\gamma}{\gamma - 1}\right)^{1/(\gamma-2)} M. \tag{3.29}$$

Here, the constant $\Gamma * u^q$ on $\partial\Omega$ may be different according as the components of Ω. Conversely, if (3.29) holds, then $u - \Gamma * u^q$ is harmonic in Ω and is a constant on the boundary of each component of Ω. This property implies the first equation of (3.29), and hence (3.29) is equivalent to (3.28).

If $u = u(x)$ satisfies

$$-\Delta u = u^q, \ u > 0 \quad \text{in } B, \qquad u = 0 \quad \text{on } \partial B \tag{3.30}$$

for $B = B(0, R)$, then $u = u(|x|)$ follows from the general theory, and, therefore, the first line of (3.29) arises with $\Omega = B$. It is known, on the other hand, that (3.30) has a solution if and only if $1 < q < (n + 2)/(n - 2)$, that is $2n/(n + 2) < \gamma < 2$.

We have the self-similarity in (3.30). If $u = u(x)$ solves

$$-\Delta u = u^q,$$

then so does $u_\mu = u_\mu(x)$ for $\mu > 0$, where

$$u_\mu(x) = \mu^{2/(q-1)}u(\mu x). \tag{3.31}$$

Thus it holds that

$$u_R(x) = R^{-2/(q-1)}U(R^{-1}x) \tag{3.32}$$

for the unique solution $U = U(y)$ to (3.30) with $B = B(0, 1)$, that is $U = u_1$.

We have

$$\int_{B(0,R)} u_R^q(x)dx = R^{n-\frac{2q}{q-1}} \int_{B(0,1)} U(y)^q dy,$$

and, therefore, if $q \neq n/(n-2)$ and $1 < q < (n+2)/(n-2)$, then any $M > 0$ admits $R > 0$ such that (3.29) has a unique solution for $\Omega = B(0,R)$. Taking zero extension outside $\Omega = B$, we obtain the equilibrium $(\rho, v) = (\bar{\rho}, 0)$ to (3.24) from this $u = u(x) \geq 0$ where

$$\bar{\rho} = \left(\frac{A\gamma}{\gamma - 1} \right)^{1/(2-\gamma)} u^q. \tag{3.33}$$

Therefore, if $\gamma \neq 2 - 2/n$ and $2n/(n+2) < \gamma < 2$, then (3.24) admits an equilibrium $(\rho, v) = (\bar{\rho}, 0)$ for each prescribed total mass $M > 0$. This $\rho = \rho(x) \geq 0$ is radially symmetric and has a compact support. Also, its total energy is defined by

$$E = \int_{\mathbf{R}^n} \frac{A\bar{\rho}^\gamma}{\gamma - 1} dx - \frac{1}{2(n-2)\omega_{n-1}} \iint_{\mathbf{R}^n \times \mathbf{R}^n} \frac{\bar{\rho}(x)\bar{\rho}(x')}{|x - x'|^{n-2}} \, dx dx'$$

$$= aR^{-\frac{4}{q-1}+n-2} \int_{\mathbf{R}^n} U(y)^{q+1} dy$$

$$-bR^{-\frac{4}{q-1}+n-2} \iint_{\mathbf{R}^n \times \mathbf{R}^n} \frac{U(y)^q U(y')^q}{|y - y'|^{n-2}} \, dy dy'$$

for

$$a = \frac{Ac^\gamma}{\gamma - 1}, \quad b = \frac{c^2}{2(n-2)\omega_{n-1}}, \quad c = \left(\frac{A\gamma}{\gamma - 1} \right)^{\frac{1}{2-\gamma}}.$$

However,

$$\bar{w} = \bar{\rho}^{(\gamma-1)/2} \approx u^{1/2}$$

does not belong to $H^1(\mathbf{R}^n)$ because this u extended 0 outside $B = B(0,R)$ has a derivative gap on ∂B by the Hopf lemma.

In the case of $\gamma = 2 - 2/n$, on the other hand, there is $\lambda_* > 0$ such that if $\lambda = \lambda^*$ then the problem (3.29) admits a unique solution for $\Omega = B(0,R)$ with $R > 0$ arbitrary, and in the other case of $\lambda \neq \lambda^*$, there is no radially symmetric solution to (3.29), where

$$\lambda = \left(\frac{A\gamma}{\gamma - 1} \right)^{1/(\gamma-2)} M.$$

If $q \geq (n+2)/(n-2)$, then

$$-\Delta v = v^q, \quad v > 0 \quad \text{in } \Omega, \qquad v = 0 \quad \text{on } \partial\Omega \tag{3.34}$$

admits no solution if Ω is star-shaped. Radially symmetric solution, on the other hand, exists for any $q > 1$ if $\Omega = \{c < |x| < d\}$. This solution, however, does not satisfy (3.29) because the values of $\Gamma * u^q$ on $|x| = c$ and $|x| = d$ are different.

There is, on the other hand, no solution to

$$-\Delta u = u^q, \quad u > 0 \quad \text{in } \mathbf{R}^n \tag{3.35}$$

if $1 < q < (n + 2)/(n - 2)$. Equation (3.35), on the contrary, has a unique radially symmetric solution $u = u(|x|)$ if $q \geq (n + 2)/(n - 2)$. In the case of $q > (n + 2)/(n - 2)$, it holds that

$$|x|^{\frac{2}{q-1}} u(|x|) \to L = \left[\frac{2}{q-1} \left(n - 2 - \frac{2}{q-1} \right) \right]^{\frac{1}{q-1}}, \quad |x| \to \infty. \tag{3.36}$$

This relation implies

$$\int_{\mathbf{R}^n} u^q = +\infty$$

and, therefore, (3.24) admits no radially symmetric equilibrium with finite total mass, in the case of $1 < \gamma < 2n/(n + 2)$.

The exponent $q = (n+2)/(n-2)$, finally, is associated with the Sobolev imbedding theorem. The solution to (3.35) is classified in this case, as it is radially symmetric with respect to some point, provided with the scaling (3.31). The solution $v = v(x)$, thus, takes the form of

$$v_{x_0,\mu}(x) = \frac{[n(n-2)\mu^2]^{(n-2)/4}}{\left(\mu^2 + |x - x_0|^2 \right)^{(n-2)/2}}, \quad x_0 \in \mathbf{R}^n, \ \mu > 0. \tag{3.37}$$

Then we see that the total mass defined by this solution is not finite. In particular, there is no radially symmetric equilibrium of (3.24) with finite total mass, also in the case of $\gamma = 2n/(n + 2)$. In this way, the radially symmetric equilibrium of the Euler-Poisson equation (3.24) has different profiles according to $\gamma \in (1, 2n/(n + 2))$, $\gamma = 2n/(n + 2)$, $\gamma \in (2n/(n + 2), 2) \setminus \{2 - 2/n\}$, and $\gamma = 2 - 2/n$.

3.2 Plasma Confinements

The equilibrium of (3.24) is reformulated by the calculus of variation. First, the total energy is reduced to

$$\mathcal{F}(\rho) = \int_{\mathbf{R}^n} \frac{p}{\gamma - 1} \, dx - \frac{1}{2} \langle \Gamma * \rho, \rho \rangle \tag{3.38}$$

in the case of $v = 0$, and this variational equilibrium is defined by $v = 0$ and $\delta\mathcal{F}(\rho) = 0$, under the constraint of

$$\rho \geq 0, \qquad \int_{\mathbf{R}^3} \rho \, dx = M.$$

We obtain the semi-unfolding

$$E(v, \rho) \geq \mathcal{F}(\rho)$$

by the definition, and, therefore, the infinitesimal stability of the variational equilibrium implies its dynamical stability by the general theory.

The functional (3.38) takes the form

$$\mathcal{F} = \frac{A\gamma}{\gamma - 1} F^* - G^*$$

for

$$F^*(\rho) = \frac{1}{\gamma} \int_{\mathbf{R}^n} \rho^\gamma \, dx, \quad \rho \geq 0$$

$$G^*(\rho) = \frac{1}{2(n-2)\omega_{n-1}} \iint_{\mathbf{R}^n \times \mathbf{R}^n} \frac{\rho(x)\rho(x')}{|x - x'|^{n-2}} \, dx dx', \quad \int_{\mathbf{R}^3} \rho \, dx = M,$$

where the Toland duality is observed.

First, we realize these functionals F^* and G^* as the Legendre transformation of proper, convex, lower semi-continuous functionals on a Banach space. To this end, we take $X = \dot{H}^1(\mathbf{R}^n) \oplus \mathbf{R}$ and define

$$G(\mu) = \frac{1}{2} \|\nabla\xi\|_2^2 + Mc, \quad \mu = \xi \oplus c \in X$$

where

$$\dot{H}^1(\mathbf{R}^n) = \left\{ \mu \in L^{\frac{2n}{n-2}}(\mathbf{R}^n) \mid \nabla\mu \in L^2(\mathbf{R}^n) \right\}.$$

We see that this functional is proper, convex, lower semi-continuous, and

$$\rho \in \partial G(\mu) \quad \Leftrightarrow \quad -\Delta\mu = \rho \text{ in } \mathbf{R}^n, \quad \int_{\mathbf{R}^n} \rho \, dx = M$$

for $(\mu, \rho) \in X \times X^*$. Furthermore, it holds that

$$G^*(\rho) = \sup_{\mu \in X} \{\langle \mu, \rho \rangle - G(\mu)\}$$

$$= \sup_{\xi \in \dot{H}^1(\mathbf{R}^n),\, c \in \mathbf{R}} \left\{ \langle \xi, \rho \rangle - \frac{1}{2} \|\nabla\xi\|_2^2 + c\langle 1, \rho \rangle - Mc \right\}$$

$$= \frac{1}{2} \langle (-\Delta)^{-1}\rho, \rho \rangle + \chi_{\{\langle \rho, 1 \rangle = M\}}, \quad \rho \in X^*,$$

under the agreement that $\Delta : \dot{H}^1(\mathbf{R}^n) \to \dot{H}^1(\mathbf{R}^n)^*$ is an isomorphism and $X^* \hookrightarrow \dot{H}^1(\mathbf{R}^n)^*$ by $\dot{H}^1(\mathbf{R}^n) \hookrightarrow X$.

Next, we put

$$F(\mu) = \frac{\gamma - 1}{\gamma} \int_{\mathbf{R}^n} \xi_+^{\frac{\gamma}{\gamma-1}} \, dx, \quad \mu = \xi \oplus c \in X,$$

which is also proper, convex, lower semi-continuous. Then we obtain

$$\rho \in \partial F(\mu) \quad \Leftrightarrow \quad \xi_+ = \rho^{\gamma-1}, \quad (\mu, \rho) \in X \times X^*, \ \mu = \xi \oplus c,$$

and, in particular, $0 \le \rho \in L^\gamma(\mathbf{R}^n)$ follows in this case. These results are summarized by

$$F^*(\rho) = \sup_{\mu \in X} \left\{ \langle \mu, \rho \rangle - F(\mu) \right\} = \begin{cases} \dfrac{1}{\gamma} \displaystyle\int_{\mathbf{R}^n} \rho^\gamma dx, \ \rho \ge 0 \\ +\infty, \qquad\qquad \text{otherwise} \end{cases}, \quad \rho \in X^*.$$

Variational equilibrium is now formulated by

$$\frac{A\gamma}{\gamma - 1} \partial F^*(\rho) \cap \partial G^*(\rho) \ne \emptyset, \quad \rho \in X^*,$$

and thus it holds that

$$-\Delta\mu = \rho \ \text{in} \ \mathbf{R}^n, \quad \rho = \left[\frac{\gamma-1}{A\gamma} \xi \right]_+^{\frac{1}{\gamma-1}}, \quad \int_{\mathbf{R}^n} \rho \, dx = M$$

for

$$\mu = \xi \oplus c \in \frac{A\gamma}{\gamma-1} \partial F^*(\rho) \cap \partial G^*(\rho).$$

This relation implies

$$-\Delta\xi = \left[\frac{\gamma-1}{A\gamma} \xi \right]_+^q \ \text{in} \ \mathbf{R}^n, \quad \int_{\mathbf{R}^n} \left[\frac{\gamma-1}{A\gamma} \xi \right]_+^q \, dx = M \qquad (3.39)$$

with $q = 1/(\gamma - 1)$, and, therefore, the variational equilibrium is defined by

$$\rho = \left[\frac{\gamma-1}{A\gamma} \xi \right]_+^q,$$

using the solution $\xi = \xi(x)$ to (3.39).

Putting $\xi = \alpha u$ and $\alpha = \left(\frac{\gamma-1}{A\gamma} \right)^{\frac{1}{2-\gamma}}$, we obtain

$$-\Delta u = u_+^q \ \text{in} \ \mathbf{R}^n, \quad \int_{\mathbf{R}^n} u_+^q \, dx = \tilde{M} \equiv \left(\frac{A\gamma}{\gamma-1} \right)^{\frac{1}{\gamma-2}} M \qquad (3.40)$$

and hence (3.29) for $\Omega = \{x \in \mathbf{R}^n \mid u(x) > 0\}$.

Thus, a variational equilibrium is a classical equilibrium. The above $\xi = \xi(x)$, however, is defined on the whole space, has the $C^{2,\theta}$-regularity, and can be negative somewhere. Therefore, the construction of the solution $\xi = \xi(x)$ to (3.39) from that of $\rho = \rho(x)$ of (3.28), or $u = u(x)$ of (3.29), is not trivial.

Motion of self-gravitating fluid enclosed in a bounded domain $\Omega \subset \mathbf{R}^n$ with smooth boundary $\partial\Omega$ may be modeled by

$$\rho_t + \nabla \cdot (\rho v) = 0, \quad \rho(v_t + (v \cdot \nabla)v) + \nabla p + \rho\nabla\Phi = 0$$

$$\Delta\Phi = \rho, \quad p = A\rho^\gamma \quad \text{in } \Omega \times (0,T), \quad (\nu \cdot v, \Phi)|_{\partial\Omega} = 0. \quad (3.41)$$

There arises the non-negativity, the total mass conservation, and the total energy conservation of the solution, that is,

$$\rho \geq 0, \quad \int_\Omega \rho \, dx = M$$

$$E = \int_\Omega \frac{\rho}{2}|v|^2 + \frac{p}{\gamma - 1} \, dx - \frac{1}{2}\left\langle(-\Delta)^{-1}\rho, \rho\right\rangle, \quad (3.42)$$

similarly. Then the variational equilibrium is formulated by $v = 0$ and $\delta\mathcal{F}(\rho) = 0$, where

$$\mathcal{F}(\rho) = \int_\Omega \frac{p}{\gamma - 1} \, dx - \frac{1}{2}\left\langle(-\Delta)^{-1}\rho, \rho\right\rangle \quad (3.43)$$

defined for

$$\rho \geq 0, \quad \int_\Omega \rho = M. \quad (3.44)$$

This functional is nothing but the Berestycki-Brezis functional concerning the plasma confinement. There is a Toland duality in this free boundary problem between the above described formulation and that of Temam, with the Nehari principle involved.

Thus we obtain a problem relative to (3.40) which arises also in plasma physics in the equilibrium of self-gravitating fluid, that is

$$-\Delta v = v_+^q \quad \text{in } \Omega, \quad v = \text{constant on } \Gamma = \partial\Omega$$

$$\int_\Omega v_+^q = \lambda \quad (3.45)$$

where $\lambda > 0$ is a constant. Here, $v = v(x)$ is a scalar function in (3.45). The associated variational functions are defined by

$$J(v) = \frac{1}{2}\|\nabla v\|_2^2 - \frac{1}{q+1}\int_\Omega v_+^{q+1} \, dx + \lambda v_\Gamma, \quad v \in H_c^1(\Omega), \int_\Omega v_+^q = \lambda$$

$$J^*(u) = \frac{q}{q+1}\int_\Omega u^{\frac{q+1}{q}} \, dx - \frac{1}{2}\left\langle(-\Delta)^{-1}u, u\right\rangle, \quad u \geq 0, \|u\|_1 = \lambda, \quad (3.46)$$

where
$$H_c^1(\Omega) = \left\{ v \in H^1(\Omega) \mid v = \text{constant on } \Gamma \right\}.$$
Henceforth, we describe the sub-critical or critical case, $1 < q \leq \frac{n+2}{(n-2)_+}$. Then, the solution $v = v(x) \in H_c^1(\Omega)$ to (3.45) is $C^{2,\theta}$ on $\overline{\Omega}$ from the elliptic regularity. The Toland duality is examined similarly to the whole space case described in §3.1. In fact, the stationary problem with $q = n/(n-2)$, $n \geq 3$, is scaling invariant, which results in the quantized blowup mechanism.

First, we define the field variational functional, putting
$$G(v) = \frac{1}{2} \|\nabla v\|_2^2 + \lambda v_\Gamma, \quad F(v) = \frac{1}{q+1} \int_\Omega v_+^{q+1} dx, \quad v \in X = H_c^1(\Omega).$$
Then it holds that
$$u \in \partial G(v) \Leftrightarrow (\nabla v, \nabla w) + \lambda w_\Gamma = \langle w, u \rangle, \quad \forall w \in X$$
$$\Leftrightarrow \langle 1, u \rangle = \lambda, \quad v - v_\Gamma = (-\Delta)^{-1} u \qquad (3.47)$$
and
$$u \in \partial F(v) \quad \Leftrightarrow \quad u = v_+^q \qquad (3.48)$$
for $(u, v) \in X \times X^*$. Thus, the problem (3.45) is equivalent to $u \in \partial F(v) \cap \partial G(v)$, or
$$\delta J(v) = 0, \quad u = v_+^q,$$
because
$$u = v_+^q, \ v \in X \quad \Rightarrow \quad 0 \leq u \in L^{\frac{q+1}{q}} \hookrightarrow X^*.$$
It holds that $D(J) = X$ and
$$J(v) = G(v) - F(v) = \frac{1}{2} \|\nabla v\|_2^2 + \lambda v_\Gamma - \frac{1}{q+1} \int_\Omega v_+^{q+1} dx,$$
and, therefore, the constraint $\int_\Omega v_+^q = \lambda$ is superfluous to derive (3.45). This property arises because the Nehari principle is involved in the original formulation (3.46).

The free energy to which the particle density $u \in X^*$ is subject is defined by the Legendre transformations. In fact, we have
$$G^*(u) = \sup_{v \in X} \left\{ \langle v, u \rangle - G(v) \right\}$$
$$= \sup_{v \in X} \left\{ \langle v - v_\Gamma, u \rangle + v_\Gamma \langle 1, u \rangle - \frac{1}{2} \|\nabla (v - v_\Gamma)\|_2^2 - \lambda v_\Gamma \right\}$$
$$= \chi_{\{\langle 1, u \rangle = \lambda\}} + \sup_{v \in H_0^1(\Omega)} \left\{ \langle v, u \rangle - \frac{1}{2} \|\nabla v\|_2^2 \right\}$$
$$= \chi_{\{\langle 1, u \rangle = \lambda\}} + \frac{1}{2} \left\langle (-\Delta)^{-1} u, u \right\rangle$$

and

$$F^*(u) = \sup_{v \in X} \{\langle v, u \rangle - F(v)\} = \begin{cases} \dfrac{q}{q+1} \displaystyle\int_\Omega u^{\frac{q+1}{q}}\, dx, & 0 \le u \in L^{\frac{q+1}{q}}(\Omega) \\ +\infty, & \text{otherwise} \end{cases}$$

because

$$u \in \partial F(v) \quad \Leftrightarrow \quad v \in \partial F^*(u) \quad \Leftrightarrow \quad u = v_+^q$$

for $(u, v) \in X \times X^*$. Hence the free energy is defined by

$$D(J^*) = D(G^*) \cap D(F^*)$$
$$= \left\{ u \in X^* \mid 0 \le u \in L^{\frac{q+1}{q}}(\Omega),\ \|u\|_1 = \lambda \right\}$$

and

$$J^*(u) = \begin{cases} \dfrac{q}{q+1} \displaystyle\int_\Omega u^{\frac{q+1}{q}}\, dx - \dfrac{1}{2}\langle (-\Delta)^{-1} u, u \rangle, & u \in D(J^*) \\ +\infty, & \text{otherwise,} \end{cases}$$

and this variational problem is nothing but Berestycki-Brezis' formulation to (3.45).

3.3   Maximum Entropy Production

The Smoluchowski-Poisson equation (1.9) with (1.12) arises also in non-equilibrium statistical mechanics in accordance with the Boltzmann entropy. A relative model is the degenerate parabolic equation of which stationary state is provided with the quantized blowup mechanism. In fact, the principle of maximum entropy production, used in §2.3 of this monograph, induces the parabolic-elliptic system

$$\mu_t = \nabla[D_* \cdot (\nabla p + \mu \nabla \varphi)], \quad \Delta \varphi = \mu \qquad \text{in } \Omega \times (0, T) \qquad (3.49)$$

as a hydrodynamical limit of self-gravitating particles. Here, $\mu = \mu(x, t) \ge 0$ is the function describing particle density at $(x, t) \in \Omega \times (0, T)$, $\Omega \subset \mathbf{R}^n$, $n \ge 2$, a domain, $\varphi = \varphi(x, t)$ the gravitational potential generated by μ, and $p \ge 0$ the pressure. We assume that this fluid is barotropic, and this pressure is determined by the state equation, i.e., density-pressure relation

$$p = p(\mu, \theta). \qquad (3.50)$$

If Ω has the boundary $\partial \Omega$, the null-flux boundary condition

$$(\nabla p + \mu \nabla \varphi) \cdot \nu = 0$$

is imposed with ν denoting the outer unit normal vector so that the total mass

$$\lambda = \int_\Omega \mu(x,t)dx$$

is conserved during the evolution.

More precisely, the density of particles at $(x,t) \in \Omega \times (0,T)$ moving at the velocity v is denoted by $0 \leq f = f(x,v,t)$. It is subject to the transport equation

$$f_t + v \cdot \nabla_x f - \nabla\varphi \cdot \nabla_v f = -\nabla_v \cdot j \qquad (3.51)$$

with the general dissipation flux term $-\nabla_v \cdot j$. This flux term is determined by the maximum entropy production principle, so that f maximizes the local entropy

$$S = \int_{\mathbf{R}^n} s(f(x,v,t))dv$$

under the constraint

$$\mu(x,t) = \int_{\mathbf{R}^n} f(x,v,t)dv, \quad p(x,t) = \frac{1}{n}\int_{\mathbf{R}^n} |v|^2 f(x,v,t)dv.$$

Averaging f over the velocities $v \in \mathbf{R}^n$, and then the passage to the limit of large friction or large time lead to the first equation of (3.49) in the (x,t) space. We have, thus, several mean field equations according to the entropy function $s(f)$ subject to the law of partition of macroscopic states of particles into mezoscopic states, that is the entropies of Boltzmann, Fermi-Dirac, Bose-Einstein, and so forth.

System (3.49) with (3.50) is still under-determined, and there are several theories to prescribe the temperature θ. In the cannonical statistics one takes the iso-thermal setting, and hence the temperature $\theta > 0$ is a constant. In the micro-cannonical statistics, on the other hand, $\theta = \theta(t) > 0$ is the function of t, where

$$E = \frac{n}{2}\int_\Omega pdx + \frac{1}{2}\int_\Omega \mu\varphi\, dx$$

is the prescribed total energy independent of t.

First, the transport equation (3.51) is obtained by putting $\gamma = 0$ and $F = \nabla_x\varphi$ in the Kramers equation (1.168). The diffusion coefficient D, furthermore, vanishes here, which is compensated by the general dissipation flux term $-\nabla_v \cdot j$. Turning to the physical quantities, the total mass is given by

$$M = \int \rho\, dx, \quad \rho = \int f\, dv. \qquad (3.52)$$

The total energy is the sum of kinetic and potential energies, that is

$$E = \frac{1}{2} \iint f|v|^2 \, dxdv + \frac{1}{2} \int \rho\varphi \, dx = \frac{1}{2} \iint f(|v|^2 + \varphi) \, dxdv. \quad (3.53)$$

The entropy is defined, using the entropy density $s = s(f)$, by

$$S = \iint s(f) \, dxdv. \quad (3.54)$$

Putting $U = \begin{pmatrix} v \\ -\nabla\varphi \end{pmatrix}$, $\nabla = \begin{pmatrix} \nabla_x \\ \nabla_v \end{pmatrix}$, we obtain the energy production as

$$
\begin{aligned}
\dot{E} = \frac{dE}{dt} &= \frac{1}{2} \iint f_t(|v|^2 + \varphi) + f\varphi_t \, dxdv \\
&= \frac{1}{2} \iint (-\nabla_v \cdot j)(|v|^2 + \varphi) \, dxdv - \frac{1}{2} \iint (U \cdot \nabla f)(|v|^2 + \varphi) - f\varphi_t \, dxdv \\
&= \frac{1}{2} \iint (-\nabla_v \cdot j)|v|^2 - (v \cdot \nabla_x f)\varphi + (\nabla_x\varphi \cdot \nabla_v f)|v|^2 + f\varphi_t \, dxdv
\end{aligned}
$$

by (3.53). Since

$$
\begin{aligned}
(v \cdot \nabla_x f)\varphi - (\nabla_x\varphi \cdot \nabla_v f)|v|^2 &= \nabla_x \cdot (vf\varphi) - \nabla_x\varphi \cdot (vf + |v|^2 \nabla_v f) \\
&= \nabla_x \cdot (vf\varphi) - \nabla_x\varphi \cdot \nabla_v \cdot ((v \otimes v)f)
\end{aligned}
$$

it holds that

$$\dot{E} = \iint j \cdot v + \frac{1}{2} f\varphi_t \, dxdv.$$

Similarly, (3.54) implies the entropy production,

$$
\begin{aligned}
\dot{S} = \frac{dS}{dt} &= \iint s'(f)f_t \, dxdv = \iint s'(f)[-v \cdot \nabla_x f + \nabla_x\varphi \cdot \nabla_v f - \nabla_v \cdot j]dxdv \\
&= \iint s''(f)j \cdot f - v \cdot \nabla_x s(f) + \nabla_x\varphi \cdot \nabla_v s(f) \, dxdv \\
&= \iint s''(f)j \cdot f \, dxdv.
\end{aligned}
$$

The maximum entropy production principle says that j is selected to maximize $\dot{S}$ under the constraint of $\dot{E} = 0$. Here we impose the boundedness of $\frac{|j|^2}{2f}$, furthermore, to get the Lagrange multiplies $\beta(t)$ and D such that

$$\delta\dot{S} - \beta(t)\delta\dot{E} - \iint \frac{1}{D} \cdot \delta\left(\frac{|j|^2}{2f}\right) \, dxdv = 0. \quad (3.55)$$

Equation (3.55) means $j = D\left[fs''(f)\nabla_v f - \beta(t)fv\right]$ and then it follows that

$$f_t + v \cdot \nabla_x f - \nabla_x \varphi \cdot \nabla_v f = \nabla_v \cdot D\left[-fs''(f)\nabla_v f + \beta(t)fv\right] \qquad (3.56)$$

from (3.51).

Taking the canonical ensemble to see the mean field of particles, we just put $\beta(t)$ in (3.56) to be a constant. Thus it is actually a function of t if the micro-canonical ensemble is used, which is to be determined by (3.56) and (3.53) with given E. The Kramers equation (1.168) is nothing but (3.56) in the canonical setting, using the Boltzmann entropy

$$s(f) = -f(\log f - 1).$$

Then the fundamental equations of fluid dynamics are obtained by several moments of the general Kramers equation (3.56) in the canonical setting.

First, the zero moment equation is derive just by an integration in v. Using the particle density ρ defined by the second equality of (3.52) and the particle velocity $u = \dfrac{1}{\rho} \displaystyle\int fv \, dv$, then we obtain the *equation of continuity*,

$$\frac{\partial \rho}{\partial t} + \nabla \cdot \rho u = 0. \qquad (3.57)$$

Next, the first moment arises with the relative velocity $w = u - v$ and stress tensor $P_{ij} = \displaystyle\int f w_i w_j \, dv$. In fact, we multiply v to (3.56) get

$$\frac{\partial}{\partial t}(\rho u_i) + \frac{\partial}{\partial x_j}(\rho u_i u_j) + \frac{\partial}{\partial x_j}P_{ij} + \rho\frac{\partial \varphi}{\partial x_i}$$
$$= D\int fs''(f)\frac{\partial f}{\partial v_i} + \beta f v_i \, dv. \qquad (3.58)$$

Using $A'(f) = fs''(f)$, here we obtain

$$\int fs''(f)\frac{\partial f}{\partial v_i} \, dv = \int \frac{\partial}{\partial v_i} A(f) \, dv = 0,$$

and, therefore, (3.58) is reduced to the *equation of motion*,

$$\frac{\partial}{\partial t}(\rho u_i) + \frac{\partial}{\partial x_j}(\rho u_i u_j) + \frac{\partial}{\partial x_j}P_{ij} + \rho\frac{\partial \varphi}{\partial x_i} = -D\beta\rho u_i. \qquad (3.59)$$

If the gasses are under the thermal equilibrium, the right-hand side on (3.59) vanishes. If the fluid is incompressible and isotropic, furthermore, then ρ is a constant and it holds that

$$P_{ij} = p\delta_{ij}. \qquad (3.60)$$

By (3.57) and (3.60) we obtain (2.2), the Euler equation of motion. Then *equation of energy balance* arises with the second moment, that is

$$\rho\frac{\partial e}{\partial t} + p\nabla \cdot u = 3D\rho, \quad p = \frac{2}{3}\rho e = \frac{1}{3}\int f|w|^2 \, dv,$$

see [Chavanis, Sommeria, and Robert (1996)].

If ρ is unknown, we require one more equation other than (3.57) and (3.59). To this end we use the free energy minimum principle at each time. For the moment we drop the variable t. Here we use the Helmholz free energy valid to the canonical ensemble,

$$F = E - TS = \iint \frac{1}{2}(|v|^2 + \varphi)f - Ts(f) \, dxdv \qquad (3.61)$$

where T denotes the temperature. Comparing (3.61) with (3.55), we regard $\beta = T^{-1}$ as the inverse temperature. At the local thermal equilibrium, the minimum of (3.61) is attained under the constraint of

$$\iint f \, dxdv = \lambda.$$

Hence this f satisfies the Euler-Lagrange equation

$$s'(f) = \beta\left(\frac{|v|^2}{2} + \lambda(x)\right)$$

with the Lagrangian multiplier $\lambda(x)$. Consequently, it follows that

$$f = A\left(\beta\left(\frac{|v|^2}{2} + \lambda(x)\right)\right) \qquad (3.62)$$

where $f = A(\sigma)$ denotes the inverse function of $\sigma = s'(f)$.

The density ρ and the pressure p satisfy

$$\rho = \int f \, dv, \quad p = \frac{1}{3}\int f|w|^2 \, dv, \quad u = \frac{1}{\rho}\int fv \, dv$$

with $w = u - v$, that is

$$\rho = \int f \, dv, \quad p = \frac{1}{3}\int f|v|^2 \, dv - \frac{1}{3}\left|\int fv \, dv\right|^2. \qquad (3.63)$$

From (3.62) and (3.63) we obtain a functional relation between ρ and p, that is the state equation indicated by

$$p = p(\rho). \qquad (3.64)$$

If the case of isotropic (3.60), equations (3.57)–(3.59) are reduced to

$$\frac{\partial\rho}{\partial t} + \nabla \cdot \rho u = 0, \quad \rho\frac{Du}{Dt} = -\nabla p - \rho\nabla\varphi - \xi\rho u \qquad (3.65)$$

where $\xi = D\beta$ and $\dfrac{D}{Dt} = \dfrac{\partial}{\partial t} + u \cdot \nabla$. In the thermal equilibrium, we obtain

$$\frac{\partial \rho}{\partial t} + \nabla \cdot \rho u = 0, \quad \rho \frac{Du}{Dt} = -\nabla p - \rho \nabla \varphi, \quad p = p(\rho) \tag{3.66}$$

by (3.65) and (3.64). System (3.66) is nothing but the *compressible Euler equation*. In the high friction limit the right-hand side on the second equation (3.65) is zero, which results in the generalized Smoluchowski equation

$$\frac{\partial \rho}{\partial t} = \frac{1}{\xi} \nabla \cdot [\nabla p + \rho \nabla \varphi]. \tag{3.67}$$

The state equation (3.64) is determined by (3.62)–(3.63), depending on the choice of the entropy density $s = s(f)$. It may be the entropies of Boltzman, Fermi-Dirac, Bose-Einstein, and so forth. In the case of the Boltzmann entropy

$$s(f) = -kf(\log f - 1) \tag{3.68}$$

equation (3.64) results in the Boyle-Charles' law

$$p = k\rho T, \tag{3.69}$$

see [Chavanis (2004)]. Then the standard Smoluchowski equation arises with (3.67) and (3.69), that is

$$\frac{\partial \rho}{\partial t} = \nabla \cdot (\nabla \rho + \rho \nabla \varphi)$$

if all the physical constants are 1. The Tsallis entropy is the q-analogue of the Boltzmann entropy, given by $s(f) = \dfrac{-1}{q-1}(f^q - f)$. Particles composing a polytropic celestial body are subject to this entropy. Then the state equation (3.64) becomes

$$p = K\rho^{1+\gamma}, \quad \frac{1}{\gamma} = \frac{1}{q-1} + \frac{n}{2} \tag{3.70}$$

where K is a constant and $n = 3$ is the space dimension. From (3.70) and (3.67) we obtain

$$\frac{\partial \rho}{\partial t} = \nabla \cdot (\nabla \rho^m + \rho \nabla \varphi), \quad \frac{1}{m-1} = \frac{1}{q-1} + \frac{n}{2}, \tag{3.71}$$

putting all the physical constants to be 1.

3.4 Degenerate Parabolic Equations

We take (3.71) on the whole space $\Omega = \mathbf{R}^n$, provided with the gravitational potential φ for the self-interaction, that is $\Delta\varphi = \rho$. Normalizing constants again, we arrive at the degenerate parabolic equation

$$u_t = \frac{m-1}{m}\Delta u^m - \nabla \cdot u\nabla\Gamma * u, \quad u \geq 0 \quad \text{in } \mathbf{R}^n \times (0,T) \qquad (3.72)$$

where

$$\Gamma(x) = \frac{1}{\omega_n(n-2)|x|^{n-2}} \qquad (3.73)$$

with ω_n denoting the $(n-1)$ dimensional volume of the boundary of the unit ball in $\mathbf{R}^n$ if $n \geq 3$ and

$$\Gamma(x) = \frac{1}{2\pi}\log\frac{1}{|x|} \qquad (3.74)$$

if $n = 2$. Thus, $\Gamma(x)$ is the fundamental solution to $-\Delta$ and

$$\Gamma * u(x,t) = \int_{\mathbf{R}^n} \Gamma(x-x')u(x',t)\,dx'.$$

If $n = 3$ and $q = \frac{5}{3}$ in (3.71), we have $m = \frac{4}{3}$. Equation (3.72) of this exponent, $m = 2 - \frac{2}{n}$ is regarded as a higher-dimensional analogue of the Smoluchowski-Poisson equation associated with the Boltzmann entropy in two-space dimensions, that is, (1.13) given as

$$u_t = \Delta u - \nabla \cdot u\nabla\Gamma * u, \quad u \geq 0 \quad \text{in } \mathbf{R}^2 \times (0,T) \qquad (3.75)$$

for $\Gamma = \Gamma(x)$ defined by (3.74). Similarly to (1.1), the simplified system of chemotaxis, equation (3.75) is provided with the total mass conservation (1.14):

$$\frac{d}{dt}\int_{\mathbf{R}^2} u\,dx = 0 \qquad (3.76)$$

and the decrease of the total free energy (1.22):

$$\frac{d}{dt}\mathcal{F}(u) = -\int_{\mathbf{R}^2} u|\nabla(\log u - \Gamma * u)|^2\,dx \leq 0$$

$$\mathcal{F}(u) = \int_{\mathbf{R}^2} u(\log u - 1)\,dx - \frac{1}{2}\langle\Gamma * u, u\rangle. \qquad (3.77)$$

Equation (3.72) is also a model (B) equation associated with the *free energy*

$$\mathcal{F}(u) = \int_{\mathbf{R}^n} \frac{u^m}{m}\,dx - \frac{1}{2}\langle\Gamma * u, u\rangle. \qquad (3.78)$$

It is the functional that formulates the equilibrium of the Euler-Poisson equation in §3.1. In fact, we have

$$\delta \mathcal{F}(u)[v] = \frac{d}{ds} \mathcal{F}(u + sv) \Big|_{s=0} = \langle v, u^{m-1} - \Gamma * u \rangle,$$

where $\langle \ , \ \rangle$ denotes the L^2-inner product. Identifying $\mathcal{F}(u)$ with $u^{m-1} - \Gamma * u$, we can write (3.72) as

$$u_t = \nabla \cdot \left(\frac{m-1}{m} \nabla u^m - u \nabla \Gamma * u \right)$$
$$= \nabla \cdot u \nabla \delta \mathcal{F}(u) \qquad \text{in } \mathbf{R}^n \times (0, T). \qquad (3.79)$$

From this form of (3.79), it is easy to infer, at least formally, the total mass conservation (3.76),

$$\|u(t)\|_1 = \|u_0\|_1 = \lambda \qquad (3.80)$$

and the decrease of the total free energy (3.77),

$$\frac{d}{dt} \mathcal{F}(u) = - \int_{\mathbf{R}^n} u \left| \nabla \delta \mathcal{F}(u) \right|^2 dx$$
$$= - \int_{\mathbf{R}^n} u \left| \nabla (u^{m-1} - \Gamma * u) \right|^2 dx \leq 0. \qquad (3.81)$$

We have rigorous proof for these properties using weak solutions.

Regarding (3.80)–(3.81), we formulate the stationary state by

$$u^{m-1} - \Gamma * u = \text{constant in } \{u > 0\}, \qquad \int_{\mathbf{R}^n} u = \lambda. \qquad (3.82)$$

Then we assume $u = v_+^{\frac{1}{m-1}}$ for $v = v(x)$ satisfying

$$-\Delta v = v_+^q \quad \text{in } \mathbf{R}^n, \qquad \int_{\mathbf{R}^n} v_+^q \, dx = \lambda, \qquad (3.83)$$

where $m = 1 + \frac{1}{q}$. We obtain, thus, a relative to (3.45) for the bounded domain case studied in §3.2.

Problem (3.83) is invariant under the scaling transformation

$$v(x) \mapsto v_\mu(x) = \mu^\gamma v(\mu x) \qquad (3.84)$$

if and only if $\gamma = n - 2$ and $q = \frac{1}{m-1} = \frac{n}{n-2}$, that is $m = 2 - \frac{2}{n}$, where $\mu > 0$ is a constant. For this exponent, problem (3.84) admits a family of solutions, each of which is necessarily radially symmetric and has compact support. Then, we recall the results in §3.1 to define the normalized solution $v_* = v_*(x)$ to (3.83) which determines the quantized mass $\lambda_* > 0$ by

$$-\Delta v_* = v_{*+}^q, \ v_* \leq v_*(0) = 1 \quad \text{in } \mathbf{R}^n, \qquad \lambda_* = \int_{\mathbf{R}^n} v_{*+}^q \, dx. \qquad (3.85)$$

The above solution $v_* = v_*(x)$ is unique and radially symmetric. The constant $\lambda_* = \lambda_*(n) > 0$ defined by (3.85) is the best constant of the following inequality, regarded as a dual form of the Trudinger-Moser inequality

$$\inf\left\{\mathcal{F}(u) \mid 0 \le u \in L^m(\mathbf{R}^n), \int_{\mathbf{R}^n} u\, dx = \lambda_*\right\} > -\infty \qquad (3.86)$$

where $m = 2 - \frac{2}{n}$.

Two-dimensional analogue of (3.85) is the Boltzmann-Poisson equation

$$-\Delta v = e^v \quad \text{in } \mathbf{R}^2, \qquad \int_{\mathbf{R}^2} e^v dx < +\infty \qquad (3.87)$$

is studied in Chapter 1.11. Different from (3.87), v_+^q in (3.85) has a compact support. The other difference is the scaling property of the free energy

$$\mathcal{F}(u_\mu) = \mu^{n-2}\mathcal{F}(u) \qquad (3.88)$$

which refines (3.86) as

$$\inf\{\mathcal{F}(u) \mid 0 \le u \in L^m(\mathbf{R}^n), \int_{\mathbf{R}^n} u\, dx = \lambda_*\} = 0. \qquad (3.89)$$

The non-stationary problem (3.72), $m = 2 - \frac{2}{n}$, is also provided with the L^1-preserving self-similar transformation

$$u_\mu(x, t) = \mu^n u(\mu x, \mu^n t). \qquad (3.90)$$

This transformation induces the backward self-similar transformation

$$v(y, s) = (T - t)u(x, t), \ y = (x - x_0)/(T - t)^{1/n}, \ s = -\log(T - t) \qquad (3.91)$$

and the rescaled equation

$$v_s = \frac{m-1}{m}\Delta v^m - \nabla \cdot v\nabla(\Gamma * v + \frac{|y|^2}{2n})$$
$$v \ge 0 \quad \text{in } \mathbf{R}^n \times (-\log T, +\infty). \qquad (3.92)$$

The main difference between 2D Smoluchowski-Poisson equation and higher dimensional degenerate parabolic equation with critical exponent lies on the potential of self-interaction, that is, $x \cdot \nabla\Gamma = -\frac{1}{2\pi}$ and $x \cdot \nabla\Gamma = -(n-2)\Gamma$ for $\Gamma(x)$ defined by (3.74) and (3.73), respectively.

Since (3.72) is degenerate, the solution to (3.72) which we handle with is the weak solution. First, given the initial value

$$0 \le u_0 \in L^1(\mathbf{R}^n) \cap L^\infty(\mathbf{R}^n), \quad u_0^m \in H^1(\mathbf{R}^n), \qquad (3.93)$$

we take the approximate solution $u_\varepsilon = u_\varepsilon(x, t)$ satisfying

$$u_{\varepsilon t} = \frac{m-1}{m}\Delta(u_\varepsilon + \varepsilon)^m - \nabla \cdot (u_\varepsilon \nabla\Gamma * u_\varepsilon) \quad \text{in } \mathbf{R}^n \times (0, T)$$
$$u|_{t=0} = u_{0\varepsilon}, \quad 0 < \varepsilon \ll 1,$$

where

$$0 \le u_{0\varepsilon} \in L^1 \cap W^{2,p}(\mathbf{R}^n), \; p \in [\frac{n}{n-1}, n+3]$$
$$\|u_{0\varepsilon}\|_p \le \|u_0\|_p, \quad p \in [1, \infty]$$
$$\|\nabla u_{0\varepsilon}^m\|_2 \le \|\nabla u_0^m\|_2$$

and

$$u_{0\varepsilon} \to u_0, \; \varepsilon \downarrow 0 \quad \text{in } L^p(\mathbf{R}^n)$$

for some $p \in [\frac{n}{n-1}, \infty)$. Then we obtain the following theorem, passing through a subsequence of $\varepsilon \downarrow 0$.

Theorem 3.1. *Assuming (3.93), we have $0 < T \ll 1$ such that (3.72) has a weak solution in the sense that*

$$\iint_{\mathbf{R}^n \times [0,T]} \frac{m-1}{m} \nabla u^m \cdot \nabla \xi - u \nabla \Gamma * u \cdot \nabla \xi - u \xi_t \, dx dt = \int_{\mathbf{R}^n} u_0 \xi \, dx$$

provided with the properties $u \in C_*([0,T), L^p(\mathbf{R}^n))$, $1 < p \le \infty$, regarding $L^p(\mathbf{R}^n) = L^{p'}(\mathbf{R}^n)'$, $\frac{1}{p'} + \frac{1}{p} = 1$,

$$u \in L^\infty([0,T]; L^1(\mathbf{R}^n)) \cap L^\infty_{loc}([0,T); L^\infty(\mathbf{R}^n))$$
$$\nabla u^m \in L^\infty(0,T; L^2(\mathbf{R}^n))$$
$$\partial_t u^{\frac{m+1}{2}} \in L^2(0,T; L^2(\mathbf{R}^n))$$
$$\nabla \Gamma * u \in L^\infty_{loc}([0,T); L^2(\mathbf{R}^n)), \tag{3.94}$$

and

$$\|u(\cdot,t)\|_1 = \|u_0\|_1, \quad t \in [0,T), \tag{3.95}$$

where $\xi \in H^1(0,T; L^2(\mathbf{R}^n)) \cap L^2(0,T; H^1(\mathbf{R}^n))$ is the test function satisfying $\xi(\cdot,t) = 0$ for $0 < T - t \ll 1$. Furthermore, it holds that

$$u_\varepsilon \rightharpoonup u \; * \text{ weakly in } L^\infty(0,T; L^q(\mathbf{R}^n)), \quad \forall q \in (1, \infty] \tag{3.96}$$

in the sense of $L^\infty(0,T; L^q(\mathbf{R}^n)) = L^1(0,T; L^{q'}(\mathbf{R}^n))'$, $\frac{1}{q'} + \frac{1}{q} = 1$, passing through a subsequence of the above approximate solutions. If the existence time of the weak solution u denoted by $T = T_{\max} \in (0, +\infty]$ is finite, then

$$\lim_{t \uparrow T} \|u(\cdot,t)\|_\infty = +\infty. \tag{3.97}$$

The above approximate solution $u_\varepsilon = u_\varepsilon(x,t)$ is extended as far as $\|u_\varepsilon(\cdot,t)\|_\infty$ is bounded, while

$$B \equiv \sup_{t\in(0,T)} \|u_\varepsilon(\cdot,t)\|_\infty \leq \frac{\|u_0\|_\infty}{1 - T\|u_0\|_\infty}, \qquad 0 < T < \|u_0\|_\infty^{-1} \qquad (3.98)$$

holds by an energy method. Then we derive several estimates of $u_\varepsilon(\cdot,t)$ using B uniform in $0 < \varepsilon \ll 1$ and $0 \leq t \leq T$ to take the process of the passage to the limit. This argument guarantees that the existence time of the weak solution $u = u(\cdot,t)$ is bounded from below by $\|u_0\|_\infty$, and, consequently, it extends in t as far as $\|u(\cdot,t)\|_\infty$ is bounded. With the lack of sufficient decay at the infinity of Γ, we use

$$\|\nabla(\Gamma * u)\|_\infty \leq C(n,q)(\|u\|_1 + \|u\|_q), \quad n < q \leq \infty$$

derived from the decomposition

$$\Gamma * u = v_1 + v_2, \quad v_1 = [\Gamma \cdot \chi_{\mathbf{R}^n \setminus B(0,1)}] * u, \quad v_2 = [\Gamma \cdot \chi_{B(0,1)}] * u \quad (3.99)$$

and also the Calderón-Zygmund estimate

$$\|\nabla^2(\Gamma * u)\|_p \leq C(n,p)\|u\|_p, \quad 1 < p < \infty.$$

There may be the case that this solution exists after the classical solution expires, which we call the incomplete blowup. Henceforth $T = T_{\max} \in (0,+\infty]$ denotes the existence time of the weak solution $u = u(\cdot,t)$. We take the case

$$\int_{\mathbf{R}^n} |x|^2 u_0 \, dx < +\infty \qquad (3.100)$$

to control the behavior of the solution at $x = \infty$. The next theorem assures the existence of the threshold of $\lambda = \|u_0\|_1$ for $T = +\infty$. The threshold value λ_* may be prescribed by the best constant $C(n)$ of the Hardy-Littlewood-Sobolev inequality

$$|\langle f, \Gamma * f\rangle| \leq C(n)\|f\|_m^m\|f\|_1^{2/n}, \quad m = 2 - \frac{2}{n}. \qquad (3.101)$$

It is actually equal to the one defined by (3.85) for $q = \frac{n}{n-2}$.

Theorem 3.2. *If $u_0 = u_0(x)$ is the initial value satisfying (3.93), (3.100), and $\|u_0\|_1 < \lambda_*$, then $T = +\infty$ holds in (3.72) for $m = 2 - \frac{2}{n}$, $n \geq 3$. Each $\lambda > \lambda_*$, on the other hand, takes $u_0 = u_0(x)$ such that (3.93), (3.100), $\|u_0\|_1 = \lambda$, and $T < +\infty$.*

The arguments described above and below are justified using the approximate solution. For the proof of Theorem 3.2, first, if $\|u_0\|_1 = \lambda < \lambda_*$ is the case then

$$\limsup_{t\uparrow T} \|u(\cdot, t)\|_m < +\infty \qquad (3.102)$$

holds by (3.81) and (3.89). Now we set up Moser's iteration scheme which guarantees

$$\limsup_{t\uparrow T} \|u(\cdot, t)\|_\infty < +\infty$$

and hence $T = +\infty$ by (3.97).

From $|x|^2 u_0 \in L^1(\mathbf{R}^n)$, next, the function

$$t \in [0, T) \mapsto \int_{\mathbf{R}^n} |x|^2 u(x, t) \, dx \in [0, +\infty)$$

is locally absolutely continuous and it holds that

$$\frac{d}{dt} \int_{\mathbf{R}^n} |x|^2 u \, dx = \frac{m-1}{m} \cdot 2n \int_{\mathbf{R}^n} u^m \, dx - (n-2) \langle \Gamma * u, u \rangle$$

$$= 2(n-2)\mathcal{F}(u), \qquad (3.103)$$

which implies $T < +\infty$ in case $\mathcal{F}(u_0) < 0$. Since (3.89) is sharp and

$$\inf\{\mathcal{F}(u) \mid 0 \le u \in L^m(\mathbf{R}^n), \int_{\mathbf{R}^n} u \, dx = \lambda\} = -\infty$$

for each $\lambda > \lambda_*$, we obtain the initial value $u_0 = u_0(x) \ge 0$ with compact support such that $\mathcal{F}(u_0) < 0$ and $\|u_0\|_1 = \lambda$.

3.5 Structure of the Blowup Set

Since (3.97) arises if $T = T_{\max} < +\infty$, the blowup set is then defined by $\mathcal{S} = \mathbf{R}^n \setminus \mathcal{B}$,

$$\mathcal{B} = \{x_0 \in \mathbf{R}^n \mid \exists r > 0 \text{ such that } \limsup_{t\uparrow T} \|u(\cdot, t)\|_{L^\infty(B(x_0, r))} < +\infty\}.$$

To detect the blowup rate, we write (3.72) as

$$u_t = \frac{m-1}{m}\Delta u^m - \nabla u \cdot \nabla\Gamma * u + u^2,$$

to take the ODE part

$$\dot{\zeta} = \zeta^2.$$

Since

$$\zeta(t) = (T-t)^{-1} \qquad (3.104)$$

we define the type I blowup rate by

$$\|u(\cdot, t)\|_\infty = O((T - t)^{-1}).$$

Then we say that $x_0 \in \mathcal{S}$ is of type I if

$$\liminf_{t \uparrow T}(T - t)\|u(\cdot, t)\|_{L^\infty(B(x_0, r_0))} < +\infty, \quad \exists r_0 > 0$$

and type II in the other case.

We obtain finiteness of the type II blowup points.

Theorem 3.3. *Let* $u_0 = u_0(x)$ *be the initial value satisfying (3.93) and (3.100), and assume* $T < +\infty$ *for the above described weak solution* $u = u(x, t)$ *to (3.72) with* $m = 2 - \frac{2}{n}$. *Then,* $\mathcal{S}$ *is bounded and* $\mathcal{S}_{II}$ *is finite, where*

$$\mathcal{S}_{II} = \left\{ x_0 \in \mathcal{S} \mid \lim_{t \uparrow T}(T - t)\|u(\cdot, t)\|_{L^\infty(B(x_0, r_0))} = +\infty, \quad \forall r_0 > 0 \right\}.$$

The first step to prove Theorem 3.3 is the ε-regularity which is a localization of the anti-blowup criterion $\|u_0\|_1 < \lambda_*$.

Theorem 3.4. *We have* $\varepsilon_0 > 0$ *and* $C > 0$ *independent of* $x_0 \in \mathbf{R}^n$ *and* $0 < R \ll 1$ *such that*

$$\sup_{t \in (0,T)} \|u(\cdot, t)\|_{L^1(B(x_0, R))} < \varepsilon_0 \quad \Rightarrow \quad \sup_{t \in (0,T)} \|u(\cdot, t)\|_{L^\infty(B(x_0, R/2))} \le C$$

$$(3.105)$$

where $u = u(x, t)$ *is a weak solution to (3.72).*

For the proof of Theorem 3.4 we use

$$\|v_1\|_{W^{1,\infty}(\mathbf{R}^n)} + \|v_2\|_{W^{1,q}(\mathbf{R}^n)} \le C(n, q)\|u\|_1, \quad 1 \le q < \frac{n}{n-1}$$

valid in the decomposition (3.99), where $v = \Gamma * u$. Let G denote the Bessel potential

$$G(x) = \int_0^\infty (4\pi s)^{-n/2} \exp\left(-\frac{|x|^2}{4s} - s\right) ds$$

which satisfies $(-\Delta + 1)G = \delta$. Then, it follows that $(-\Delta + 1)v = v + u$ which implies

$$v = G * v + G * u$$

because $(-\Delta + 1)w = 0$, $w \in L^1(\mathbf{R}^n) + L^\infty(\mathbf{R}^n)$ implies $w = 0$. The exponentially decay at infinity of $G = G(x)$ now justifies the calculations used in the localization mentioned before Theorem 3.4.

A direct consequence of Theorem 3.4 is the boundedness of the blowup set $\mathcal{S}$. Here we use (3.100).

Lemma 3.1. *It holds that*

$$\limsup_{t \uparrow T} \|u(\cdot, t)\|_{L^\infty(\mathbf{R}^n \setminus B(0,R))} < +\infty$$

for $R \gg 1$.

Proof. We have

$$\int_{\mathbf{R}^n} |x|^2 u \, dx \leq C(T, u_0) \tag{3.106}$$

for

$$C(T, u_0) = 2(n-2)T\mathcal{F}(u_0) + \int_{\mathbf{R}^n} |x|^2 u_0 \, dx$$

and hence

$$\limsup_{t \uparrow T} \int_{|x| > R} u(x, t) \, dx \leq R^{-2} C(T, u_0).$$

Taking $R \gg 1$ as $C(T, u_0)R^{-2} < \varepsilon_0$, we obtain $\mathcal{S} \subset \mathbf{R}^n \setminus B(0, R)$ by Theorem 3.4. $\qquad\square$

The constant C in (3.105) is involved by the initial value. This inconvenience, however, is removed by the parabolic regularity concerning local norms of the solution as in §1.2 concerning 2D Smocluchowski-Poisson equation.

Lemma 3.2. *Each $r \in [2, \infty)$ admits $0 < \varepsilon_r \ll 1$ such that*

$$\sup_{t \in (0,1)} \|u(\cdot, t)\|_{L^1(B(x_0, 2R))} < \varepsilon_r \quad \Rightarrow \quad \|u(\cdot, t)\|_{L^r(B(x_0, R))} \leq t^{-1}$$

for $0 < t \leq 1$, where $u = u(\cdot, t)$ is the weak solution to (3.72), $x_0 \in \mathbf{R}^n$, and $R > 0$.

Given $x_0 \in \mathcal{S}$ and $0 < R \ll 1$, we take $0 \leq \varphi = \varphi_{x_0, R}(x) \in C_0^\infty(\mathbf{R}^n)$ satisfying supp $\varphi \subset \overline{B(x_0, 2R)}$ and $\varphi = 1$ on $B(x_0, R)$ and put

$$A(t) = \int_{\mathbf{R}^n} \varphi(x)u(x, t) \, dx.$$

First, it holds that

$$\left| \frac{d}{dt} \int_{\mathbf{R}^n} \varphi u \, dx \right|^2 = \left| \int_{\mathbf{R}^n} u\nabla(u^{m-1} - \Gamma * u) \cdot \nabla\varphi \, dx \right|^2$$

$$\leq \int_{\mathbf{R}^n} u \left| \nabla(u^{m-1} - \Gamma * u) \right|^2 dx \cdot \int_{\mathbf{R}^n} u \left| \nabla\varphi \right|^2 dx$$

$$\leq -\lambda \|\nabla\varphi\|_\infty^2 \frac{d}{dt}\mathcal{F}(u) \tag{3.107}$$

which means

$$(A')^2 \leq -\frac{\|\nabla\varphi\|_\infty^2 \lambda}{2(n-2)} H'', \quad H(t) = \int_{\mathbf{R}^n} |x|^2 u(x,t) dx. \qquad (3.108)$$

If

$$\lim_{t\uparrow T} \mathcal{F}(u(t)) > -\infty \qquad (3.109)$$

is the case, therefore, it follows that

$$\int_0^T \left| \frac{d}{dt} \int_{\mathbf{R}^n} \varphi u \, dx \right| dt \leq T^{1/2} \left\{ \int_0^T \left| \frac{d}{dt} \int_{\mathbf{R}^n} \varphi u \, dx \right|^2 dt \right\}^{1/2} < +\infty$$

and hence

$$\lim_{t\uparrow T} A(t) = \lim_{t\uparrow T} \int_{\mathbf{R}^n} \varphi(x) u(x,t) \, dx \qquad (3.110)$$

exists. Since Lemma 3.1 guarantees

$$\liminf_{t\uparrow T} A(t) = \limsup_{t\uparrow T} A(t) \geq \limsup_{t\uparrow T} \|u(t)\|_{L^1(B(x_0,R))} \geq \varepsilon_0,$$

we obtain

$$\lim_{R\downarrow 0} \liminf_{t\uparrow T} \|u(\cdot,t)\|_{L^1(B(x_0,2R))} \geq \varepsilon_0$$

for any $x_0 \in \mathcal{S}$, and hence the finiteness of $\mathcal{S}$ by the total mass conservation.

In the other case of

$$\lim_{t\uparrow T} \mathcal{F}(u(\cdot,t)) = -\infty, \qquad (3.111)$$

we have $\mathcal{F}(u(t_0)) < 0$ for some $t_0 \in [0,T)$. We may assume $t_0 = 0$ without loss of generality. Equality (3.103) then implies

$$\frac{dH}{dt} < 0 \qquad (3.112)$$

and hence there is $H(T) = \lim_{t\uparrow T} H(t) \geq 0$. If $H(T) = 0$ is the case, then

$$\lim_{t\uparrow T} \int_{|x|>\varepsilon} u(x,t) \, dx = 0$$

for any $\varepsilon > 0$ which implies $\mathcal{S} \subset \{0\}$ by Theorem 3.4. Thus we may assume $H(T) > 0$ furthermore.

Lemma 3.3. *It holds that*

$$\sup_{t'\in[t,\frac{t+T}{2}]} A(t') \leq A(t) + C(H(t) - H(T))^{1/2}. \qquad (3.113)$$

Proof. Inequality (3.108) implies

$$\int_t^{t'} (t' - s)A'(s)^2 ds \leq \frac{\|\nabla\varphi\|_\infty^2 \lambda}{2(n-2)}(H(t) - H(t'))$$

for $0 \leq t \leq t' < T$ by $H'(t) \leq 0$. Therefore, it holds that

$$\left| A(\frac{t+t'}{2}) - A(t) \right|^2 = \left| \int_t^{\frac{t+t'}{2}} A'(s) ds \right|^2$$

$$\leq \int_t^{\frac{t+t'}{2}} (t' - s)^{-1} ds \cdot \int_t^{t'} (t' - s)A'(s)^2 ds$$

$$\leq \frac{\log 2}{2} \cdot \frac{\|\nabla\varphi\|_\infty^2}{n-2} \cdot \lambda \cdot (H(t) - H(t'))$$

$$\leq \frac{\log 2}{2} \cdot \frac{\|\nabla\varphi\|_\infty^2}{n-2} \cdot \lambda \cdot (H(t) - H(T))$$

for $t' \in [t, T)$. This inequality implies

$$A(\frac{t+t'}{2}) \leq A(t) + C(H(t) - H(T))^{1/2}$$

for $t' \in [t, T)$ and hence (3.113). $\qquad\square$

Now we use the scaling property (3.90).

Lemma 3.4. *Each $r_0 > 0$ admits $t_0 \in [0, T)$ and $C > 0$ such that*

$$\|u(\cdot, t_1)\|_{L^1(B(x_0, r_0))} < \varepsilon_0/2 \tag{3.114}$$

implies

$$\sup_{t \in (t_1 + \frac{1}{8}(T - t_1), t_1 + \frac{3}{8}(T - t_1))} (T - t)\|u(\cdot, t)\|_{L^\infty(B(x_0, (T - t_1)^{1/n}))} \leq C \tag{3.115}$$

where $x_0 \in \mathbf{R}^n$ and $t_1 \in [t_0, T)$.

Proof. We have $A(t_1) < \varepsilon_0/2$ for

$$A(t) = \int_{\mathbf{R}^n} \varphi_{x_0, r_0}(x) u(x, t) \, dx$$

by (3.114). Hence it holds that

$$\sup_{t' \in [t_1, \frac{T+t_1}{2}]} A(t') < \varepsilon_0, \quad t_1 \in [t_0, T) \tag{3.116}$$

for $0 < T - t_0 \ll 1$ by Lemma 3.3. Here we use the scaling property (3.90) and take $\mu > 0$ and $\tilde{u}(x, t)$ by

$$\tilde{u}(x, t) = \mu^n u(\mu x + x_0, \mu^n t + t_1), \quad \mu^n + t_1 = \frac{T + t_1}{2}.$$

It holds that $\mu^n = \frac{T-t_1}{2}$ and

$$\tilde{u}_t = \frac{m-1}{m}\Delta\tilde{u}^m - \nabla\cdot(\tilde{u}\nabla\Gamma*\tilde{u}), \ \tilde{u}\geq 0 \quad \text{in } \mathbf{R}^n\times(0,1)$$

$$\sup_{t\in(0,1)}\|\tilde{u}(\cdot,t)\|_{L^1(B(0,r_0\mu^{-1}))} < \varepsilon_0 \qquad (3.117)$$

by (3.116). Now we use Lemma 3.2 and then Moser's iteration scheme applied to the proof of Theorem 3.4. We obtain

$$\sup_{t\in[1/4,3/4]}\|\tilde{u}(\cdot,t)\|_{L^\infty(B(0,1))} \leq C_1 \qquad (3.118)$$

similarly, because $r_0\mu^{-1}\geq 2$ holds for $0 < T-t_1 \ll 1$.

Inequality (3.118) implies

$$\sup_{t\in(t_1+\frac{1}{8}(T-t_1),t_1+\frac{3}{8}(T-t_1))}(T-t_1)\|u(t)\|_{L^\infty(B(x_0,(T-t_1)^{1/n}))} \leq C_1.$$

Then (3.115) follows for $C = \frac{3}{4}C_1$. $\qquad\qquad\qquad\qquad\qquad\qquad \square$

Proof of Theorem 3.3. This proof is complete with

$$\inf_{x_0\in\mathcal{S}_{II}}\liminf_{r\downarrow 0}\liminf_{t\uparrow T}\|u(\cdot,t)\|_{L^1(B(x_0,r))} \geq \varepsilon_0/2. \qquad (3.119)$$

In fact, inequality (3.119) implies $\sharp\mathcal{S}_{II} < +\infty$ from the total mass conservation of $u = u(\cdot,t)$. Assuming the contrary, we have $x_0\in\mathcal{S}_{II}$, $r_0 > 0$, and $t_j\uparrow T$ such that

$$\|u(\cdot,t_j)\|_{L^1(B(x_0,2r_0))} < \varepsilon_0/2, \quad j = 1,2,\cdots.$$

Then we obtain

$$\sup_{y\in B(x_0,r_0)}\|u(\cdot,t_j)\|_{L^1(B(y,r_0))} < \varepsilon_0/2,$$

and, therefore,

$$\sup_{t\in(t_j+\frac{1}{8}(T-t_j),t_j+\frac{3}{8}(T-t_j))}(T-t)\|u(\cdot,t)\|_{L^\infty(B(y,(T-t_j)^{1/n}))} \leq C \qquad (3.120)$$

by Lemma 3.4, where $y\in B(x_0,r_0)$ is arbitrary.

Inequality (3.120) implies

$$\sup_{t\in(t_j+\frac{1}{8}(T-t_j),t_j+\frac{3}{8}(T-t_j))}(T-t)\|u(\cdot,t)\|_{L^\infty(B(x_0,r_0))} \leq C$$

and hence

$$\liminf_{t\uparrow T}(T-t)\|u(\cdot,t)\|_{L^\infty(B(x_0,r_0))} < +\infty,$$

that is $x_0\in\mathcal{S}_I = \mathcal{S}\setminus\mathcal{S}_{II}$, a contradiction. $\qquad\qquad\qquad \square$

3.6 Stationary States

Putting $w = v + \log \lambda - \log \int_\Omega e^v dx$ in the Boltzmann-Poisson equation (2.27),

$$-\Delta v = \frac{\lambda e^v}{\int_\Omega e^v dx} \quad \text{in } \Omega, \qquad v = 0 \quad \text{on } \Gamma = \partial\Omega, \tag{3.121}$$

we obtain

$$-\Delta w = e^w \quad \text{in } \Omega, \quad w = w_\Gamma \text{ on } \Gamma = \partial\Omega, \quad \int_\Omega e^w \, dx = \lambda \tag{3.122}$$

for $w_\Gamma = \log \lambda - \log \int_\Omega e^v \in \mathbf{R}$. If $w = w(x)$ is a solution to (3.122), conversely, $v = w - w_\Gamma$ solves (3.121). In (3.122), $w_\Gamma \in \mathbf{R}$ is an unknown constant to be determined by the third equation for given λ. Equation (3.45) is a reduced form of the Grad-Shafranov equation, where $\Omega_p = \{x \in \Omega \mid w(x) > 0\}$ stands for the plasma region. It is a free boundary problem in the sense that the plasma regions Ω_p is determined by the solution w.

Theorem 2.15 in §2.14 is written for (3.122) in the following form, where $H_\ell = H_\ell(x)$, $x = (x_1, \cdots, x_\ell)$, denotes the ℓ-th Hamiltonian defined by (2.175).

Theorem 3.5. *Let $\Omega \subset \mathbf{R}^2$ be a bounded domain with smooth boundary $\partial\Omega$. Let $\{(\lambda_k, w_k)\}$ be a family of solutions to (3.122) for $\lambda = \lambda_k$ and $w = w_k$ satisfying $\lambda_k \to \lambda_0$. Then we obtain the following alternatives, up to a subsequence.*

(1) $\{w_k\}$ is uniformly bounded on $\overline{\Omega}$.

(2) $w_k \to -\infty$ locally uniformly in Ω.

(3) $\lambda_0 = 8\pi\ell$ for some $\ell \in \mathbf{N}$. There is a critical point $x^* = (x_1^*, \cdots, x_\ell^*) \in \Omega^\ell$, $x_i^* \neq x_j^*$, $i \neq j$, of the ℓ-th Hamiltonian $H_\ell = H_\ell(x)$, $x = (x_1, \cdots, x_\ell)$, defined by (2.175), and also there is a local maximizer $x = x_k^j \to x_j^*$ of $w_k = w_k(x)$, such that $w_k(x_k^j) \to +\infty$. It holds that $w_k \to -\infty$ locally uniformly in $\overline{\Omega} \setminus \{x_1^*, \ldots, x_\ell^*\}$ and

$$e^{w_k} dx \rightharpoonup \sum_{j=1}^{\ell} 8\pi\delta_{x_j^*}(dx) \quad \text{in } \mathcal{M}(\overline{\Omega}).$$

Hence $\mathcal{S} = \{x_1^*, \cdots, x_\ell^*\}$ is the blowup set of $\{w_k\}$.

Proof. From Theorems 2.13 and 2.15 and the proof, we have the following alternatives on the behavior of $\{w_k\}$ up to the subsequence:

(1) locally uniformly bounded in Ω.

(2) $w_k \to -\infty$ locally uniformly in Ω.

(3) $w_k \to -\infty$ locally uniformly in $\Omega \setminus \mathcal{S}$, where $\mathcal{S} \subset \Omega$ is a finite set. Each $x_0 \in \mathcal{S}$ admits a local maximizer $x_k \to x_0$ of $w_k(x)$ such that $w_k(x_k) \to +\infty$. There is $m(x_0) \in 8\pi\mathbf{N}$ such that

$$e^{w_k}dx \rightharpoonup \sum_{x_0 \in \mathcal{S}} m(x_0)\delta_{x_0}(dx) \quad \text{in } \mathcal{M}(\overline{\Omega}).$$

We have, on the other hand,

$$v_k = w_k - w_{k\Gamma} \geq 0 \quad \text{in } \Omega \tag{3.123}$$

by the maximum principle. Furthermore, there is an open set ω satisfying $\partial\Omega \subset \omega$ and

$$\|v_k\|_{L^\infty(\Omega \cap \omega)} \leq C \tag{3.124}$$

from the proof of Theorem 2.11. By (3.123)–(3.124) and the above alternatives, we have either $w_{k\Gamma} \to -\infty$ or $w_{k\Gamma} = O(1)$, passing through a subsequence.

We have $w_{k\Gamma} = O(1)$ with the first alternative, and then $\{v_k\}$ is uniformly bounded on $\overline{\Omega}$ by (3.124). This property is the first case of this theorem. If the case $w_{k\Gamma} \to -\infty$ arises with the uniform boundedness of $\{v_k\}$ on $\overline{\Omega}$, then the second case of this theorem occurs. Otherwise, we have $\|v_k\|_\infty \to +\infty$ up to a subsequence, and hence the second alternative of Theorem 2.10, that is, the case treated by Theorem 2.11. Then it holds that

$$-(\Delta v_k)dx = -(\Delta w_k)dx = e^{w_k}dx \rightharpoonup \sum_j 8\pi\delta_{x_j^*}(dx) \quad \text{in } \mathcal{M}(\overline{\Omega}),$$

the third alternative of this theorem. $\square$

Analogous result to Theorem 3.5 is observed to (3.45),

$$-\Delta w = w_+^q \text{ in } \Omega, \quad w = \text{constant on } \Gamma, \quad \int_\Omega w_+^q \, dx = \lambda \tag{3.125}$$

in higher space dimensions, where $\Omega \subset \mathbf{R}^n$, $n \geq 3$, is a bounded domain with smooth boundary $\partial\Omega = \Gamma$ and $q = \frac{n}{n-2}$ is the critical exponent. Equation (3.125) arises as the stationary state of self-graviating fluid in §3.1, where $q = \frac{n+2}{n-2}$ and $q = \frac{n}{n-2}$ stand for the critical exponents. As we have observed, the critical mass to (3.125) for $q = \frac{n}{n-2}$ is defined by the solution $U = U(x)$ to

$$-\Delta U = U^q, \quad U > 0 \quad \text{in } B, \qquad U = 0 \quad \text{on } \partial B, \tag{3.126}$$

that is,

$$m_* = \int_B U^q \, dx \tag{3.127}$$

for $B = B(0, R)$. First, we have $U = U(|x|)$ and (3.126) is reduced to

$$U_{rr} + \frac{n-1}{r} U_r + U^q = 0, \; U_r(0) = 0$$

$$U(r) > U(R) = 0, \; 0 \le r < R. \tag{3.128}$$

Then a unique solution $U = U(r)$ to (3.128) exists for each $R > 0$, while the value m_* in (3.127) is independent of $R > 0$, because of the scaling invariance of the first equation of (3.126) formulated by

$$U_\mu(x) = \mu^{\frac{2}{q-1}} U(\mu x), \quad \mu > 0. \tag{3.129}$$

Theorem 3.6. *Let $\Omega \subset \mathbf{R}^n$ be a bounded open set. Assume $n \ge 3$, and put $q = \frac{n}{n-2}$. Let $\{(\lambda_k, w_k)\}$ be a family of solutions to*

$$-\Delta w = w_+^q \quad in \; \Omega, \qquad \int_\Omega w_+^q \, dx = \lambda$$

for $w = w_k$, $\lambda = \lambda_k$, satisfying $\lambda_k = O(1)$. Then the following alternatives hold up to a subsequence:

(1) $\{w_k\}$ is locally uniformly bounded in Ω.
(2) $w_k \to -\infty$ locally uniformly in Ω.
(3) For $\ell \in \mathbf{N}$, $x_j^* \in \Omega$, $j = 1, \cdots, \ell$, and a local maximizer $x_k^j \to x_j^*$ of $w_k(x)$ it holds that $w_k(x_k^j) \to +\infty$, $w_k \to -\infty$ locally uniformly in $\Omega \setminus \{x_1^*, \cdots, x_\ell^*\}$, and

$$w_k(x)_+^q dx \rightharpoonup \sum_{j=1}^\ell m_* n_j \delta_{x_*^j}(dx) \quad in \; \mathcal{M}(\Omega),$$

where $n_j \in \mathbf{N}$.

The structure of the proof of Theorem 3.6 is the same as that of Theorem 2.15 on two space dimensions. We use ε-regularity analogous to Theorem 2.13, scaling (3.129), classification of the entire solution as in Theorem 2.14, and sup + inf inequality. There are two differences between (3.122) concerning two space dimensions. First, without algebraic and geometric structures, ε-regularity and sup + inf inequality are proven by the blowup analysis. This argument guarantees even boundary ε-regularity. Thus, there are $\varepsilon_0 > 0$ and C_0 such that

$$-\Delta w = w_+^q \text{ in } \Omega \cap B, \quad w = c \in (-\infty, 0] \text{ on } \partial\Omega \cap B, \quad \int_{\Omega \cap B} w_+^q \, dx < \varepsilon_0$$

implies

$$\|w\|_{L^\infty(B/2)} \le C_0,$$

where $q = \frac{n}{n-2}$, $B = B(0,1) \subset \mathbf{R}^n$, and $B/2 = B(0, 1/2)$. Second, differently from (2.225), the solution to

$$-\Delta w = w_+^q, \quad w \le w(0) = 1 \quad \text{in } \mathbf{R}^n, \qquad \int_{\mathbf{R}^n} w_+^q \, dx < +\infty \qquad (3.130)$$

has a compact support.

Let $G = G(x, x')$ be the Green function defined by

$$-\Delta G(\cdot, x') = \delta_{x'}(dx), \quad G(\cdot, x')|_{\partial\Omega} = 0$$

for $x' \in \Omega$. We recall the fundamental solution $\Gamma = \Gamma(x)$ of $-\Delta$ given by (3.23), to define the Robin function

$$R(x) = [G(x, x') - \Gamma(x - x')]_{x'=x}$$

and the ℓ-th Hamiltonian (2.175),

$$H_\ell(x_1, \cdots, x_\ell) = \frac{1}{2}\sum_{j=1}^{\ell} R(x_j) + \sum_{1 \le i < j \le \ell} G(x_i, x_j).$$

We begin with the exclusion of the boundary blowup.

Theorem 3.7. *Let $\{(\lambda_k, w_k)\}$ be a family of solutions to (3.125) for $\lambda = \lambda_k > 0$, $w = w_k(x)$, satisfying*

$$\lambda_k \to \lambda_0, \quad \|w_{k+}\|_\infty \to +\infty.$$

Then, passing to a subsequence we have $\sharp \mathcal{S} < +\infty$, $c_k \to -\infty$, and

$$w_{k+}^q \, dx \rightharpoonup \sum_{x_0 \in \mathcal{S}} m(x_0)\delta_{x_0}(dx) \quad \text{in } \mathcal{M}(\overline{\Omega})$$

$$w_k \to -\infty \quad \text{locally uniformly in } \overline{\Omega} \setminus \mathcal{S}, \qquad (3.131)$$

where

$$\mathcal{S} = \{x_0 \in \overline{\Omega} \mid \text{there exists } x_k \to x_0$$
$$\text{such that } w_k(x_k) \to +\infty\}$$

is the blowup set and $c_k = w_k|_{\partial\Omega} \in \mathbf{R}$. Furthermore, it holds that $\mathcal{S} \subset \Omega$.

Proof. If $c_k \geq -C$, the maximum principle guarantees $w_k \geq -C$ in Ω, which contradicts the second relation of (3.131). Then the boundary ε-regularity is applicable besides the interior one, which implies $\mathcal{S} < +\infty$, and therefore, (3.131) up to a subsequence.

To show $\mathcal{S} \subset \Omega$, assume the contrary, $x_0 \in \partial\Omega \cap \mathcal{S}$. For $0 < R \ll 1$ it holds that $\overline{w} \cap (\mathcal{S} \setminus \{x_0\}) = \emptyset$ and

$$w_k - c_k \;\to\; \sum_{x_0 \in \mathcal{S}} m(x_0) G(\cdot, x_0) \tag{3.132}$$

locally uniformly in $\overline{w} \setminus \{x_0\}$ including their derivatives of any order.

To apply the Kazdan-Warner identity

$$\int_\omega \Delta u \, \nabla u \, dx = \int_{\partial\omega} \frac{\partial u}{\partial\nu} \nabla u - \frac{1}{2} |\nabla u|^2 \, dS, \tag{3.133}$$

let

$$f(w) = w_+^q, \quad F(w) = \frac{1}{q+1} w_+^{q+1},$$

in (3.130). Then it holds that

$$-\int_\omega \Delta w \nabla w \, dx = \int_\omega f(w) \nabla w \, dx = \int_\omega \nabla F(w) \, dx = \int_{\partial\omega} \nu \cdot F(w) \, dS. \tag{3.134}$$

By (3.133) and $c_k \to -\infty$, it holds that

$$\int_{\partial\omega} -\frac{\partial w_k}{\partial\nu} \nabla w_k + \frac{1}{2} |\nabla w_k|^2 \nu \, dS = 0$$

for k sufficiently large. Then we obtain

$$\int_{\partial\Omega \cap \partial\omega} \frac{1}{2} |\nabla w_k|^2 \nu \, dS = O(1)$$

because (3.132) implies

$$\int_{\Omega \cap \partial\omega} |\nabla w_k|^2 \, dS = O(1).$$

It thus follows that

$$\int_{\partial\Omega \cap \partial\omega} |\nabla w_k|^2 \, dS = O(1)$$

and hence

$$\int_{\partial\Omega \cap \partial\omega} \left(\frac{\partial G}{\partial\nu}(\cdot, x_0) \right)^2 \, dS < +\infty. \tag{3.135}$$

Since

$$\left(\frac{\partial}{\partial\nu} G(x, x_0) \right)^2 \approx |x - x_0|^{-2n+2}$$

as $x \to x_0$, inequality (3.135) is impossible. $\qquad\square$

Here we use the local second moment based on duality to prove the following theorem. Combined with Theorem 3.7, it is a higher dimensional version of Theorem 2.11.

Theorem 3.8. *Let $\Omega \subset \mathbf{R}^n$, $n \geq 3$, be a bounded domain with smooth boundary $\partial\Omega$, and let $\{(\lambda_k, w_k)\}$ be a family of solutions to (3.125), $q = \frac{n}{n-2}$, for $(\lambda, w) = (\lambda_k, w_k)$ satisfying $\lambda_k \to \lambda_0$. Assume the third alternative in Theorem 3.6. Then, it holds that $n_j = 1$ for $1 \leq j \leq \ell$ and $x_* = (x_1^*, \cdots, x_\ell^*)$ is a critical point of the ℓ-th Hamiltonian $H_\ell = H_\ell(x)$, $x = (x_1, \cdots, x_\ell)$.*

Proof. It holds that

$$\lim_{k \to \infty} c_k = -\infty, \quad c_k = w_k|_{\partial\Omega} \tag{3.136}$$

by the previous theorem. For the moment we drop the suffix k.

Using

$$u = w_+^q, \quad q = \frac{n}{n-2}, \tag{3.137}$$

we transform (3.125) to

$$w - w_\Gamma = \int_\Omega G(\cdot, x')u(x')\,dx', \quad \int_\Omega u = \lambda, \tag{3.138}$$

which implies

$$\nabla w(x) = \int_\Omega \nabla_x G(x, x')u(x')\,dx'.$$

Given $\psi \in C^1(\overline{\Omega})^n$, we have

$$\int_\Omega (\psi \cdot \nabla w)u = \iint_{\Omega \times \Omega} \psi(x) \cdot [\nabla_x G(x, x')]u(x)u(x')\,dx dx'. \tag{3.139}$$

Then, from (3.136) the left-hand side on (3.139) is equal to

$$\int_\Omega (\psi \cdot \nabla w)u\,dx = \frac{1}{q+1}\int_\Omega \psi \cdot \nabla w_+^{q+1}\,dx = -\frac{1}{q+1}\int_\Omega w_+^{q+1}\nabla \cdot \psi\,dx \tag{3.140}$$

for k sufficiently large.

Let $0 \leq \varphi = \varphi_{x_0, R} \leq 1$ be the smooth cut-off function with the support in $\overline{B(x_0, R)}$ and is equal to 1 on $\overline{B(x_0, R/2)}$. We put $\mathcal{S} = \{x_1^*, \cdots, x_\ell^*\} \subset \Omega$, recalling Theorem 3.7. Let $x_0 \in \mathcal{S}$ and take $0 < R \ll 1$ satisfying $B(x_0, 2R) \subset \Omega$ and $B(x_0, 2R) \cap \mathcal{S} = \{x_0\}$, to put

$$\psi(x) = (x - a)\varphi(x), \quad a \in \mathbf{R}^n, \quad \varphi = \varphi_{x_0, R}.$$

Since (3.140) and

$$\nabla \cdot \psi = n\varphi + (x - a) \cdot \nabla\varphi$$

it holds that

$$\int_\Omega (\psi \cdot \nabla w)u \, dx = -\frac{n}{q+1} \int_\Omega w_+^{q+1} \varphi \, dx + o(1),$$

which implies

$$\frac{n}{q+1} \int_\Omega w_+^{q+1} \varphi + \iint_{\Omega \times \Omega} \psi(x) \cdot [\nabla_x G(x, x')]u(x)u(x') \, dx dx'$$

$$= o(1), \quad k \to \infty \tag{3.141}$$

by (3.138).

Let $\hat{\varphi} = \varphi_{x_0, 2R}$. Then the second term on the left-hand side on (3.141) is equal to

$$\iint_{\Omega \times \Omega} \psi(x) \cdot [\nabla_x G(x, x')]u(x)u(x') \, dx dx'$$

$$= \iint_{\Omega \times \Omega} \psi(x) \cdot [\nabla_x G(x, x')]u(x)\hat{\varphi}(x)u(x') \, dx dx'$$

$$= \iint_{\Omega \times \Omega} \psi(x) \cdot [\nabla_x G(x, x')]u(x)\hat{\varphi}(x)u(x')\hat{\varphi}(x') \, dx dx'$$

$$+ \iint_{\Omega \times \Omega} \psi(x) \cdot [\nabla_x G(x, x')]u(x)\hat{\varphi}(x)u(x')(1 - \hat{\varphi}(x')) \, dx dx'. \tag{3.142}$$

The second term of the right-hand side on this (3.142), next, is equal to

$$\int\int_{\Omega \times \Omega} \psi(x) \cdot [\nabla_x G(x, x')]u(x)\hat{\varphi}(x)u(x')(1 - \hat{\varphi}(x')) \, dx dx'$$

$$= m(x_0)(x_0 - a) \cdot \sum_{x_0' \in \mathcal{S} \setminus \{x_0\}} m(x_0')\nabla_x G(x_0, x_0') + o(1),$$

while the method of symmetrization is applied to the first term of the right-hand side on (3.142). In fact, since

$$K(x, x') = G(x, x') - \Gamma(x - x')$$

is smooth in $(\overline{\Omega} \times \Omega) \bigcup (\Omega \times \overline{\Omega})$, with $u^0 = u\hat{\varphi}$, $\rho_\psi^0(x, x') = (\psi(x) - \psi(x')) \cdot \nabla\Gamma(x - x')$ this term is equal to

$$\iint_{\Omega \times \Omega} \psi(x) \cdot [\nabla_x G(x, x')]u(x)\hat{\varphi}(x)u(x')\hat{\varphi}(x') \, dx dx'$$

$$= \frac{1}{2} \iint_{\Omega \times \Omega} \rho_\psi^0(x, x')u^0(x)u^0(x') \, dx dx'$$

$$+ \iint_{\Omega \times \Omega} \psi(x) \cdot [\nabla_x K(x, x')]u^0(x)u^0(x') \, dx dx'. \tag{3.143}$$

Here, for the second term on the right-hand side of this (3.143) we have

$$\iint_{\Omega \times \Omega} \psi(x) \cdot [\nabla_x K(x, x')] u^0(x) u^0(x') \, dx dx'$$
$$= m(x_0)^2 (x_0 - a) \cdot \nabla_x K(x_0, x_0) + o(1), \quad k \to \infty, \qquad (3.144)$$

while

$$\rho_\psi^0(x, x') = -(n-2)\Gamma(x - x') \quad \text{in } B(x_0, R/2) \times B(x_0, R/2)$$

holds in the first term. Therefore, putting

$$\tilde{u}^0 = u\tilde{\varphi}, \quad \tilde{\varphi} = \varphi_{x_0, R/2},$$

we obtain

$$\rho_\psi^0(x, x') u^0(x) u^0(x') = -(n-2)\Gamma(x - x')\tilde{u}^0(x)\tilde{u}^0(x')$$
$$+ \rho_\psi^0(x, x')(1 - \tilde{\varphi}(x))\tilde{\varphi}(x') u^0(x) u^0(x')$$
$$+ \rho_\psi^0(x, x')(1 - \tilde{\varphi}(x')) u^0(x) u^0(x'). \qquad (3.145)$$

The second and the third terms of the right-hand side on (3.145) are symmetric with respect to x and x', and therefore, these two terms are treated similarly. Concerning the third term, we use

$$\left| \rho_\psi^0(x, x') \right| \le C\Gamma(x - x').$$

It holds also that

$$0 \le \iint_{\Omega \times \Omega} \Gamma(x - x')(1 - \tilde{\varphi}(x')) u^0(x) u^0(x') \, dx dx'$$
$$= \langle \Gamma * u^0, (1 - \tilde{\varphi}) u^0 \rangle$$

and

$$\left\| (1 - \tilde{\varphi}) u^0 \right\|_\infty = o(1), \quad \left\| \Gamma * u^0 \right\|_1 \le C \left\| u \right\|_1 = O(1), \quad k \to \infty,$$

which implies

$$\iint_{\Omega \times \Omega} \Gamma(x - x')(1 - \tilde{\varphi}(x')) u^0(x) u^0(x') \, dx dx' = o(1). \qquad (3.146)$$

Then, by (3.145)–(3.146) we obtain

$$\frac{1}{2} \iint_{\Omega \times \Omega} \rho_\psi^0(x, x') u^0(x) u^0(x') \, dx dx'$$
$$= -\frac{n-2}{2} \iint_{\Omega \times \Omega} \Gamma(x - x')\tilde{u}^0(x)\tilde{u}^0(x') \, dx dx' + o(1). \qquad (3.147)$$

From (3.142)–(3.144) and (3.147), equality (3.141) is reduced to

$$\frac{n}{q+1} \int_\Omega w_+^{q+1} \varphi - \frac{n-2}{2} \iint_{\Omega \times \Omega} \Gamma(x-x') \tilde{u}^0(x) \tilde{u}^0(x') \, dx dx'$$

$$+ m(x_0)(x_0 - a) \cdot \sum_{x_0' \in \mathcal{S} \setminus \{x_0\}} m(x_0') \nabla_x G(x_0, x_0')$$

$$+ m(x_0)^2 (x_0 - a) \cdot \nabla_x K(x_0, x_0) = o(1), \quad k \to \infty. \tag{3.148}$$

For simplicity we take zero extension of $u = u(x)$ in (3.137) where it is not defined and put

$$\mathcal{F}_0(u) = \frac{1}{\gamma} \int_{\mathbf{R}^n} u^\gamma \, dx - \frac{1}{2} \langle \Gamma * u, u \rangle$$

for $\gamma = 1 + \frac{1}{q} = 2 - \frac{2}{n}$. From

$$\frac{n}{q+1} \int_\Omega w_+^{q+1} \varphi \, dx = \frac{n-2}{\gamma} \int_\Omega u^\gamma \varphi \, dx = \frac{n-2}{\gamma} \int_\Omega (\tilde{u}^0)^\gamma + o(1)$$

in (3.148), it follows that

$$o(1) = (n-2)\mathcal{F}_0(\tilde{\varphi} u) + m(x_0)(x_0 - a)$$

$$\cdot \left[\sum_{x_0' \in \mathcal{S} \setminus \{x_0\}} m(x_0') \nabla_x G(x_0, x_0') + m(x_0) \nabla_x K(x_0, x_0) \right]. \tag{3.149}$$

Since $a \in \mathbf{R}^n$ is arbitrary, the second term on the right-hand side of (3.149) vanishes. Hence each $x_0 \in \mathcal{S}$ admits the relation

$$\frac{m(x_0)}{2} \nabla R(x_0) + \sum_{x_0' \in \mathcal{S} \setminus \{x_0\}} m(x_0') \nabla_x G(x_0, x_0') = 0 \tag{3.150}$$

and also

$$\mathcal{F}_0(\tilde{\varphi} u) = o(1). \tag{3.151}$$

Having (3.150), we shall show the simplicity of the collapse indicated by $m(x_0) = m_*$. The scaling (3.129) of $w(x)$ to (3.125) is translated as

$$u_\mu(x) = \mu^n u(\mu x + x_0), \quad \mu > 0$$

for $u(x)$ satisfying (3.137), and then it holds that

$$\mathcal{F}_0(u_\mu) = \mu^{n-2} \mathcal{F}_0(u).$$

To execute the blowup analysis below, we write the suffix k explicitly again.

First, from the proof of Theorem 3.6, each $x_0 \in \mathcal{S}$ admits a local maximizer $x = x_k^0 \to x_0$ of $u_k = u_k(x)$ in $x \in B(x_0, 2R)$, for k sufficiently large. Under the scaling

$$\tilde{u}_k(x) = \mu_k^n u_k(\mu_k x + x_k^0), \quad \mu_k = u_k(x_k^0)^{-1/n} \to 0$$
$$\tilde{w}_k(x) = \mu_k^{n-2} w_k(\mu_k x + x_k^0)$$

there is a subsequence denoted by the same symbol such that

$$\tilde{w}_k \to \tilde{w} \quad \text{locally uniformly in } \mathbf{R}^n \tag{3.152}$$

with $\tilde{w} = \tilde{w}(x)$ satisfying

$$-\Delta\tilde{w} = \tilde{w}_+^{\frac{n}{n-2}}, \quad \tilde{w} \le \tilde{w}(0) = 1 \quad \text{in } \mathbf{R}^n, \quad \int_{\mathbf{R}^n} \tilde{w}_+^{\frac{n}{n-2}} \, dx < +\infty. \tag{3.153}$$

From the classification of entire solutions, equation (3.153) implies

$$\tilde{w} = \tilde{w}(|x|), \quad \text{supp } \tilde{w} \subset \overline{B}, \quad \int_{\mathbf{R}^n} \tilde{w}_+^{\frac{n}{n-2}} \, dx = m_* \tag{3.154}$$

for $B = B(0, L)$ with $L \gg 1$.

Henceforth, we put

$$\hat{u}_k(x) = \mu_k^n(\tilde{\varphi}u_k)(\mu_k x + x_k^0). \tag{3.155}$$

First, by (3.151) it holds that

$$\mathcal{F}_0(\hat{u}_k) = \mu_k^{n-2}\mathcal{F}_0(\tilde{\varphi}u_k) \to 0. \tag{3.156}$$

Second, we apply Theorem 3.6 to $\hat{w}_k(x) = \tilde{w}_k(x)$, $x \in \mu_k^{-1}(B(x_0, R/4) - \{x_0\})$. Similarly to (3.152), we have

$$\hat{w}_k(x) \to \tilde{w} \quad \text{locally uniformly in } \mathbf{R}^n, \tag{3.157}$$

passing through a subsequence. From (3.137) and (3.157) it follows that

$$\hat{u}_k \to \tilde{u} \equiv \tilde{w}_+^{\frac{n}{n-2}} \quad \text{locally uniformly in } \mathbf{R}^n,$$

and hence $\nabla\tilde{w} = \nabla\Gamma * \tilde{u}$ by Liouville's theorem. This relation implies

$$\frac{x}{q+1} \cdot \nabla w_+^{q+1} = x \cdot \tilde{u}\nabla\Gamma * \tilde{u}$$

and hence

$$\frac{n}{q+1} \int_{\mathbf{R}^n} \tilde{w}_+^{q+1} dx + \frac{1}{2} \iint_{\mathbf{R}^n \times \mathbf{R}^n} (x - x') \cdot \nabla\Gamma(x - x')\tilde{u} \otimes \tilde{u} \, dx dx' = 0,$$

which means

$$\mathcal{F}_0(\tilde{u}) = \frac{1}{\gamma} \int_{\mathbf{R}^n} \tilde{u}^\gamma \, dx - \frac{1}{2} \langle \Gamma * \tilde{u}, \tilde{u} \rangle = 0 \tag{3.158}$$

by $q = \frac{n}{n-2}$ and $x \cdot \nabla \Gamma(x) = -(n-2)\Gamma(x)$.

We have, on the other hand,

$$\|\hat{u}_k\|_{L^p(\mathbf{R}^n)} \leq C, \quad p = 1, \infty \tag{3.159}$$

and hence

$$\|\Gamma * \hat{u}_k\|_{W^{2,p}(\mathbf{R}^n)} \leq C, \quad 1 < p < \infty. \tag{3.160}$$

By (3.159)–(3.160), a subsequence denoted by the same symbol satisfies

$$\lim_{k \to \infty} \langle \Gamma * \hat{u}_k, \hat{u}_k \rangle = \langle \Gamma * \tilde{u}, \tilde{u} \rangle. \tag{3.161}$$

Relations (3.156), (3.158), and (3.161) imply

$$\lim_{k \to \infty} \int_{\mathbf{R}^n} \hat{u}_k^\gamma \, dx = \int_{\mathbf{R}^n} \tilde{u}^\gamma \, dx,$$

and therefore, it holds that

$$\hat{u}_k \to \tilde{u} \quad \text{in } L^\gamma(\mathbf{R}^n). \tag{3.162}$$

Now we complete the proof of the simpleness of the collapse at x_0. By the proof of Theorem 3.6, condition $m(x_0) > m_*$ implies the collision of collapses. Hence we have a local maximizer $x_k^1 \neq x_k^0$ of $u_k = u_k(x)$ and $r_k^0, r_k^1 \to 0$ such that $x_k^1 \to x_0$ and

$$\lim_{k \to \infty} \int_{B(x_k^0, r_k^0)} u_k \, dx = \lim_{k \to \infty} \int_{B(x_k^1, r_k^1)} u_k \, dx = m_*$$

$$B(x_k^0, 2r_k^0) \cap B(x_k^1, 2r_k^1) = \emptyset.$$

This fact follows from the scaling (3.155). Namely, we have $L' > L$, $x_k' \in \mathbf{R}^2$, and $r_k' > 0$ such that

$$\lim_{k \to \infty} \int_{B(0,L')} \hat{u}_k \, dx = \lim_{k \to \infty} \int_{B(x_k', r_k')} \hat{u}_k \, dx = m_*$$

$$B(0, 2L') \cap B(x_k', 2r_k') = \emptyset. \tag{3.163}$$

Furthermore, $\tilde{u}_k \to 0$ locally uniformly in $B(0, 2L')^c$, and therefore, it holds that

$$\lim_{k \to \infty} |x_k'| = +\infty. \tag{3.164}$$

Here we take the second scaling $\tilde{u}_k'(x) = (\mu_k')^n \hat{u}_k(\mu_k' x + x_k')$, $\mu_k' = \hat{u}_k(x_k')^{-1/n} \geq 1$. We show

$$\lim_{k \to \infty} \tilde{\mu}_k' = +\infty, \tag{3.165}$$

passing through a subsequence. In fact, if (3.165) is not the case it holds that

$$\mu'_k = \tilde{u}_k(x'_k)^{-1/n} \approx 1. \tag{3.166}$$

Then we apply Theorem 3.6 to $\tilde{u}''_k = \tilde{u}'_k(\cdot + x'_k)$. Since $0 \leq \hat{u}_k \leq \tilde{u}_k \leq C$, a subsequence satisfies

$$\tilde{u}''_k \to \tilde{u}'' = (\tilde{w}'')_+^{\frac{n}{n-2}} \quad \text{locally uniformly in } \mathbf{R}^2$$

with $\tilde{w}'' = \tilde{w}''(x)$ satisfying

$$-\Delta \tilde{w}'' = (\tilde{w}'')_+^{\frac{n}{n-2}} \quad \text{in } \mathbf{R}^n, \quad 0 < \tilde{w}''(0) = 1 \leq \max_{\mathbf{R}^n} \tilde{w}'' < +\infty$$

$$\int_{\mathbf{R}^n} (\tilde{w}'')_+^{\frac{n}{n-2}} \, dx < +\infty.$$

This property implies $\int_{\mathbf{R}^n} u'' dx = m_*$, and therefore, relation (3.166) implies

$$r'_k \approx 1 \tag{3.167}$$

in (3.163).

We thus obtain

$$\lim_k \int_{B(0,2L)^c} \hat{u}_k^\gamma \, dx \geq \lim_k \int_{B(x'_k, 2r'_k)} (\hat{u}'_k)^\gamma \, dx > 0 \tag{3.168}$$

by (3.163), (3.164), and (3.167), which contradicts (3.162), recalling $B = B(0, L)$ in (3.154). Hence relation (3.165), which implies

$$\lim_{k \to \infty} \frac{u_k(x_k^0)}{u_k(x_k^1)} = +\infty.$$

From the same argument with changing the roles of x_k^0 and x_k^1, however, it follows that

$$\lim_{k \to \infty} \frac{u_k(x_k^0)}{u_k(x_k^1)} = 0$$

a contradiction. $\qquad\qquad\qquad\qquad\qquad\qquad\qquad\qquad\qquad\qquad\qquad\qquad$ $\square$

3.7 Notes

See Section 2.1.2 of [Suzuki and Senba (2011)] for the proof of (3.7), that is, Liouville's formula. For the state equation (3.22) in astrophysics, see [Chandrasekhar (1939)]. Several fundamental equations of self-interacting fluids are provided with the semi-unfolding-minimality. This notion is formulated in Section 4.3 of [Suzuki (2015)]. There, Chapter 7 is devoted to the self-interacting fluid, where several fundamental properties in fluid dynamics are described, such as Cauchy integral, circulation theorem, and vortex theorem. New points of view using calculus of variation, dynamical systems, and kinetic theory are also given there, such as the Hamilton and kinetic formalisms, which results in the irrotational blowup of the solution. Mathematical study on the Euler-Poisson equation (3.24), such as local-in-time well-posedness using the theory of symmetric hyperbolic system [Kato (1985)], blowup and decay of the solution via second moment of the density and kinetic-Hamilton formalisms.

Thermodynamical relation is used in confirming that the isentropic fluid is barotropic. Theory of equilibrium thermodynamics is described in Section 5.1 of [Suzuki (2015)]. First, the objects of thermodynamics are the quantity of state, independent of the history of the system. The absolute temperature T, the pressure p, and the volume V are the quantities of state of the ideal gas, subject to the state equation

$$f(p, T, V) = 0.$$

The one-form dA derived from a thermodynamical quantity of state A, therefore, is completely integrable, and it holds that

$$\int_\gamma dA = 0$$

for any closed path γ in the phase space.

The first law of thermodynamics is the energy balance described by

$$dU = d'Q - pdV,$$

where the left-hand side indicates the energy variation of the system in accordance with the volume variation in the right-hand side caused by the energy variation denoted by $d'Q$. Here, it is emphasized that work or heat is not a quantity state. If the process is reversible, then

$$\int_\gamma \frac{d'Q}{T} = 0$$

for any closed path γ, and thus we can define the entropy variation dS by

$$dS = \frac{d'Q}{T}.$$

Entropy decomposition is obtained by introducing the inside thermal energy transport $d'Q^*$ caused by the thermal energy contact, denoted by $d'Q_{ir}$, to the outer system. More precisely, the entropy variation and the inner entropy production arising at this contact are defined by

$$d_e S = \frac{d'Q_{ir}}{T}$$

and

$$d_i S = \frac{d'Q^*}{T},$$

respectively. Then Clausius-Duhem's inequality

$$dS > \frac{d'Q}{T} \tag{3.169}$$

is replaced by the equality

$$dS = d_e S + d_i S, \tag{3.170}$$

and therefore, the second law of thermodynamics is re-formulated by $d_i S \geq 0$ with the equality if and only if the process is reversible. The first and second laws of thermodynamics are thus summarized by

$$dU = d'Q - pdV, \quad dS = \frac{d'Q}{T} + d_i S. \tag{3.171}$$

Then, (3.12) follows from (3.171) and (3.170).

For existence and uniqueness of the solution $u = u_R(x)$ to (3.30) we refer to [Ni (2005)]. The equilibrium made by this solution, however, does not belong to the function space where the Cauchy problem (3.24) is locally well-posed by the theory of symmetric hyperbolic systems, that is, $H^s(\mathbf{R}^n)$, $s > 1 + n/2$ for $(\rho^{(\gamma-1)/2}, v)$, because there is a derivative gap of $\overline{w} = \overline{\rho}^{(\gamma-1)/2}$ on the interface. See Section 7.3 of [Suzuki (2015)] for more details. The fact that (3.34) admits no solution in the case that $q \geq (n+2)/(n-2)$ and Ω is star-shaped follows from the classical Pohozaev identity [Pohozaev (1965)]. Existence of radially symmetric solution for super-critical nonlinearity on annulus follows from a simple variational formulation. See [Kazdan and Warner (1975)], for example. Non-existence of the solution to (3.35) in the whole space for $1 < q < (n+2)/(n-2)$, on the other hand, is shown in [Gidas and Spruck (1981a)]. This fact guarantees a priori bounds of the solution to the elliptic and parabolic problems with sub-critical nonlinearity defined on the bounded domain [Gidas and Spruck (1981b)] through the blowup analysis. Asymptotic behavior (3.36) for the case of $q > (n+2)/(n-2)$ is due to [Gui, Ni, and Wang (1992)]. The solution to (3.35) with $p = \frac{n+2}{n-2}$ is classified by [Caffarelli, Gidas, and Spruck (1989); Chen and Li (1991)]. The classification (3.37) of the solution to (3.35), $q = (n+2)/(n-2)$ was done, first, for the case of

$$v(x) = O\left(|x|^{2-n}\right) \qquad \text{as } |x| \to \infty. \tag{3.172}$$

In the context of differential geometry, this property means that any metric conformal to the standard metric ds_0 on S^n with the same mean curvature is a pull-back of ds_0 by the conformal transformation on S^n. See [Obata (1971)]. Although radial symmetry of the general solution to (3.35) does not follow in the super-critical case $q > (n+2)/(n-2)$, the total set of entire radial solutions has several remarkable structures. See [Ni (2005)] and the references therein.

It will be natural to regard (3.39) as a different description of the equilibrium to (3.24). Actually, uniform boundedness and radial symmetry of the solution $\xi = \xi(x)$ to (3.39) with bounded Morse indices are known for $1 < q < (n+2)/(n-2)$ if $n \geq 4$ and for $2 < q < 5$ if $n = 3$ ([Harrabi and Rebhi (1998a,b)]). A close problem to the variational equilibrium in ρ variable is

$$\frac{\gamma}{\gamma-1}\nabla\rho^{\gamma-1} = \nabla\Gamma * \rho, \ \rho \geq \quad \text{in } \mathbf{R}^n. \tag{3.173}$$

For (3.173) if $0 \leq \rho^\gamma \in H^1_{loc}(\mathbf{R}^n) \cap L^1(\mathbf{R}^n) \cap L^\infty(\mathbf{R}^n)$ then there is $x_0 \in \mathbf{R}^n$ such that $\rho = \rho(\cdot - x_0)$ is radially symmetric, and $\rho(|x - x_0|)$ is non-increasing in $|x - x_0|$. This result induces the asymptotic behavior of the solution to the degenerate parabolic equation in the next section. See [Carrillo, Hittmeir, Volzone, and Yao (2017)] for the proof. The blowup mechanism of (3.24) for the critical case $\gamma = 2 - 2/n$, on the other hand, is studied in Section 7.3 of [Suzuki (2015)] including self-similar blowup found by [Straumann (1991)]. The Berestycki-Brezis functional (3.43)–(3.44) is introduced by [Berestycki, H. and Brezis, H. (1980)], while Temam's variational functional for (3.45) was used in [Temam (1977)]. Damlamian [Damlamian (1978)] observed the Toland duality between these formulations. See [Nehari (1960, 1961)] for the Nehari principle involved in the Temam formulation. See [Temam (1975)] for (3.45) formulated in plasma physics. From the general theory of dual variation, any infinitesimally stable stationary state is dynamically stable in (3.41). Neumann boundary condition, Navier-Stokes-Poisson equation, and Murakami-Nishihara-Hanawa equation [Murakami, Nishihara, and Hanawa (2004)] can be also treated under this framework. See Sections 7.4 and 7.5 of [Suzuki (2015)] for these topics.

Kinetic transport theory consistent to thermodynamics is described in [Chavanis (2004)]. There, the aximum entropy production induces several models in canonical and micro-canonical settings, applied to the transport equation. See Section 6.2 of [Suzuki and Senba (2011)] for the transport equation. State equation (3.71) is derived in [Chavanis (2004); Chavanis and Sire (2004)]. Quantized blowup mechanism for stationary solutions to such equations in higher space dimensions is noticed by [Suzuki and Takahashi (2008)].

Theorem 3.1, construction of the solution provided with the fundamental properties (3.80)–(3.81), is done by [Suzuki and Takahashi (2009a)]. The scheme is due to [Sugiyama (2006, 2015); Sugiyama and Kunii (2006)] where the Bessel potential is used instead of the Newton potential $\Gamma(x)$. The stationary state, problem (3.84) with $m = 2 - \frac{2}{n}$, $n \geq 3$, is studied in [Wang and Ye (2003)]. Particularly, there is a family of solutions, each of which is necessarily radially symmetric and has compact support. See [Wang and Ye (2003); Suzuki and Takahashi (2009a)] for the proof of inequality (3.86). The argument used for the proof of inequality (3.98) is called the L^∞-energy method in [Ôtani (2004)]. We may suspect the profile of incomplete blowup in an example by [Li and Zhang (2010)] concerning $n = 1$, $m = 3$, and the Bessel potential for Γ. Thus given the regular non-negative initial value $u_0 = u_0(x)$, $u_0 \not\equiv 0$ with compact support, we have a unique classical solution local-in-time which breaks down in a finite time T_c satisfying

$$\lim_{t \uparrow T_c} \|\partial_x u(\cdot, t)\|_\infty = +\infty, \quad \limsup_{t \uparrow T_c} \|u(\cdot, t)\|_p < +\infty, \quad 1 \leq p \leq \infty.$$

Theorem 3.2 is proven by [Blanchet, Carrillo, and Laurençot (2009); Suzuki and Takahashi (2009b)], using λ_* defined differently, which however coincides because of the Toland duality.

Theorems 3.4 and 3.3 are proven by [Suzuki and Takahashi (2012)] and [Suzuki and Takahashi (2010)], respectively. Section 3.5 of the present monograph is based on Section 11.4 of [Suzuki (2015)]. Other properties of the solution are described in Section 11.5 of [Suzuki (2015)] in accordance with the behavior of the pre-scaled and rescaled free energies, and several formal solutions for the higher-dimensional Smoluchowski-Poisson equation obtained by the method of mached expansions.

The Grad-Shafranov equation is described in Section 8.2 of [Suzuki (2015)]. See also Temam [Temam (1975)]. Theorem 3.6 is due to [Wang and Ye (2003)], where ε-regularity, sup + inf inequality, and classification of the entire solution are provided. Analogous results for the other exponents with different constraints are shown by [Suzuki and Takahashi (2017)]. Theorem 3.8 is given in [Suzuki and Takahashi (2008)]. Related results are also described there. Boundary ε-regularity to (3.125) follows from the interior ε-regularity shown in [Suzuki and Takahashi (2017)]. Duality argument used for the proof of Theorem 3.8 is developed by [Ohtsuka and Suzuki (2003)] for (3.121) in two space dimensions.

The Kazdan-Warner identity is used in [Kazdan and Warner (1975)]. Here we give the proof for completeness.

Proof of (3.133). Let $u_j = \dfrac{\partial u}{\partial x_j}$ and $u_{ij} = \dfrac{\partial^2 u}{\partial x_i \partial x_j}$ for simplicity. It holds that

$$I_{ij} \equiv \int_\omega u_{ii} u_j \, dx = \int_{\partial\omega} \nu_i u_i u_j \, dS - \int_\omega u_i u_{ij} \, dx$$

$$= \int_{\partial\omega} \nu_i u_i u_j - \nu_i u u_{ij} \, dS + \int_\omega u u_{iij} \, dx$$

$$= \int_{\partial\omega} \nu_i u_i u_j - \nu_i u u_{ij} + \nu_j u u_{ii} \, dS - I_{ij}$$

and hence

$$I_{ij} = \frac{1}{2} \int_{\partial\omega} \nu_i u_i u_j - \nu_i u u_{ij} + \nu_j u u_{ii} \, dS. \tag{3.174}$$

Summing up i and j in (3.174), we obtain (3.133) as

$$\int_\omega \Delta u \, \nabla u \, dx = \frac{1}{2} \int_{\partial\omega} \frac{\partial u}{\partial \nu} \nabla u - u \frac{\partial}{\partial \nu}(\nabla u) + \nu u \Delta u \, dS$$

$$= \int_{\partial\omega} \frac{\partial u}{\partial \nu} \nabla u - \frac{1}{2} \frac{\partial}{\partial \nu}(u \nabla u) + \frac{1}{2}\nu u \Delta u \, dS$$

$$= \int_{\partial\omega} \frac{\partial u}{\partial \nu} \nabla u - \frac{1}{2} \int_\omega \Delta(u \nabla u) - \nabla(u \Delta u) \, dx$$

$$= \int_{\partial\omega} \frac{\partial u}{\partial \nu} \nabla u \, dS - \frac{1}{2} \int_\omega \nabla(|\nabla u|^2) \, dx$$

$$= \int_{\partial\omega} \frac{\partial u}{\partial \nu} \nabla u - \frac{1}{2}|\nabla u|^2 \nu \, dS.$$

$\square$

Concerning the concept of dual variation, particularly, notions of dynamical and infinitesimal stabilities are developed in Section 4.3 of [Suzuki

(2015)]. See also more detailed theories. The general theory of Toland duality is developed in Chapter 3 of [Suzuki (2015)]. Here, we state the fundamental part.

First, let X be a Banach space over $\mathbf{R}$. Its dual space and the duality paring are denoted by X^* and $\langle\,,\,\rangle = \langle\,,\,\rangle_{X,X^*}$, respectively.

Given $F : X \to [-\infty, +\infty]$, we define its Legendre transformation $F^* : X^* \to [-\infty, +\infty]$ by

$$F^*(p) = \sup_{x \in X} \{\langle x, p \rangle - F(x)\}, \qquad p \in X^*.$$

Then, Fenchel-Moreau's theorem guarantees that if

$$F : X \to (-\infty, +\infty]$$

is proper, convex, lower semi-continuous, then so is

$$F^* : X^* \to (-\infty, +\infty],$$

and the second Legendre transformation defined by

$$F^{**}(x) = \sup_{p \in X^*} \{\langle x, p \rangle - F^*(p)\}, \qquad x \in X$$

is equal to $F(x)$, see [Ekeland and Temam (1976); Brezis (1983)]. Here and henceforth, we say that $F : X \to (-\infty, +\infty]$ is proper if its effective domain defined by

$$D(F) = \{x \in X \mid F(x) \in \mathbf{R}\}$$

is not empty; *convex* if

$$F(\theta x + (1 - \theta)y) \leq \theta F(x) + (1 - \theta)F(y)$$

for any $x, y \in X$ and $0 < \theta < 1$; and lower semi-continuous if

$$F(x) \leq \liminf_k F(x_k),$$

provided that $x_k \to x$ in X.

Let $F, G : X \to (-\infty, +\infty]$ be proper, convex, lower semi-continuous, with the effective domains $D(F)$ and $D(G)$, respectively. Let

$$D(F) = \{x \in X \mid F(x) < +\infty\}$$
$$D(G) = \{x \in X \mid G(x) < +\infty\}$$

and

$$\varphi(x, y) = F(x + y) - G(x) \tag{3.175}$$

for $x \in D(G)$. Then,

$$y \in X \mapsto \varphi(x, y) \in (-\infty, +\infty]$$

is proper, convex, lower semi-continuous, and its Legendre transformation is defined by

$$L(x,p) = \sup_{y \in X} \{ \langle y, p \rangle - \varphi(x,y) \} \tag{3.176}$$

for $p \in X^*$. Thus

$$L(x, \cdot) : X^* \to (-\infty, +\infty]$$

is proper, convex, lower semi-continuous.

If $(x,p) \in D(G) \times X^*$, then it holds that

$$L(x,p) = \sup_{y \in X} \{ \langle y + x, p \rangle - F(x+y) + G(x) - \langle x, p \rangle \}$$
$$= F^*(p) + G(x) - \langle x, p \rangle. \tag{3.177}$$

Putting $L(x,p) = +\infty$ for $x \notin D(G)$, on the other hand, we obtain (3.177) for any $(x,p) \in X \times X^*$. Thus we define

$$J^*(p) = \begin{cases} F^*(p) - G^*(p), & p \in D(F^*) \\ +\infty, & \text{otherwise} \end{cases} \tag{3.178}$$

for $p \in X^*$. Then, it holds that

$$\inf_{x \in X} L(x,p) = F^*(p) - \sup_{x \in X} \{ \langle x, p \rangle - G(x) \}$$
$$= F^*(p) - G^*(p) = J^*(p)$$

for $p \in D(F^*)$. Let us note that this relation is valid even to $p \notin D(F^*)$ by (3.177) and (3.178).

Similarly, we define

$$J(x) = \begin{cases} G(x) - F(x), & x \in D(G) \\ +\infty, & \text{otherwise} \end{cases} \tag{3.179}$$

for $x \in X$, and obtain

$$\inf_{p \in X^*} L(x,p) = G(x) - \sup_{p \in X^*} \{ \langle x, p \rangle - F^*(p) \}$$
$$= G(x) - F^{**}(x) = J(x)$$

for $x \in D(G)$, which is valid even to $x \notin D(G)$ by (3.177) and (3.179). Thus, we have

$$D(J) = \{ x \in X \mid J(x) \neq \pm\infty \} = D(G) \cap D(F)$$
$$D(J^*) = \{ p \in X^* \mid J^*(p) \neq \pm\infty \} = D(G^*) \cap D(F^*)$$

and

$$\inf_{x \in X} L(x,p) = J^*(p), \, p \in X^*$$
$$\inf_{p \in X^*} L(x,p) = J(x), \, x \in X. \tag{3.180}$$

Relation (3.180) implies

$$\inf_{(x,p)\in X\times X^*} L(x,p) = \inf_{p\in X^*} J^*(p) = \inf_{x\in X} J(x), \qquad (3.181)$$

called the Toland duality [Toland (1978, 1979)]. Here, we call $\langle v, u\rangle$ of (3.177) the hook term. The functional

$$\varphi(x,y) = F(x+y) - G(x) : X \times X \to [-\infty, +\infty]$$

of (3.175) is called the cost function, which is convex in y-component. Above mentioned global theory can be localized by sub-differentials. First, given $F : X \to [-\infty, +\infty]$, $x \in X$, and $p \in X^*$, we define

$$p \in \partial F(x) \quad \Leftrightarrow \quad F(y) \geq F(x) + \langle y - x, p\rangle, \ \forall y \in X$$

$$x \in \partial F^*(p) \quad \Leftrightarrow \quad F^*(q) \geq F^*(p) + \langle x, q - p\rangle, \ \forall q \in X^*.$$

It is obvious that $\partial F(x) \neq \emptyset$ implies $x \in D(F)$, but if $F : X \to (-\infty, +\infty]$ is proper, convex, lower semi-continuous, then

$$x \in \partial F^*(p) \quad \Leftrightarrow \quad p \in \partial F(x), \qquad (3.182)$$

and Fenchel-Moreau's identity

$$F(x) + F^*(p) = \langle x, p\rangle \qquad (3.183)$$

follows [Ekeland and Temam (1976)]. If $p \in \partial F(x)$, then $x \in \partial F^*(p)$, and, therefore,

$$F^*(p) - \langle x, p\rangle = -F(x).$$

This relation implies

$$L\big|_{p\in\partial F(x)} = J, \qquad (3.184)$$

where $L(x,p) = F^*(p) + G(x) - \langle x, p\rangle$ for

$$J(x) = G(x) - F(x), \quad J^*(p) = F^*(p) - G^*(p).$$

Similarly, we obtain the unfolding, indicated by

$$L\big|_{x\in\partial G^*(p)} = J^*. \qquad (3.185)$$

The second important notion of minimality follows immediately from (3.181). More precisely, we have

$$L(x,p) \geq J(x), \quad L(x,p) \geq J^*(p),$$

and, therefore,

$$L(x,p) \geq \max\{J(x), J^*(p)\} \qquad (3.186)$$

for any $(x,p) \in X \times X^*$. Then we obtain three fundamental theorems on this dual variation; spectral equivalence, dynamical equivalence, and dynamical stability. See Theorems 3.2.1, 3.2.2, and 3.2.3 of [Suzuki (2015)].

Chapter 4

Reaction

This chapter deals with reaction diffusion systems. Some of these systems contain the variational structure due to the stationary state, realized as Lyapunov functions in the context of dynamical theory. There may be the other structure derived from the ODE part, formulated by the Hamilton system. For the system of two components, if a level set of this Hamiltonian is a Jordan curve on which no stationary solution lies, then the solution on this orbit becomes periodic-in-time. Added to the diffusion terms, this Hamiltonian acts as a Lyapunov function, and the PDE orbits are absorbed into the spatially homogeneous parts. This phenomenon is observed in several models associated with thermodynamically isolated systems. In this chapter, first, we describe the method of mathematical modeling for chemical reactions. Then mathematical analysis using variational structures and positivity of the solution is exposed. Applications to biology are taken on the cell populations and echological models.

4.1 Mathematical Modeling of Reaction

A fundamental idea of mathematical modeling is to balance the amount, which is assumed to be non-negative. If $u = u(t)$ denotes a quantity of some material, two models of ordinary differential equations,

$$\frac{du}{dt} = a, \quad u(0) = u_0 > 0 \tag{4.1}$$

and

$$\frac{du}{dt} = au, \quad u(0) = u_0 > 0 \tag{4.2}$$

describe different laws of its time variation, where a is a constant. The first model can stand for supply or consumption according as $a > 0$ and $a < 0$, respectively. In the latter case of (4.1) the solution

$$u(t) = at + u_0$$

becomes eventually negative in spite of $u_0 > 0$. This feature has a strong contrast with this case $a < 0$ of (4.2), where the solution

$$u(t) = u_0 e^{at}$$

keeps positive even if it decays fast as $t \uparrow +\infty$.

The logistic equation

$$\frac{du}{dt} = u(1 - u), \quad u(0) = u_0 > 0$$

admits a global-in-time solution $u = u(t) >$ converging to 1 as $t \uparrow +\infty$. This feature is associated with the stability of two equilibria, $u = 0$ and $u = 1$. There may arise the blowup of the solution in accordance with the nonlinearity of the right-hand side. A typical example is

$$\frac{du}{dt} = u^2, \quad u(0) = u_0 > 0$$

of which solution

$$u(t) = (T - t)^{-1}, \quad T = u_0^{-1},$$

is usually thought to expire at the blowup time $t = T$. Activation and inhibition may be modeled by

$$u_t = \alpha v, \quad v_t = -\beta u,$$

using positive constants α and β. General autonomous system

$$\frac{du}{dt} = f(u), \quad u(0) = u_0 \in \mathbf{R}_+^N$$

can generate a local-in-time orbit, provided that $f = f(u)$ is locally Lipschitz continuous, where

$$\mathbf{R}_+^N = \{ u = (u_i) \in \mathbf{R}^N \mid u_i > 0, \ 1 \le i \le N \}$$

denotes the positive cone. The solutions keep the positivity $u = u(t) \in \mathbf{R}_+^N$, if this $f = f(u)$ is quasi-positive,

$$f_i(\hat{u}_i) \ge 0, \quad u \in \mathbf{R}_+^N, \ 1 \le i \le N, \tag{4.3}$$

where $f = (f_i)$ and

$$\hat{u}_i = (u_1, \cdots, u_{i-1}, 0, u_{i+1}, \cdots, u_N)$$

for $u = (u_i) \in \mathbf{R}_+^N$.

Fundamental process of the chemical reaction

$$A + B \;\to\; AB \; (k), \quad AB \;\to\; A + B \; (\ell) \qquad (4.4)$$

is formulated by the quadratic process using mass action law. Here we assume, first, that the collision of two particles causes reaction with a definite rate. Second, the collision rate is proportional to the product of concentrations of particles. Hence k and ℓ in (4.4) denote the attachment and detachment reaction rates, and if $[X]$ stands for the concentration of X-molecules there arises

$$\frac{d}{dt}[A] = -k[A][B] + \ell[AB]$$

$$\frac{d}{dt}[B] = -k[A][B] + \ell[AB]$$

$$\frac{d}{dt}[AB] = k[A][B] - \ell[AB]. \qquad (4.5)$$

This (4.5) implies the total mass conservations of A and B molecules,

$$\frac{d}{dt}([A] + [AB]) = 0, \quad \frac{d}{dt}([B] + [AB]) = 0.$$

If B is always homo-dymerized as in BB, reaction law (4.4) implies

$$A + BB \;\to\; ABB \; (2k), \quad ABB \;\to\; A + BB \; (\ell)$$

and then it follows that

$$\frac{d}{dt}[A] = -2k[A][BB] + \ell[ABB]$$

$$\frac{d}{dt}[BB] = -2k[A][BB] + \ell[ABB]$$

$$\frac{d}{dt}[ABB] = 2k[A][BB] - \ell[ABB]$$

from the above described mass action law. Similarly, if B is always hetero-dymerized as in AB, we obtain

$$A + AB \;\to\; ABA \; (k), \quad ABA \;\to\; A + AB \; (2\ell)$$

which results in

$$\frac{d}{dt}[A] = -k[A][AB] + 2\ell[ABA]$$

$$\frac{d}{dt}[AB] = -k[A][AB] + 2\ell[ABA]$$

$$\frac{d}{dt}[ABA] = k[A][AB] - 2\ell[ABA].$$

To formulate the polymerization $A + A \to AA$, we take the case that this A is masked as in AB and define the reaction rates k_* and ℓ_* by

$$A + AB \ \to \ AAB \ (k_*), \quad AAB \ \to A + AB \ (\ell_*).$$

If N_A denotes the number of the particle A, then the number of the collision of two A-particles is $\dfrac{1}{2} N_A (N_A - 1) \approx \dfrac{N_A^2}{2}$. Hence the above mass action law induces

$$A + A \ \to \ AA \ (k_*/2), \quad AA \ \to \ A + A \ (\ell_*).$$

Regarding the symmetry of the process, we reach

$$\frac{d}{dt}[A] = 2\left(-\frac{k_*}{2}[A]^2 + \ell_*[AA]\right)$$

$$\frac{d}{dt}[AA] = \frac{k_*}{2}[A]^2 - \ell_*[AA].$$

We can construct a reaction network, compiling the fundamental processes. Some aspects of its mathematical structure are described in the next chapter of this monograph. We have several data basis on chemical reaction rates. In some cases the application of dimension analysis is useful to suspect unknown coefficients.

Models using compartments or cell automata are actually useful, but partial differential equations (PDE) are essential in realizing the averaged movement of particles interacting the environment continuously distributed in space and time. Gradient of the scalar field is the fundamental notion to describe the action through medium. If Ω_t denotes the domain moving with the velocity $v = v(\cdot, t)$ and $\rho = \rho(x, t)$ is the smooth scalar field varying in t, then Liouville's theorem guarantees

$$\frac{d}{dt}\int_{\Omega_t} \rho \ dx = \int_{\Omega_t} \rho_t + \nabla \cdot \rho v \ dx. \tag{4.6}$$

Mass conservation has two expressions. The differential form

$$\rho_t = -\nabla \cdot j \tag{4.7}$$

comes from the Euler coordinate, that is,

$$\frac{d}{dt}\int_{\omega} \rho \ dx = -\int_{\partial \omega} \nu \cdot j \ dS$$

and hence $j = j(x, t)$ stands for the flux of this flow. The integral form,

$$\frac{d}{dt}\int_{\Omega_t} \rho_t \ dx = 0,$$

on the other hand, is subject to the Lagrange coordinate, and hence

$$\rho_t + \nabla \cdot \rho v = 0 \tag{4.8}$$

by (4.6). Adjusting (4.7) and (4.8), we obtain

$$j = \rho v.$$

If mass is changing under this flow, (4.7) is replaced by

$$\rho_t + \nabla \cdot \rho v = \rho X,$$

where X stands for the rate of mass change.

As is described in §1.1, the diffusion is the case of (4.7) with $j = -\nabla u$, if the diffusion constant is one. Then, Keller-Segel system (1.1) arises as a typical multi-scale model. Semilinear parabolic equation

$$\frac{\partial u}{\partial t} - \Delta u = f(u), \quad \frac{\partial u}{\partial \nu}\bigg|_{\partial\Omega} = 0, \quad u|_{t=0} = u_0(x) \tag{4.9}$$

casts a model in many fields. The Neumann boundary condition in (4.9) can be replaced by the Dirichlet boundary condition and others. If $f = f(u)$ is locally Lipschitz continuous it is transformed into the integral equation in a function space, say $X = L^\infty(\Omega)$,

$$u(t) = e^{t\Delta}u_0 + \int_0^t e^{(t-s)\Delta} f(u(s)) \, ds, \tag{4.10}$$

where $-\Delta$ is provided with the Neumann boundary condition. Equation (4.10) is solved by the fixed point argument in the space $C([0,T];X)$. Since the existence time T is estimated below by $\|u_0\|_\infty$, it holds that

$$T < +\infty \quad \Rightarrow \quad \lim_{t\uparrow T} \|u(\cdot,t)\|_\infty = +\infty. \tag{4.11}$$

Thus the orbit of (4.9) generated by the smooth initial value u_0 is classified into compact global-in-time, blowup in finite time, blowup in infinite time indicated by (1.106):

$$T = +\infty, \quad \limsup_{t\uparrow+\infty} \|u(\cdot,t)\|_\infty = +\infty.$$

Using the semi-group estimate

$$\|e^{t\Delta}u\|_r \le C_7(q,r) \max\left\{1, t^{-\frac{n}{2}\left(\frac{1}{q}-\frac{1}{r}\right)}\right\} \|u\|_q, \quad 1 \le q \le r \le \infty, \tag{4.12}$$

this theory is extended to $X = L^p(\Omega)$, $1 \le p \le \infty$ according as the growth rate of $f = f(u)$.

4.2   Thermodynamics of Reaction Diffusion

Reaction diffusion equation in two species is formulated by

$$u_t = \varepsilon^2 \Delta u + f(u,v), \quad \tau v_t = D\Delta v + g(u,v) \qquad \text{in } \Omega \times (0,T)$$

$$\left.\frac{\partial}{\partial \nu}(u,v)\right|_{\partial \Omega} = 0, \qquad (u,v)|_{t=0} = (u_0(x), v_0(x)) > 0,$$

where the quasi-positivity (4.3) is reduced to

$$f(0,v) \geq 0, \ g(u,0) \geq 0, \quad u,v \geq 0.$$

It casts the prey-predator system if

$$f(u,v) = u(a - bv), \quad g(u,v) = v(-c + du)$$

where $a, b, c, d > 0$ are constants. The ODE part

$$\frac{du}{dt} = f(u,v), \quad \tau \frac{dv}{dt} = g(u,v)$$

takes the unique stationary state $u_* = c/d$, $v_* = a/b$, and otherwise, the orbit becomes periodic-in-time. This fact follows because

$$\frac{du}{dt} = u(a - bv), \ \tau\frac{dv}{dt} = v(-c + du), \quad (u,v)|_{t=0} = (u_0, v_0) > 0 \quad (4.13)$$

has a first integral

$$a \log v - bv + c \log u - du = \text{constant}, \tag{4.14}$$

and this (4.14) determines a Jordan curve in the first quadrant in uv space except for the stationary solution.

To derive (4.14), we use the growth rates

$$\xi = \log u, \quad \eta = \log v \tag{4.15}$$

to write (4.13) as

$$\frac{d\xi}{dt} = a - be^\eta, \quad \frac{d\eta}{dt} = -\tau^{-1}c + \tau^{-1}de^\xi. \tag{4.16}$$

The right-hand side on (4.16) does not depend on the variables in the left-hand side explicitly, and therefore, it casts the form of Hamilton system

$$\xi_t = -H_\eta, \quad \eta_t = H_\xi$$

with the Hamiltonian

$$H(\xi, \eta) = -a\eta + be^\eta - \tau^{-1}c\xi + \tau^{-1}de^\xi. \tag{4.17}$$

Then we obtain (4.14) by

$$\frac{dH}{dt} = 0.$$

Under (4.15), the prey-predator system with diffusion

$$u_t = \varepsilon^2 \Delta u + u(a - bv), \quad \tau v_t = D\Delta v + v(-c + du)$$

$$\frac{\partial}{\partial \nu}(u, v)\Big|_{\partial\Omega} = 0, \quad (u, v)|_{t=0} = (u_0(x), v_0(x)) > 0 \qquad (4.18)$$

takes the form

$$\xi_t = \varepsilon^2 e^{-\xi} \Delta e^\xi - H_\eta, \quad \eta_t = \tau^{-1} D e^{-\eta} \Delta e^\eta + H_\xi, \quad \frac{\partial}{\partial \nu}(\xi, \eta)\Big|_{\partial\Omega} = 0.$$

Then we obtain

$$\frac{\partial}{\partial t} H(\xi, \eta) = H_\xi \varepsilon^2 e^{-\xi} \Delta e^\xi + H_\eta \tau^{-1} D e^{-\eta} \Delta e^\eta$$

$$= -\tau^{-1}(c\varepsilon^2 e^{-\xi} \Delta e^\xi + aD e^{-\eta} \Delta e^\eta)$$

$$+\tau^{-1}(\varepsilon^2 d\Delta e^\xi + Db\Delta e^\eta),$$

and therefore,

$$\frac{d}{dt} \int_\Omega H(\xi, \eta) = -\tau^{-1} \int_\Omega c\varepsilon^2 |\nabla \xi|^2 + aD|\nabla \eta|^2 \, dx \leq 0. \qquad (4.19)$$

Thus $\int_\Omega H(\xi, \eta)$ casts the Lyapunov function to (4.18), and then (4.19) implies the a priori estimate

$$\int_\Omega e^\xi + e^\eta \, dx = \|u(\cdot, t)\|_1 + \|v(\cdot, t)\|_1 \leq C. \qquad (4.20)$$

We combine this (4.20) and

$$u_t \leq \varepsilon^2 \Delta u + au, \ u > 0, \quad \frac{\partial u}{\partial \nu}\Big|_{\partial\Omega} = 0$$

derived from the first equation of (4.18). Repeating the semi-group estimates (4.12), we obtain

$$\limsup_{t\uparrow T} \|u(\cdot, t)\|_\infty < +\infty. \qquad (4.21)$$

Then we write

$$\tau v_t = D\Delta v + v(-c + du), \quad \frac{\partial v}{\partial \nu}\Big|_{\partial\Omega} = 0,$$

for the second equation of (4.18). By (4.21) and the linear theory, it holds that

$$\limsup_{t\uparrow T} \|v(\cdot, t)\|_\infty < +\infty$$

and hence $T = +\infty$. Then the parabolic regularity guarantees that the orbit $\mathcal{O} = \{(u(\cdot,t), v(\cdot,t)\}_{t\geq 0}$ is compact in $C(\overline{\Omega})^2$.

From the theory of dynamical systems, the ω-limit set

$$\omega(u_0, v_0) = \{(u_\infty, v_\infty) \mid \exists t_k \uparrow +\infty \text{ s.t.}$$

$$\lim_{k\to\infty} \|u(\cdot, t_k) - u_0, v(\cdot, t_k) - v_0\|_{C^2} = 0\}$$

induced by the initial value $(u_0, v_0) = (u, v)|_{t=0}$ is non-empty, connected, and compact set. This $\omega(u_0, v_0)$ is invariant under the flow induced by (4.18), and the Lyapunov function

$$\int_\Omega H(\xi, \eta) = \int_\Omega -a \log v + bv - \tau^{-1} c \log u + \tau^{-1} du\ dx \qquad (4.22)$$

on it.

More precisely, although each $(u_\infty, v_\infty) \in \omega(u_0, v_0)$ is non-negative, but is not positive definite, inequality (4.19) implies

$$\int_\Omega \log u_\infty + \log v_\infty\ dx > -\infty$$

by Fatou's lemma. Combining this inequality with (4.20), we see that the right-hand side on (4.22) converges absolutely as $t \uparrow +\infty$. It holds also that u_∞ and v_∞ are not identically zero, and hence the strong maximum principle of the parabolic equation is applicable. Thus, the solution $(\tilde{u}(\cdot,t), \tilde{v}(\cdot,t))$ to (4.18) with the initial value (u_∞, v_∞) is positive definite for $t > 0$, and the quantity defined by the right-hand side on (4.22),

$$L(\tilde{u}(\cdot,t), \tilde{v}(\cdot,t)) = \int_\Omega -a \log \tilde{v} + b\tilde{v} - \tau^{-1} c \log \tilde{u} + \tau^{-1} d\tilde{u}\ dx$$

is differentiable in $t > 0$. Putting $\tilde{\xi} = \log \tilde{u}$ and $\tilde{\eta} = \log \tilde{v}$, therefore, we obtain

$$\frac{d}{dt} \int_\Omega H(\tilde{\xi}, \tilde{\eta}) = -\tau^{-1} \int_\Omega c\varepsilon^2 |\nabla \tilde{\xi}|^2 + aD|\nabla \tilde{\eta}|^2\ dx = 0, \quad t > 0.$$

It holds that $\nabla \tilde{\xi}(\cdot, t) = \nabla \tilde{\eta}(\cdot, t) = 0$ for $t > 0$, and hence any $(u_\infty, v_\infty) \in \omega(u_0, v_0)$ is spatially homogeneous.

By the invariance of $\omega(u_0, v_0)$ under the flow (4.18) and the above described classification of ODE orbits we obtain the following theorem.

Theorem 4.1. *It holds always that $T = +\infty$ in (4.18). Each initial value (u_0, v_0) admits the ODE orbit $\mathcal{O} \subset \mathbf{R}^2$ such that*

$$\lim_{t\uparrow +\infty} dist_{C^2}((u(\cdot,t), v(\cdot,t)), \mathcal{O}) = 0.$$

If this $\mathcal{O}$ is not a singleton, therefore, there is $\ell > 0$ such that

$$\lim_{t\uparrow +\infty} \|u(\cdot, t+\ell) - u(\cdot,t), v(\cdot, t+\ell) - v(\cdot,t)\|_{C^2} = 0.$$

Inequality (4.20) is independent of D and hence so is (4.21). Then we obtain

$$\sup_{0 \le t < T} \|v(\cdot, t)\|_\infty \le C_T$$

for each $T > 0$ by the second equation of (4.18), with $C_T > 0$ independent of D. The parabolic regularity now guarantees the convergence of $(u, v) = (u_D(\cdot, t), v_D(\cdot, t))$, the solution to (4.18), to $(U, V) = (U(\cdot, t), V(t))$, the solution to

$$U_t = \varepsilon^2 \Delta U + U(a - bV), \quad \frac{\partial u}{\partial \nu}\bigg|_{\partial\Omega} = 0, \quad U|_{t=0} = u_0(x)$$

$$\tau \frac{dV}{dt} = \frac{V}{|\Omega|} \int_\Omega -c + dU \; dx, \quad V|_{t=0} = \bar{v}_0 \equiv \frac{1}{|\Omega|} \int_\Omega v_0(x) \quad (4.23)$$

locally uniformly in $\overline{\Omega} \times [0, \infty)$ as $D \uparrow +\infty$. This (4.23) is called the shadow system of (4.18).

The Hamiltonian $H(\xi, \eta)$ in (4.17) is independent of D, which induces

$$\frac{d}{dt} \int_\Omega H(\hat{\xi}, \hat{\eta}) \; dx = -\tau^{-1} \int_\Omega c\varepsilon^2 |\nabla \hat{\xi}|^2 \; dx \le 0$$

for (4.19). The above argument now guarantees the convergence of $(U, V) = (U(\cdot, t), V(t))$ to a solution to the ODE part as $t \uparrow +\infty$. The periods and phases of these limits of (u, v) and (U, V) are different. The shadow system (4.23) actually reproduces the transient profile of (4.18) for $D \gg 1$.

ODE part of the Gierer-Meinhard equation in morphogenesis is described by

$$\frac{du}{dt} = -u + \frac{u^p}{v^q}, \quad \tau \frac{dv}{dt} = -v + \frac{u^r}{v^s}. \quad (4.24)$$

It has the same property as that of the prey-predator system (4.18), if

$$\tau = \frac{s+1}{p-1}. \quad (4.25)$$

Thus there is a unique stationary solution and the other orbit is periodic-in-time in the first quadrant in the uv plane. There is actually a first integral of (4.24) for (4.25). Here we derive this fact from the Hamilton formalism as in (4.18).

For this purpose we rewrite (4.24) as

$$u^{-p}(u_t + u) = v^{-q}, \quad v^s(v_t + \tau^{-1}v) = \tau^{-1}u^r. \quad (4.26)$$

Then, regarding the first terms of the left-hand side on (4.26), we put

$$\xi = \frac{u^{-p+1}}{p-1}, \quad \eta = \frac{v^{s+1}}{s+1},$$

to obtain

$$\xi_t = u^{-p+1} - v^{-q} = (p-1)\xi - \{(s+1)\eta\}^{-\frac{q}{s+1}}$$

$$\eta_t = -\frac{v^{s+1}}{\tau} + \frac{u^r}{\tau} = -\tau^{-1}(s+1)\eta + \tau^{-1}\{(p-1)\xi\}^{-\frac{r}{p-1}}. \quad (4.27)$$

The cross terms of the right-hand side on (4.27) are linear in ξ and η, which are contained in the Hamiltonian under the assumption of (4.25). Then, (4.27) is reduced to

$$\frac{d\xi}{dt} = H_\eta, \quad \frac{d\eta}{dt} = -H_\xi$$

for

$$H(\xi,\eta) = (p-1)\xi\eta + \left(\frac{r}{p-1}-1\right)^{-1} A(\xi) + \left(\frac{q}{s+1}-1\right)^{-1} B(\eta)$$

$$A(\xi) = \tau^{-1}(p-1)^{-\frac{r}{p-1}}\xi^{1-\frac{r}{p-1}}, \quad B(\eta) = (s+1)^{-\frac{q}{s+1}}\eta^{1-\frac{q}{s+1}}.$$

Then a similar result to Theorem 4.1 is obtained under some more restrictions on the parameter.

Even in the case that the Hamilton structure is not expected, some algebraic structures can make the system integrable. For example, a Lotka-Volterra system different from (4.13),

$$\tau_1\frac{du_1}{dt} = (u_2-u_3)u_1, \quad \tau_2\frac{du_2}{dt} = (u_3-u_1)u_2, \quad \tau_3\frac{du_3}{dt} = (u_1-u_2)u_3 \quad (4.28)$$

can take the form

$$\tau_1\frac{d\xi_1}{dt} = e^{\xi_2} - e^{\xi_3}, \quad \tau_2\frac{d\xi_2}{dt} = e^{\xi_3} - e^{\xi_1}, \quad \tau_3\frac{d\xi_3}{dt} = e^{\xi_1} - e^{\xi_2}, \quad (4.29)$$

in the variables $\xi_i = \log u_i$ for $i = 1,2,3$. If $\times$ denotes the outer product in $\mathbf{R}^3$, this (4.29) means

$$\mathcal{M}\frac{d\xi}{dt} = H(\xi) \times a, \quad (4.30)$$

where

$$\xi = \begin{pmatrix} \xi_1 \\ \xi_2 \\ \xi_3 \end{pmatrix}, \quad a = \begin{pmatrix} 1 \\ 1 \\ 1 \end{pmatrix}, \quad H(\xi) = \begin{pmatrix} e^{\xi_1} \\ e^{\xi_2} \\ e^{\xi_3} \end{pmatrix}$$

and $\mathcal{M} = \text{diag}(\tau_1,\tau_2,\tau_3)$. Then, it is easy to see that (4.30) takes the conservation laws

$$0 = \frac{d}{dt}\mathcal{M}\xi \cdot a = \frac{d}{dt}(\tau_1\xi_1 + \tau_2\xi_2 + \tau_3\xi_3)$$

$$0 = \frac{d}{dt}\mathcal{M}\xi \cdot H(\xi) = \frac{d}{dt}(\tau_1 e^{\xi_1} + \tau_2 e^{\xi_2} + \tau_3 e^{\xi_3}). \quad (4.31)$$

In the reaction diffusion system

$$\tau_1 \frac{\partial u_1}{\partial t} = d_1 \Delta u_1 + (u_2 - u_3)u_1, \quad \left.\frac{\partial u_1}{\partial \nu}\right|_{\partial\Omega} = 0$$

$$\tau_2 \frac{\partial u_2}{\partial t} = d_2 \Delta u_2 + (u_3 - u_1)u_2, \quad \left.\frac{\partial u_2}{\partial \nu}\right|_{\partial\Omega} = 0$$

$$\tau_3 \frac{\partial u_3}{\partial t} = d_3 \Delta u_3 + (u_1 - u_2)u_3, \quad \left.\frac{\partial u_3}{\partial \nu}\right|_{\partial\Omega} = 0 \qquad (4.32)$$

the quantities

$$\int_\Omega \sum_i \tau_i \log u_i \, dx, \quad \int_\Omega \sum_i \tau_i u_i \, dx$$

associated with (4.31) cast the total entropy and the total mass of the system, which reflects the structure of the solution set. Thus, an analogous result to Theorem 4.1 is valid to (4.32) if $n \leq 2$, where n denotes the space dimension.

4.3 Critical Dimension of Lotka-Volterra Systems

Lotka-Volterra system describes the population dynamics in echology. For the case of N species with diffusion, it is described by

$$\tau_j u_{jt} = d_j \Delta u_j + (-e_j + \sum_k a_{jk} u_k)u_j \qquad \text{in } \Omega \times (0, T)$$

$$\left.\frac{\partial u_j}{\partial \nu}\right|_{\partial\Omega} = 0, \quad u_j|_{t=0} = u_{j0}(x) \geq 0, \quad 1 \leq j \leq N, \qquad (4.33)$$

where $\Omega \subset \mathbf{R}^n$ denotes a bounded domain with smooth boundary $\partial\Omega$, ν is the outer unit normal vector, τ_j, $d_j > 0$, e_j, and $a_{jk} \in \mathbf{R}$ are constants, and $u_0 = (u_{j0})$ is a smooth initial value. There arises local-in-time solution positive in $\overline{\Omega} \times (0, T)$ unless $u_{j0} \equiv 0$. System (4.32) is the case of $N = 3$, $e_1 = e_2 = e_3 = 0$, and

$$A \equiv (a_{jk}) = \begin{pmatrix} 0 & 1 & -1 \\ -1 & 0 & 1 \\ 1 & -1 & 0 \end{pmatrix}.$$

Let

$$A = (a_{jk})_{j:1\downarrow N; k:1\to N} \in M_N(\mathbf{R}), \quad u = (u_j)_{j:1\downarrow N} \in \mathbf{R}^N$$

$$e = (e_j)_{j:1\downarrow N} \in \mathbf{R}^N, \quad \tau = (\tau_j)_{j:1\downarrow N} \in \mathbf{R}^N, \quad d = (d_j)_{j:1\downarrow N} \in \mathbf{R}^N,$$

and

$$|u| = \left(\sum_{i=1}^{N} u_j^2 \right)^{1/2}. \tag{4.34}$$

We assume

$$^t A + A \le 0, \quad e \ge 0. \tag{4.35}$$

Here, $^t A + A \le 0$ and $e = (e_j)_{j:1\downarrow N} \ge 0$ mean the non-positive definiteness of the symmetric matrix $^t A + A$ and $e_j \ge 0$ for any $1 \le j \le N$, respectively.

To examine thermodynamical structure derived from (4.35), we take the ODE part of (4.33),

$$\tau_j \frac{dv_j}{dt} = (-e_j + \sum_k a_{jk} v_k) v_j, \quad v_j|_{t=0} = v_{j0} > 0. \tag{4.36}$$

It holds that $v_j(t) > 0$ as far as the solution exists. Then

$$M = \tau \cdot v = \sum_j \tau_j v_j$$

stands for the total mass of the species, where $v = (v_j)_{j:1\downarrow N}$. We obtain

$$\frac{dM}{dt} = \tau \cdot \frac{dv}{dt} = \sum_j \left(-e_j + \sum_k a_{jk} v_k \right) v_j \le -e \cdot v \le 0 \tag{4.37}$$

by (4.35). Since M is bounded, it holds that $T = +\infty$ in (4.36), and furthermore,

$$0 \le v_j(t) \le C_1, \quad t \ge 0, \ 1 \le j \le N. \tag{4.38}$$

We assume, without loss of generality,

$$e_1, \cdots, e_\ell = 0 < e_{\ell+1}, \cdots, e_N \tag{4.39}$$

for some $\ell = 0, \cdots, N$, and put

$$c_0 = \frac{\min_{j=\ell+1,\cdots,N} e_j}{\max_{j=\ell+1,N} \tau_j} > 0, \quad \tilde{M} = \sum_{j=\ell+1}^{N} \tau_j v_j.$$

Inequality (4.37) implies

$$M(t) \le M(0) - \int_0^t e \cdot v(t') dt'$$

and hence

$$\tilde{M}(t) \le M(0) - c_0 \int_0^t \tilde{M}(t') dt', \quad t \ge 0. \tag{4.40}$$

This (4.40) means

$$f'(t) \leq M(0) - c_0 f(t), \quad t \geq 0$$

for $f(t) = \int_0^t \tilde{M}(t')dt'$. Therefore, it follows that $f(t) \leq C_2$, that is,

$$\int_0^\infty \tilde{M}(t')dt' < +\infty.$$

Then there is $t_k \uparrow +\infty$ such that

$$\tilde{M}(t_k) \to 0.$$

We have, furthermore,

$$\frac{d\tilde{M}}{dt} \leq \sum_{j=\ell+1}^N \sum_{k=1}^N a_{jk}v_k v_j \leq C_3 \tilde{M}$$

by (4.38), which implies

$$\tilde{M}(t) \leq \tilde{M}(t_k) + C_3 \int_{t_k}^t \tilde{M}(t')dt', \quad t \geq t_k. \tag{4.41}$$

Making $t \uparrow +\infty$ and then $k \to \infty$ in (4.41), we obtain

$$\limsup_{t\uparrow+\infty} \tilde{M}(t) \leq 0, \tag{4.42}$$

which means

$$\lim_{t\uparrow+\infty} v_j(t) = 0, \quad \ell+1 \leq j \leq N.$$

Here, the following structure to (4.36) assures the entropy increasing, that is, the existence of $0 < b \in \mathbf{R}^\ell$ such that

$${}^t\tilde{A}b \geq 0, \quad \tilde{A} = \left(\tau_j^{-1}a_{jk}\right)_{j:1\downarrow\ell;\ k:1\to\ell}. \tag{4.43}$$

In fact, then,

$$H = \sum_{j=1}^\ell b_j \log v_j, \quad b = (b_j)_{j:1\downarrow\ell}$$

casts the entropy in (4.39) as

$$\frac{dH}{dt} = \sum_{j=1}^\ell b_j v_j^{-1}\frac{dv_j}{dt} = \sum_{j=1}^\ell \sum_{k=1}^N b_j \tau_j^{-1}a_{jk}v_k \geq 0. \tag{4.44}$$

Inequality (4.44) implies

$$\liminf_{t\uparrow+\infty} v_j(t) > 0, \ 1 \leq j \leq \ell.$$

The ω-limit set

$$\omega(v_0) = \{v_* = (v_{*j})_{j:1\downarrow N} \in \mathbf{R}^N \mid \exists t_k \uparrow +\infty \text{ s.t. } v(t_k) \to v_*\}$$

defined for the initial value $v_0 = (v_{0j})_{j:1\downarrow N}$ of (4.36) is non-empty, connected, and compact. Furthermore, it holds that

$$v_{\ell+1}^* = \cdots = v_N^* = 0 < v_1^*, \cdots, v_\ell^*, \quad b \cdot (\log v_j^*)_{j:1\downarrow \ell} \geq H(0)$$

for any $v_* = (v_{*j})_{j:1\downarrow N}$.

As we shall confirm later, these structures of ODE part induce those for (4.33) together with the homogenization in space. Also, condition (4.43) will be clarified according as the algebraic structure of $A = (a_{jk})$ in the next chapter.

Here we take a preliminary study on (4.33) with (4.35). The fundamental property is

$$\frac{dM}{dt} = \int_\Omega \tau \cdot u_t \, dx \leq - \int_\Omega e \cdot u \, dx \leq 0 \tag{4.45}$$

for

$$M = \int_\Omega \tau \cdot u \, dx. \tag{4.46}$$

The a priori estimate

$$\sup_{t \geq 0} \|u(\cdot, t)\|_1 \leq C_4 \tag{4.47}$$

follows from this (4.45), and then there arises

$$\sup_{t \geq 0} \|u(\cdot, t)\|_\infty \leq C_5 \tag{4.48}$$

and hence $T = +\infty$ in the case of $n = 1$. This is shown by the proof of the following theorem, global-in-time ε-regularity for the case of $n = 2$.

Theorem 4.2. *If $n = 2$ and (4.35) in (4.33), there is $\varepsilon_0 > 0$ such that*

$$\lambda \equiv \int_\Omega \tau \cdot u_0 \, dx = \sum_j \int_\Omega \tau_j u_{j0} \, dx < \varepsilon_0 \tag{4.49}$$

implies

$$T = +\infty, \quad \sup_{t \geq 0} \|u(\cdot, t)\|_\infty < +\infty. \tag{4.50}$$

Theorem 4.2 is a relative of Lemma 1.1 in §1.2. To understand the situation in more details, let $\ell = N$ in (4.39). This assumption means $e_j = 0$ for $j = 1, \cdots, N$, and then, (4.33) admits the self-similar transformation

$$u_j^\mu(x,t) = \mu^2 u_j(\mu x, \mu^2 t), \quad \mu > 0, \ 1 \le j \le N. \tag{4.51}$$

In this case, the quantity λ in (4.49) is invariant in $\mu > 0$, if and only if $n = 2$. Thermodynamical structure of the total mass and entropy controls are also similar to the Smoluchowski-Poisson equation (1.13). An analogous form of monotonicity formula (1.25) even arises, which admits the blowup analysis as in Chapter 1. Below we combine this method with the notion of weak solution to remove the assumption (4.49) in Theorem 4.2. Here we shall show this theorem by the classical argument.

Proof of Theorem 4.2. First, we have (4.47) by (4.45). Second, it holds that

$$\tau_j u_{jt} \le d_j \Delta u_j + a_0 \sum_k u_k u_j, \ u_j \ge 0 \quad \left. \frac{\partial u_j}{\partial \nu} \right|_{\partial\Omega} = 0, \quad 1 \le j \le N \tag{4.52}$$

where

$$a_0 = \max_{j,k}(a_{jk})_+, \quad (a_{jk})_+ = \max\{a_{jk}, 0\}.$$

Then, Young's inequality implies

$$\frac{\tau_j}{2} \frac{d}{dt} \|u_j\|_2^2 + d_j \|\nabla u_j\|_2^2 \le a_0 \sum_k \int_\Omega u_k u_j^2 \, dx$$

$$\le \frac{2Na_0}{3} \|u_j\|_3^3 + \frac{a_0}{3} \sum_k \|u_k\|_3^3. \tag{4.53}$$

For the right-hand side on (4.53) we apply the Gagliardo-Nirenberg inequality valid to $n = 2$,

$$\|w\|_3^3 \le C_6 \|w\|_{H^1}^2 \|w\|_1, \quad w \in H^1(\Omega), \tag{4.54}$$

and also Poincaré-Wirtinger's inequality

$$\mu_2 \|w - \overline{w}\|_2^2 \le \|\nabla w\|_2^2, \quad \overline{w} = \frac{1}{|\Omega|} \int_\Omega w \, dx \tag{4.55}$$

for $w = u_j$. Since (4.47) it holds that

$$\frac{\tau_j}{2} \frac{d}{dt} \|u_j\|_2^2 + d_j \|\nabla u_j\|_2^2 \le C_7 \lambda \|\nabla u\|_2^2 + C_7,$$

and hence

$$\frac{1}{2} \frac{d}{dt} \sum_j \tau_j \|u_j\|_2^2 + d_0 \|\nabla u\|_2^2 \le NC_7 \lambda \|\nabla u\|_2^2 + NC_7 \tag{4.56}$$

for $d_0 = \min_j d_j$, where $|\nabla u|^2 = \sum_j |\nabla u_j|^2$.

If $\lambda < d_0/(2NC_2) \equiv \varepsilon_0$ in (4.56) we have

$$\frac{d}{dt} \sum_j \tau_j \|u_j\|_2^2 + d_0 \|\nabla u\|_2^2 \leq C_8$$

and then it holds that

$$\frac{d}{dt}[u]_2^2 + \mu_2 \tau_0^{-1} d_0 [u]_2^2 \leq C_9$$

for (4.55), (4.47), and

$$\tau_0 = \max_j \tau_j, \quad [u]_2^2 = \sum_j \tau_j \|u_j\|_2^2.$$

We thus obtain

$$\|u(\cdot, t)\|_2 \leq C_{10}. \tag{4.57}$$

Based on (4.57), we apply the comparison theorem. Regarding

$$\tau_j u_{jt} \leq d_j \Delta u_j + \frac{a_0}{2}(|u|^2 + Nu_j^2), \ u_j \geq 0, \ \left.\frac{\partial u_j}{\partial \nu}\right|_{\partial\Omega} = 0, \quad 1 \leq j \leq N,$$

we take

$$g_j = \mu u_j + \frac{a_0}{2}(|u|^2 + Nu_j^2)$$

for $\mu \gg 1$. Let $\bar{u}_j = \bar{u}_j(x, t)$, $1 \leq j < N$, be the solution to

$$\tau_j \bar{u}_{jt} = d_j \Delta \bar{u}_j - \mu \bar{u}_j + g_j(\cdot, t)$$
$$\left.\frac{\partial \bar{u}_j}{\partial \nu}\right|_{\partial\Omega} = 0, \quad \bar{u}_j|_{t=0} = u_{j0}(x) \geq 0. \tag{4.58}$$

Then we obtain

$$0 \leq u_j \leq \bar{u}_j \tag{4.59}$$

by the comparison theorem. Here we have

$$\|g_j(\cdot, t)\|_1 \leq C_{11}$$

by (4.57).

Now we use the semi-group estimate (4.12). Then, $L_j = -d_j \Delta + \mu$, $1 \leq q \leq r \leq \infty$ satisfies

$$\|e^{-tL_j}\|_{L^q \to L^r} \leq C(q, r)e^{-\mu t} \max\left\{1, t^{-\frac{n}{2}\left(\frac{1}{q} - \frac{1}{r}\right)}\right\}. \tag{4.60}$$

We apply (4.60) to

$$\overline{u}_j(\cdot, t) = e^{-t\tau_j^{-1}L_j} u_{j0} + \int_0^t e^{-(t-s)\tau_j^{-1}L_j} g_j(\cdot, s) ds \qquad (4.61)$$

derived from (4.58), repeatedly. In fact, assuming

$$\|u(\cdot, t)\|_q \leq C_{12} \qquad (4.62)$$

for some $q \geq 2$ it holds that

$$\|g_j(\cdot, t)\|_{q/2} \leq C_{13}.$$

Then, we use (4.60)–(4.61) to obtain

$$\|\overline{u}_j(\cdot, t)\|_r \leq C_{14}, \quad 1 \leq j \leq N \qquad (4.63)$$

for $\frac{n}{2}(\frac{2}{q} - \frac{1}{r}) < 1$, satisfying $r \in [q, \infty]$. Then, inequality (4.59) implies

$$\|u_j(\cdot, t)\|_r \leq C_{15} \qquad (4.64)$$

for $1 \leq r < \frac{nq}{2(n-q)_+}$ and $r = \infty$ if $q > n$ and $q \leq n$, respectively.

If $n = 2$, we have (4.63) for $1 \leq r < \infty$ by (4.62), $q = 2$. Then we apply (4.62) for $q > 2 = n$, to obtain (4.64). The proof is complete. $\qquad \square$

Let us confirm that the above proof guarantees (4.50) for $n = 1$ in (4.33) with (4.35). Generally, if the nonlinearity is quadratic growth in (4.33), condition (4.57) implies (4.48) for $n \leq 3$.

4.4 Virus Dynamics

Spatial homogenization of isolated systems are widely observed in population dynamics in biology under the presense of diffusion. Here we study the asymptotic behavior of the solution to the virus dynamics model

$$u_{1t} = d_1 \Delta u_1 + \lambda - mu_1 - \beta u_1 u_3$$
$$u_{2t} = d_2 \Delta u_2 + \beta u_1 u_3 - au_2$$
$$u_{3t} = d_3 \Delta u_3 + aru_2 - bu_3 \qquad (4.65)$$

in a bounded domain $\Omega \subset \mathbf{R}^n$ with smooth boundary $\partial\Omega$, where $u_1 = u_1(x, t)$, $u_2 = u_2(x, t)$, and $u_3 = u_3(x, t)$ stand for the population density of uninfected cells, that of infected cells, and that of viruses, respectively, at the time t and at the point $x \in \Omega$. We impose the Neumann boundary condition

$$\left. \frac{\partial u}{\partial \nu} \right|_{\partial\Omega} = 0, \qquad (4.66)$$

where ν is the outer unit normal vector on $\partial\Omega$, and $u = (u_1, u_2, u_3)$. In the initial condition

$$u(x,0) = u_0(x) \quad \text{in } \Omega, \tag{4.67}$$

we assume that $u_0 = (u_{1,0}, u_{2,0}, u_{3,0}) \in C^2(\overline{\Omega}; \mathbf{R}^3)$, $\partial u_0/\partial\nu|_{\partial\Omega} = 0$, and each component of u_0 is non-negative and not identically equal to zero.

Equation (4.65) is obtained by incorporating the diffusion terms to the ordinary differential equation (ODE) model

$$\frac{du_1}{dt} = \lambda - mu_1 - \beta u_1 u_3$$

$$\frac{du_2}{dt} = \beta u_1 u_3 - au_2$$

$$\frac{du_3}{dt} = aru_2 - bu_3 \tag{4.68}$$

for the persistent infection of HIV. Thus, uninfected cells are produced at a constant rate λ, have a death rate m, and become infected at a rate βu_3. The death rates of infected cells and viruses are denoted, respectively, by a and b. Each infected cell releases r virus particles when it bursts.

Equation (4.68) has two equilibria $u_* = (u_{*1}, 0, 0)$ and $u^* = (u_1^*, u_2^*, u_3^*)$, where

$$u_{*1} = \frac{\lambda}{m}, \; u_1^* = \frac{b}{r\beta}, \; u_2^* = \frac{bm}{ar\beta}(R_0 - 1), \; u_3^* = \frac{m}{\beta}(R_0 - 1). \tag{4.69}$$

Here, $R_0 = r\beta\lambda/(bm)$ is the basic reproduction number of Bonhoeffer. The equilibrium u_* is the disease free state. The equilibrium u^* is the interior equilibrium, whose elements are all positive, if and only if $R_0 > 1$.

Let

$$\Phi(s) = s - \log s - 1 \quad (s > 0). \tag{4.70}$$

Equation (4.68) is provided with Korobeinikov's Lyapunov function:

$$U(u) = u_{*1}\Phi\left(\frac{u_1}{u_{*1}}\right) + u_2 + \frac{1}{r}u_3 \tag{4.71}$$

for u_*, and

$$V(u) = u_1^*\Phi\left(\frac{u_1}{u_1^*}\right) + u_2^*\Phi\left(\frac{u_2}{u_2^*}\right) + \frac{u_3^*}{r}\Phi\left(\frac{u_3}{u_3^*}\right) \tag{4.72}$$

for u^*, which implies the global stability of u_* if $R_0 \leq 1$, and that of u^* if $R_0 > 1$.

To confirm this property, we put $f(u) = (f_1(u), f_2(u), f_3(u))$ and

$$f_1(u) = \lambda - mu_1 - \beta u_1 u_3$$
$$f_2(u) = \beta u_1 u_3 - au_2$$
$$f_3(u) = aru_2 - bu_3. \tag{4.73}$$

Along the solution $u = (u_1, u_2, u_3)$ to (4.68) it holds that

$$\frac{d}{dt} U(u) = \nabla U(u) \cdot f(u) = mu_{*1}\left(2 - \frac{u_1}{u_{*1}} - \frac{u_{*1}}{u_1}\right) + \frac{b}{r}(R_0 - 1)u_3 \tag{4.74}$$

and

$$\frac{d}{dt} V(u) = \nabla V(u) \cdot f(u)$$
$$= mu_1^*\left(2 - \frac{u_1}{u_1^*} - \frac{u_1^*}{u_1}\right) + au_2^*\left(3 - \frac{u_1^*}{u_1} - \frac{u_2 u_3^*}{u_2^* u_3} - \frac{u_1 u_2^* u_3}{u_1^* u_2 u_3^*}\right). \tag{4.75}$$

By the inequality on arithmetic and geometric means, the right-hand sides on (4.74) and (4.75) are non-positive if $u_1, u_{*1}, u_3 > 0$ and $R_0 \leq 1$ and if $u_1, u_2, u_3 > 0$ and $u_1^*, u_2^*, u_3^* > 0$, respectively. We shall use these relations for the study of (4.65)–(4.66). Asymptotic behavior of the solution to this system is thus clarified by the method introduced in §4.2.

First, spatially homogeneous steady solutions to (4.65) are $u(x,t) \equiv u_*$ and $u(x,t) \equiv u^*$. Second, Lyapunov functions associated with (4.71) and (4.72) guarantees that as $t \to \infty$, each solution $u(\cdot, t)$ to (4.65) converges in $C(\overline{\Omega}; \mathbf{R}^3)$ to u_* and u^* if $R_0 \leq 1$ and $R_0 > 1$, respectively.

Theorem 4.3. *Let $u = (u_1, u_2, u_3)$ be the solution to (4.65), (4.66), and (4.67). Then it holds that*

$$u(\cdot, t) \to \hat{u} \text{ in } C(\overline{\Omega}; \mathbf{R}^3), \quad t \uparrow +\infty, \tag{4.76}$$

where $\hat{u} = u_*$, the disease free spatially homogeneous equilibrium, if $R_0 \leq 1$, and $\hat{u} = u^*$, the infected spatially homogeneous equilibrium, if $R_0 > 1$.

The Lyapunov functions play two roles. First, they establish L^1 bound of the orbit. Second, they provide the ω-limit argument by the pre-compactness of the orbit in $C(\overline{\Omega}; \mathbf{R}^3)$, via the semi-group estimates and the standard parabolic regularity.

The reaction terms f_1, f_2, and f_3 defined by (4.73) are quasi-positive:

$$f_1(0, u_2, u_3) = \lambda > 0, \ f_2(u_1, 0, u_3) = \beta u_1 u_2 \geq 0, \ f_3(u_1, u_2, 0) = aru_2 \geq 0$$

for $u_1, u_2, u_3 \geq 0$, which implies $u_i(x,t) > 0$ in $\overline{\Omega} \times (0, T]$ from the assumption on the initial value.

Having assumed the smooth initial value, we have unique existence of the classical solution local-in-time, which is extended to global-in-time if we can derive a priori bound on its L^∞ norm. To this end we use

$$\mathcal{U}(u) = \mathcal{U}(u)(t) = \int_\Omega U(u(x,t))\,dx \qquad (4.77)$$

and

$$\mathcal{V}(u) = \mathcal{V}(u)(t) = \int_\Omega V(u(x,t))\,dx \qquad (4.78)$$

in accordance with (4.71) and (4.72), respectively.

First we consider the case $R_0 \leq 1$.

Lemma 4.1. *Suppose $R_0 \leq 1$, and let $u = (u_i(x,t))$ be the solution to (4.65), (4.66), and (4.67). Then there exists a positive constant C_1 such that*

$$\left\| \Phi\left(\frac{u_1(\cdot,t)}{u_{*1}} \right) \right\|_1 + \|u_2(\cdot,t)\|_1 + \|u_3(\cdot,t)\|_1 \leq C_1, \quad t > 0.$$

Proof. Along the classical solution $u = u(x,t)$ to (4.65), (4.66), and (4.67), it follows that

$$\frac{d}{dt}\int_\Omega U(u)\,dx = \int_\Omega \frac{\partial}{\partial t}\left\{ \{u_1 - u_{*1}\log u_1 - u_{*1}\} + u_2 + \frac{1}{r}u_3 \right\}\,dx$$

$$= \int_\Omega \left(1 - \frac{u_{*1}}{u_1}\right)(d_1\Delta u_1 + f_1(u)) + (d_2\Delta u_2 + f_2(u))$$

$$+ \frac{1}{r}(d_3\Delta u_3 + f_3(u))\,dx$$

$$= \int_\Omega \left(1 - \frac{u_{*1}}{u_1}\right)d_1\Delta u_1 + d_2\Delta u_1 + \frac{d_3}{r}\Delta u_3\,dx + \int_\Omega \nabla U(u)\cdot f(u)\,dx.$$

Hence it holds that

$$\frac{d}{dt}\int_\Omega U(u)\,dx = \int_\Omega -\frac{d_1 u_{*1}}{u_1^2}|\nabla u_1|^2\,dx + \nabla U(u)\cdot f(u)\,dx \qquad (4.79)$$

by Green's formula and the Neumann condition (4.66). Here, the last term of the right-hand side on (4.79) is non-positive by (4.74). We thus end up with

$$\frac{d}{dt}\int_\Omega U(u)\,dx \leq 0, \qquad (4.80)$$

and hence

$$\int_\Omega U(u)\,dx = \int_\Omega u_{*1}\Phi\left(\frac{u_1}{u_{*1}} \right) + u_2 + \frac{1}{r}u_3\,dx \leq C_2, \quad t > 0,$$

which proves the proposition because $\Phi(s) \geq 0$ and $u_2, u_3 > 0$. $\qquad\square$

For the case $R_0 > 1$, we have the following lemma.

Lemma 4.2. *Suppose $R_0 > 1$, and let $u = (u_i(x,t))$ be the solution to (4.65) with (4.66) and (4.67). Then there exists a positive constant C_3 such that*

$$\left\| \Phi\left(\frac{u_i(\cdot,t)}{u_i^*}\right) \right\|_1 \le C_3 \quad (t > 0, \ i = 1,2,3).$$

Proof. For $\mathcal{V}(u)$ defined by (4.78), it holds that

$$\frac{d}{dt}\int_\Omega V(u)\,dx = \int_\Omega \left(1 - \frac{u_1^*}{u_1}\right)(d_1\Delta u_1 + f_1(u))$$

$$+ \left(1 - \frac{u_2^*}{u_2}\right)(d_2\Delta u_2 + f_2(u)) + \frac{1}{r}\left(1 - \frac{u_3^*}{u_3}\right)(d_3\Delta u_3 + f_3(u))\,dx$$

$$= \int_\Omega \left(1 - \frac{u_1^*}{u_1}\right)d_1\Delta u_1 + \left(1 - \frac{u_2^*}{u_2}\right)d_2\Delta u_1 + \frac{1}{r}\left(1 - \frac{u_3^*}{u_3}\right)d_3\Delta u_3$$

$$+ \nabla V(u)\cdot f(u)\,dx = -\int_\Omega \frac{d_1 u_1^*}{u_1^2}|\nabla u_1|^2 + \frac{d_2 u_2^*}{u_2^2}|\nabla u_2|^2 + \frac{d_3 u_3^*}{ru_3^2}|\nabla u_3|^2\,dx$$

$$+ \int_\Omega \nabla V(u)\cdot f(u)\,dx. \tag{4.81}$$

by Green's formula and the Neumann condition (4.66). By (4.75) we obtain

$$\frac{d}{dt}\int_\Omega V(u)\,dx \le 0 \tag{4.82}$$

and hence

$$\int_\Omega V(u)\,dx = \int_\Omega u_1^*\Phi\left(\frac{u_1}{u_1^*}\right) + u_2^*\Phi\left(\frac{u_2}{u_2^*}\right) + \frac{u_3^*}{r}\Phi\left(\frac{u_3}{u_3^*}\right)\,dx \le C_4, \quad t > 0.$$

Then we obtain the result by $\Phi(s) \ge 0$. □

To complete the proof of Lemma 4.4, we note the following lemma.

Lemma 4.3. *Any function $v = v(x) > 0$ on $\overline{\Omega}$ satisfies*

$$\|v\|_1 \le 2\|\Phi(v)\|_1 + s_0|\Omega|,$$

where $|\Omega|$ denotes the volume of Ω, and s_0 is the root of $\Phi(s/2) = \log 2$ with $s > 2$.

Proof. The equation $\Phi(s/2) = \log 2$ has the unique root $s = s_0$ in $s > 2$, because $\Phi(s)$ is strictly increasing in $s \ge 1$, $\Phi(1) = 0$, and $\lim_{s\uparrow+\infty}\Phi(s) = +\infty$. By the monotonicity of $\Phi(s)$ in $s \ge 1$, $\Phi(s/2) > \log 2$ if $s > s_0$. On the other hand, from (4.70) it follows that

$$\Phi\left(\frac{s}{2}\right) - \log 2 = \Phi(s) - \frac{s}{2}.$$

Hence $\Phi(s) > s/2$ if and only if $\Phi(s/2) > \log 2$, and therefore,

$$\Phi(s) > \frac{s}{2}, \quad s > s_0. \tag{4.83}$$

Given $v = v(x) > 0$, let

$$\Omega_1 = \{x \in \Omega \mid v(x) > s_0\}, \quad \Omega_2 = \Omega \setminus \Omega_1.$$

Then (4.83) implies

$$\begin{aligned}
\|v\|_1 = \|v\|_{L^1(\Omega_1)} + \|v\|_{L^1(\Omega_2)} &\leq \|2\Phi(v)\|_{L^1(\Omega_1)} + s_0|\Omega_2| \\
&\leq 2\|\Phi(v)\|_1 + s_0|\Omega|.
\end{aligned}$$

□

Combining Lemma 4.3 with Lemmas 4.1 and 4.2, we obtain the following result.

Lemma 4.4. *The solution $u = (u_i(x,t))$ to (4.65), (4.66), and (4.67) is L^1-bounded, that is, there exists a $C_1 > 0$ such that*

$$\|u_i(\cdot,t)\|_1 \leq C_5, \quad t > 0; i = 1,2,3.$$

The above L^1 bound of the solution is improved to the L^p bound for $1 \leq p < \infty$, including its derivatives. The first observation is the following lemma.

Lemma 4.5. *Let $u = (u_i(x,t))$ be the solution to (4.65), (4.66), and (4.67). Given $1 \leq q < \infty$, assume*

$$\|u_1(\cdot,t)\|_q \leq C_6, \quad t > 0.$$

Then, for each $p \in [q, \infty)$, satisfying

$$\frac{1}{p} > \frac{1}{q} - \frac{2}{n}, \tag{4.84}$$

we have

$$\|u_1(\cdot,t)\|_p \leq C_7, \quad t > 0.$$

Proof. Let $\bar{u}_1 = \bar{u}_1(x,t)$ be the solution to

$$\frac{\partial \bar{u}_1}{\partial t} = \left(d_1\Delta - \frac{m}{2}\right)\bar{u}_1 + \lambda - \frac{m}{2}u_1, \quad \frac{\partial \bar{u}_1}{\partial \nu}\Big|_{\partial\Omega} = 0, \quad \bar{u}_1(x,0) = u_{10}(x).$$

Then, $v_1 = \bar{u}_1 - u_1$ satisfies

$$\frac{\partial v_1}{\partial t} - d_1\Delta v_1 + \frac{m}{2}v_1 > 0, \quad \frac{\partial v_1}{\partial \nu}\Big|_{\partial\Omega} = 0, \quad v_1(x,0) = 0,$$

by $u_1, u_3 > 0$, and hence it follows that $v_1(x,t) > 0$ from the maximum principle. Then we obtain

$$0 < u_1(x,t) < \bar{u}_1(x,t), \quad t > 0, \ x \in \overline{\Omega}. \tag{4.85}$$

Henceforth, Δ is provided with the Neumann boundary condition. Let $L = d_1 \Delta - m/2$. Then it holds that

$$\bar{u}_1(\cdot, t) = e^{tL} u_{10} + \int_0^t e^{(t-s)L} \left(\lambda - \frac{m}{2} u_1 \right) (\cdot, s) \, ds,$$

which yields

$$\|\bar{u}_1(\cdot,t)\|_p \leq \|e^{tL} u_{10}\|_p + \int_0^t \left\| e^{(t-s)L} \left(\lambda - \frac{m}{2} u_1 \right) (\cdot, s) \right\|_p ds \tag{4.86}$$

for $p \geq 1$. Here, $e^{tL} u_{10}$ is the solution to $\partial u / \partial u = d_1 \Delta u - (m/2) u$ with the initial value u_{10}. Hence it follows that $\|e^{tL} u_{1,0}\|_p \leq \|u_{10}\|_p$. Now we use the semi-group estimate (4.12),

$$\|e^{rL} w\|_p \leq C_8 e^{-\frac{m}{2} r} \max \left\{ 1, (d_1 r)^{-\frac{n}{2} \left(\frac{1}{q} - \frac{1}{p} \right)} \right\} \|w\|_q,$$

to deduce

$$\int_0^t \left\| e^{(t-s)L} \left(\lambda - \frac{m}{2} u_1 \right) (\cdot, s) \right\|_p ds$$

$$\leq \int_0^t C_8 e^{-\frac{m}{2}(t-s)} \max \left\{ 1, \{(t-s)d_1\}^{-\frac{n}{2} \left(\frac{1}{q} - \frac{1}{p} \right)} \right\} \left\| \left(\lambda - \frac{m}{2} u_1 \right) (\cdot, s) \right\|_q ds$$

$$\leq C_8 \sup_{s>0} \left\| \left(\lambda - \frac{m}{2} u_1 \right) (\cdot, s) \right\|_q \int_0^t e^{-\frac{m}{2} r} \max\{1, (d_1 r)^{-\frac{n}{2} \left(\frac{1}{q} - \frac{1}{p} \right)}\} \, dr. \tag{4.87}$$

If

$$-\frac{n}{2} \left(\frac{1}{q} - \frac{1}{p} \right) > -1$$

which is equivalent to (4.84), the last integral of the right-hand side on (4.87) is bounded in $t > 0$. Since $\|\lambda - (m/2) u_1\|_q \leq \lambda |\Omega|^{1/q} + (m/2) \|u_1\|_q$, the assumption, $\|u_1(\cdot,t)\|_q \leq C_6$ for $t > 0$, and inequalities (4.85) and (4.86) imply the result. $\qquad \square$

The following lemmas are immediate consequences of Lemma 4.5.

Lemma 4.6. *Assume that for an integer ℓ in $0 \leq \ell \leq n-2$, there exists a constant C_9 such that*

$$\|u_1(\cdot,t)\|_{\frac{n}{n-\ell}} \leq C_{10}, \quad t > 0.$$

If $\ell < n-2$, then it holds that

$$\|u_1(\cdot,t)\|_{\frac{n}{n-(\ell+1)}} \leq C_{11}, \quad t > 0.$$

If $\ell = n-2$, then for any p in $1 \leq p < \infty$, there holds

$$\|u_1(\cdot,t)\|_p \leq C_{12}, \quad t > 0.$$

Proof. If $\ell < n - 2$, we take

$$p = \frac{n}{n - (\ell + 1)}, \quad q = \frac{n}{n - \ell}.$$

Then it holds that

$$\frac{1}{p} = \frac{n - \ell - 1}{n} > \frac{n - \ell - 2}{n} = \frac{n - \ell}{n} - \frac{2}{n} = \frac{1}{q} - \frac{2}{n},$$

and hence, by Lemma 4.5, there holds that $\|u_1(\cdot, t)\|_p \leq C_{10}$ for $t > 0$.

If $\ell = n-2$ we take $q = n/(n-(n-2)) = n/2$, which means $1/q - 2/n = 0$. Then the conclusion of Lemma 4.5 holds for any $p \geq 1$, and we obtain the result. $\square$

Lemma 4.7. *For any $p \geq 1$, there exists a constant $C_{12} > 0$ such that*

$$\|u_1(\cdot, t)\|_p \leq C_{12}, \quad t > 0.$$

Proof. By Lemma 4.4, the assumption in Lemma 4.6 holds for $\ell = 0$. Then, repeating the use of Lemma 4.6 implies the result. $\square$

Now we turn to the estimates on u_2 and u_3.

Lemma 4.8. *Let $u = (u_i(x, t))$ be the solution to (4.65), (4.66), and (4.67), and $1 \leq q < \infty$. If there holds*

$$\|u_2(\cdot, t)\|_q \leq C_{13}, \quad t > 0, \tag{4.88}$$

then for $p \geq q$ with (4.84), i.e.,

$$\frac{1}{p} > \frac{1}{q} - \frac{2}{n},$$

we have a constant

$$\|u_3(\cdot, t)\|_p \leq C_{14}, \quad t > 0.$$

Similarly, if there holds

$$\|u_1 u_3(\cdot, t)\|_q \leq C_{15}, \quad t > 0 \tag{4.89}$$

then for $p \geq q$ with (4.84) we have

$$\|u_2(\cdot, t)\|_p \leq C_{16}, \quad t > 0.$$

Proof. The proof is similar to that of Lemma 4.5, because we have

$$u_3(\cdot, t) = e^{t(d_3 \Delta - b)} u_{30} + \int_0^t e^{(t-s)(d_3 \Delta - b)} a r u_2(\cdot, s)\, ds$$

$$u_2(\cdot, t) = e^{t(d_2 \Delta - a)} u_{20} + \int_0^t e^{(t-s)(d_2 \Delta - a)} \beta u_1 u_3(\cdot, s)\, ds \tag{4.90}$$

by (4.65), (4.66), and (4.67). $\square$

Lemma 4.9. *Assume that for an integer ℓ in $0 \le \ell \le n-2$, there exists a constant C_{17} such that*

$$\|u_2(\cdot,t)\|_{\frac{n}{n-l}} \le C_{17}, \quad \|u_3(\cdot,t)\|_{\frac{n}{n-l}} \le C_{17}, \quad t > 0.$$

If $\ell < n-2$, then there exists a constant C_{18} such that

$$\|u_2(\cdot,t)\|_{\frac{n}{n-(l+1)}} \le C_{18}, \quad \|u_3(\cdot,t)\|_{\frac{n}{n-(l+1)}} \le C_{18}, \quad t > 0.$$

If $\ell = n-2$, then for any p in $1 \le p < \infty$, there exists a constant C_{19} such that

$$\|u_2(\cdot,t)\|_p \le C_{19}, \quad \|u_3(\cdot,t)\|_p \le C_{19}, \quad t > 0.$$

Proof. If $0 \le \ell < n-2$ we have (4.88) for $q = n/(n-\ell)$. Then, by Lemma 4.8, it holds that

$$\|u_3(\cdot,t)\|_{\frac{n}{n-(l+1)}} \le C_{20}, \quad t > 0$$

as in the proof of Lemma 4.6. By Lemma 4.7, on the other hand, it holds that

$$\|u_1(\cdot,t)\|_n \le C_{22}, \quad t > 0.$$

Hence, noting

$$\frac{n-l}{n} = \frac{1}{n} + \frac{n-(l+1)}{n}$$

and the Hölder inequality, we obtain

$$\|u_1 u_3(\cdot,t)\|_{\frac{n}{n-l}} \le \|u_1(\cdot,t)\|_n \|u_3(\cdot,t)\|_{\frac{n}{n-(l+1)}} \le C_{21} C_{220}.$$

Inequality (4.89) thus arises for $q = n/(n-l)$, and, therefore, Lemma 4.8 implies

$$\|u_2(\cdot,t)\|_{\frac{n}{n-(l+1)}} \le C_{22}, \quad t > 0.$$

If $\ell = n-2$, then, by Lemma 4.8, (4.88) for $q = n/(n-\ell) = n/2$ implies that for any p in $1 \le p < \infty$, there exists a constant C_{23} such that

$$\|u_3(\cdot,t)\|_p \le C_{23}, \quad t > 0.$$

Using this inequality for $p = n$, the Hölder inequality, and Lemma 4.7, we have

$$\|u_1 u_3(\cdot,t)\|_{\frac{n}{2}} \le \|u_1(\cdot,t)\|_n \|u_3(\cdot,t)\|_n \le C_{21} C_{23}, \quad t > 0.$$

Thus, by Lemma 4.8 again, each p in $1 \le p < \infty$ admits

$$\|u_2(\cdot,t)\|_p \le C_{24}, \quad t > 0,$$

which completes the proof. $\qquad\square$

Lemma 4.10. *For any p in $1 \leq p < \infty$, there exists a constant C_{25} such that*

$$\|u_2(\cdot, t)\|_p \leq C_{25}, \quad \|u_3(\cdot, t)\|_p \leq C_{25}, \quad t > 0.$$

Proof. By Lemmas 4.4 and 4.9, the result follows exactly as in Lemma 4.7. ∎

Lemmas 4.7 and 4.10 assure the following lemma.

Lemma 4.11. *Let $u = (u_i(x, t))$ be the solution to (4.65), (4.66), and (4.67). Then for any p in $1 \leq p < \infty$, there exists a constant C_{26} such that*

$$\|u_i(\cdot, t)\|_p \leq C_{26}, \quad t > 0; \ i = 1, 2, 3.$$

Now we apply the other form of semi-group estimates to derive $W^{2,p}$ bound of the solution. Henceforth, for the vector valued function $v = (v_1, v_2, \ldots, v_n)$, its L^p norm is denoted by $\|v\|_p = \|v_1\|_p + \|v_2\|_p + \cdots + \|v_n\|_p$.

Lemma 4.12. *Let $u = (u_i(x, t))$ be the solution of (4.65), (4.66), and (4.67). Then for each p in $1 \leq p < \infty$, there exists a constant C_{27} such that*

$$\|\nabla u_i(\cdot, t)\|_p \leq C_{27}, \quad t > 0; \ i = 1, 2, 3.$$

Proof. By equation (4.65), we get

$$\|\nabla u_1(\cdot, t)\|_p \leq \|\nabla e^{t(d_1\Delta - m)} u_{10}\|_p$$
$$+ \int_0^t \|\nabla e^{(t-s)(d_1\Delta - m)} \left(\lambda - \beta u_1 u_3\right)(\cdot, s)\|_p \, ds \tag{4.91}$$

for $1 < p < \infty$. Then it holds that

$$\|\nabla e^{t(d_1\Delta - m)} u_{10}\|_p \leq C_{28}\|e^{t(d_1\Delta - m)} u_{10}\|_{W^{1,p}} \leq C_{29}\|\Delta^{1/2} e^{t(d_1\Delta - m)} u_{10}\|_p$$
$$= C_{29}\|e^{t(d_1\Delta - m)} \Delta^{1/2} u_{10}\|_p \leq C_{29}\|\Delta^{1/2} u_{10}\|_p$$
$$\leq C_{29}\|u_{10}\|_{W^{1,p}} \leq C_{30}$$

by $u_{1,0} \in C^2(\overline{\Omega})$.

Here we use the semi-group estimate

$$\|\nabla e^{r(d_1\Delta - m)} w\|_p \leq C_{31} e^{-mr} \max\{1, (d_1 r)^{-1/2}\} \|w\|_p. \tag{4.92}$$

Then it follows that

$$\int_0^t \|\nabla e^{(t-s)(d_1\Delta - m)} \left(\lambda - \beta u_1 u_3\right)(\cdot, s)\|_p \, ds$$

$$\leq C_{32} \sup_{s>0} \|(\lambda - \beta u_1 u_3)(\cdot, s)\|_p \int_0^t e^{-mr} \max\{1, (dr)^{1/2}\} \, dr.$$

Here, by the Hölder inequality we obtain

$$\|(\lambda - \beta u_1 u_3)(\cdot, s)\|_p \leq \|\lambda\|_p + |\beta| \cdot \|u_1 u_3(\cdot, s)\|_p$$
$$= |\lambda| \cdot |\Omega|^{1/p} + |\beta| \cdot \|u_1(\cdot, s)\|_{2p} \|u_3(\cdot, s)\|_{2p}.$$

Then, by Lemma 4.11 (replaced p by $2p$), (4.91) proves the result for $i = 1$. Using (4.90), we can similarly prove it for $i = 2, 3$. $\qquad \square$

Now we consider the derivatives of the second order.

Lemma 4.13. *Let* $u = (u_i(x, t))$ *be the solution to (4.65), (4.66), and (4.67). Then for each* p *in* $1 \leq p < \infty$*, there exists a constant* C_{33} *such that*

$$\|\nabla^2 u_i(\cdot, t)\|_p \leq C_{33}, \quad t > 0; \ i = 1, 2, 3. \qquad (4.93)$$

Proof. Equation (4.65) implies

$$\|\Delta u_1(\cdot, t)\|_p \leqq \|\Delta e^{t(d_1 \Delta - m)} u_{10}\|_p$$
$$+ \int_0^t \|\Delta e^{(t-s)(d_1 \Delta - m)} (\lambda - \beta u_1 u_3)(\cdot, s)\|_p \, ds$$

for $1 < p < \infty$. The first term on the right-hand side is estimated as

$$\|\Delta e^{t(d_1 \Delta - m)} u_{10}\|_p = \|e^{t(d_1 \Delta - m)} \Delta u_{10}\|_p \leq C_{34}, \quad t > 0$$

by $u_{10} \in D(\Delta)$. For the second term, we use (4.205) in the final section of this chapter to deduce

$$\|\Delta e^{(t-s)(d_1 \Delta - m)} (\lambda - \beta u_1 u_3)(\cdot, s)\|_p$$
$$= \|(-\Delta)^{1/2} e^{(t-s)(d_1 \Delta - m)} (-\Delta)^{1/2} (\lambda - \beta u_1 u_3)(\cdot, s)\|_p$$
$$\leq C_{35} e^{-m(t-s)} \max \left\{ 1, \{(t - s)d_1\}^{-\frac{1}{2}} \right\} \|(-\Delta)^{1/2} (\lambda - \beta u_1 u_3)(\cdot, t)\|_p.$$

Then, inequality (4.206) in the final section of this chapter, the Hölder inequality, and Lemmas 4.11 and 4.12, we have

$$\|(-\Delta)^{1/2} (\lambda - \beta u_1 u_3)(\cdot, t)\|_p \leq C_{36} \|\nabla(\lambda - \beta u_1 u_3)(\cdot, t)\|_p$$
$$\leq C_{36} \beta (\|u_3(\cdot, t)\|_{2p} \|\nabla u_1(\cdot, t)\|_{2p} + \|u_1(\cdot, t)\|_{2p} \|\nabla u_3(\cdot, t)\|_{2p}) \leq C_{37}.$$

This inequality implies

$$\|\Delta u_i(\cdot, t)\|_p \leq C_{38}, \quad t > 0$$

for $i = 1$. Then inequality (4.93) for u_1 follows from the elliptic estimate. The other inequalities for u_2 and u_3 are shown similarly. $\qquad \square$

To complete the proof of Theorem 4.3, we derive an L^∞ bound of the solution.

Lemma 4.14. *For each solution* $u = (u_i(x,t))$ *to (4.65), (4.66), and (4.67), there exists a constant* C_{39} *such that*

$$\|u_i(\cdot,t)\|_\infty + \|\nabla u_i(\cdot,t)\|_\infty \le C_{39}, \quad t > 0, \, i = 1,2,3.$$

Proof. By Morrey's theorem we have $W^{1,p}(\Omega) \hookrightarrow C^{1-n/p}(\overline{\Omega})$ with continuous imbedding. By Lemmas 4.11 and 4.12 for $p > n$, it holds that

$$\|u_i(\cdot,t)\|_\infty \le C_{40}, \quad t > 0, \, i = 1,2,3.$$

By Lemma 4.13, $u_i(\cdot,t)$, $i = 1,2,3$, is bounded in $W^{2,p}(\Omega)$ for any $p > 1$ by the elliptic estimate, and hence $\nabla u_i(\cdot,t)$ is bounded in $W^{1,p}(\Omega)$. Thus, by Morrey's theorem again, we obtain

$$\|\nabla u_i(\cdot,t)\|_\infty \le C_{41}, \quad t > 0, \, i = 1,2,3.$$

$\square$

From the above a priori estimate, the solution $u = (u_i(x,t))$ to (4.65), (4.66), and (4.67) is global-in-time. Now we can use the invariance principle to study its asymptotic behavior under the presense of the Lyapunov, noting the following lemma.

Lemma 4.15. *The orbit* $u = (u_i(\cdot,t))$ *created by the solution to (4.65), (4.66), and (4.67) is pre-compact in* $C(\overline{\Omega}; \mathbf{R}^3)$.

Proof. By Lemma 4.14, the family $\{u_i(\cdot,t) \mid t \ge 0\}$ is uniformly bounded and equicontinuous on $\overline{\Omega}$. Therefore, the Arzelá-Ascoli theorem implies the result. $\square$

Lemma 4.15 implies that the ω-limit set $\omega((u_{i,0}))$ is nonempty, compact, invariant under the flow defined by (4.65)–(4.66), and is connected. Furthermore, the Lyapunov function is constant on this set. We are thus ready to complete the following proof.

Proof of Theorem 4.3. First, we take the case $R_0 > 1$. Let $u = (u_i(\cdot,t))$ be the solution to (4.65), (4.66), and (4.67) and $\omega((u_{i0}))$ be its ω-limit set. Take $\tilde{u}_0 = (\tilde{u}_{i0}) \in \omega((u_{i,0}))$, and let $\tilde{u} = (\tilde{u}_i(x,t))$ be the solution to (4.65)-(4.66) with the initial value $\tilde{u}_0$. Then, it holds that $\tilde{u}(\cdot,t) \in \omega((u_{i0}))$ and hence

$$\frac{d}{dt} \int_\Omega V(\tilde{u}) \, dx = 0,$$

which means

$$\int_\Omega |\nabla \tilde{u}_i(\cdot,t)|^2\, dx = 0,\ i=1,2,3,\quad \int_\Omega \nabla V(u)\cdot f(u)\, dx = 0.$$

By the first equality, $\tilde{u}_i(x,t)$, $i=1,2,3$, is spatially homogeneous, and hence $\tilde{u}(t) = (\tilde{u}_i(x,t))$ satisfies (4.68). By the last equality and (4.75), on the other hand, we get

$$2 - \frac{\tilde{u}_1}{u_1^*} - \frac{u_1^*}{\tilde{u}_1} = 0,\quad 3 - \frac{u_1^*}{\tilde{u}_1} - \frac{\tilde{u}_2 u_3^*}{\tilde{u}_2^* \tilde{u}_3} - \frac{\tilde{u}_1 u_2^* \tilde{u}_3}{u_1^* \tilde{u}_2 u_3^*} = 0$$

for $\tilde{u}_i = \tilde{u}_i(t)$, $i=1,2,3$, which yields

$$\frac{\tilde{u}_1}{u_1^*} = \frac{u_1^*}{\tilde{u}_1},\quad \frac{u_1^*}{\tilde{u}_1} = \frac{\tilde{u}_2 u_3^*}{\tilde{u}_2^* \tilde{u}_3} = \frac{\tilde{u}_1 u_2^* \tilde{u}_3}{\tilde{u}_1^* \tilde{u}_2 u_3^*}.$$

Hence we have

$$\tilde{u}_1(t) \equiv u_1^*,\quad \frac{\tilde{u}_2(t)}{u_2^*} \equiv \frac{\tilde{u}_3(t)}{u_3^*}.$$

Since $\tilde{u}_1(t)$ is independent of t, the first equation in (4.68) shows that $\tilde{u}_3(t)$ is also a constant. Hence $\tilde{u}_2(t)$ is so. Thus we obtain $(\tilde{u}_{10}, \tilde{u}_{20}, \tilde{u}_{30}) \equiv (u_1^*, u_2^*, u_3^*)$, and therefore, $\omega((u_{10}, u_{20}, u_{30})) = \{(u_1^*, u_2^*, u_3^*)\}$. This means (4.76) for the case $R_0 > 1$.

In the other case $R_0 \le 1$, we use $\mathcal{U}$ defined by (4.77). Given $(\tilde{u}_{i0}) \in \omega((u_{i0}))$, we take the solution to (4.65)–(4.66) with the initial value $(\tilde{u}_{i0})$, denoted by $\tilde{u} = (\tilde{u}_i(\cdot,t))$. Then it holds that $\tilde{u} = \tilde{u}(\cdot,t) \in \omega((u_{i0}))$, and hence

$$\frac{d}{dt} \int_\Omega U(\tilde{u})\, dx = 0,$$

which means

$$\int_\Omega |\nabla \tilde{u}_1(\cdot,t)|^2\, dx = 0,\quad \int_\Omega \nabla U(u)\cdot f(u)\, dx = 0.$$

Thus we obtain $\nabla \tilde{u}_1(x,t) \equiv 0$ for $t > 0$, which implies, by the first equation of (4.65), $\nabla \tilde{u}_3(x,t) \equiv 0$. Hence, by the third equation of (4.65), it holds that $\nabla \tilde{u}_2(x,t) \equiv 0$. Hence $\tilde{u} = (\tilde{u}_i(\cdot,t))$ is spatially homogeneous, and satisfies (4.68).

By (4.74), on the other hand, we get $2 - \frac{\tilde{u}_1(t)}{u_{*1}} - \frac{u_{*1}}{\tilde{u}_1(t)} = 0$ which yields

$$\frac{\tilde{u}_1(t)}{u_{*1}} = \frac{u_{*1}}{\tilde{u}_1(t)}.$$

Therefore, it holds that $\tilde{u}_1(t) \equiv u_{*1}$. Since $\tilde{u}_1(t)$ is constant, $\tilde{u}_3(t)$ and $\tilde{u}_2(t)$ are so by the first and the third equations of (4.65), respectively. Thus we obtain $\omega((u_{i,0})) = \{u_*\}$ and hence (4.76) for $R_0 \le 1$. $\quad\Box$

4.5   Dissipative Systems with Quadratic Growth

Here we introduce a class of reaction diffusion systems of which weak solution exists global-in-time with relatively compact orbit in L^1. Reaction
term in this class is quasi-positive, dissipative, and up to with quadratic
growth rate. If the space dimension is less than or equal to two, the solution
is classical and uniformly bounded. Provided with the entropy structure, on
the other hand, this weak solution is asymptotically spatially homogeneous.

Let $\Omega \subset \mathbf{R}^n$ be a bounded domain with smooth boundary $\partial\Omega$, and
$\tau_j > 0$ and $d_j > 0$, $1 \leq j \leq N$, be constants. We consider the system

$$\tau_j \frac{\partial u_j}{\partial t} - d_j \Delta u_j = f_j(u) \quad \text{in } Q_T = \Omega \times (0,T), \ 1 \leq j \leq N$$

$$\left. \frac{\partial u_j}{\partial \nu} \right|_{\partial\Omega} = 0, \quad u_j|_{t=0} = u_{j0}(x) \geq 0, \tag{4.94}$$

where $u = (u_j)$ and $T > 0$.

We assume that

$$f_j : \mathbf{R}^N \to \mathbf{R} \text{ is locally Lipschitz continuous, } 1 \leq j \leq N, \tag{4.95}$$

and therefore, system (4.94) admits a unique classical solution local-in-
time if the initial value $u_0 = (u_{j0}(x))$ is sufficiently smooth. Also, the
nonlinearity is assumed to be quasi-positive, which means

$$f_j(u_1, \cdots, u_{j-1}, 0, u_{j+1}, \cdots, u_n) \geq 0, \quad 1 \leq j \leq N \tag{4.96}$$

for $0 \leq u = (u_j) \in \mathbf{R}^N$, where $u = (u_j) \geq 0$ if and only if $u_j \geq 0$ for any
$1 \leq j \leq N$. From this condition, the solution satisfies $u = (u_j(\cdot, t)) \geq 0$ as
long as it exists.

The solution which we handle with, however, is mostly weak solution.
Here we say that

$$0 \leq u = (u_j(\cdot, t)) \in L^\infty_{loc}([0,T), L^1(\Omega)^N) \cap L^1_{loc}(0,T; W^{1,1}(\Omega)^N)$$

is a weak solution to (4.94) if $f(u) \in L^1_{loc}(\overline{\Omega} \times (0,T))$,

$$\tau_j \frac{d}{dt} \int_\Omega u_j \varphi \, dx + d_j \int_\Omega \nabla u_j \cdot \nabla \varphi \, dx = \int_\Omega f_j(u)\varphi \, dx, \quad 1 \leq j \leq N$$

for any $\varphi \in W^{1,\infty}(\Omega)$ in the sense of distributions with respect to t, and

$$u_j|_{t=0} = u_{j0}(x), \quad 1 \leq j \leq N$$

in the sense of measures on $\overline{\Omega}$. Then it is shown that this weak solution
$u = (u_j(\cdot, t))$ belongs to $C((0,T), L^1(\Omega)^N)$ and it holds that

$$u_j(\cdot, t) = e^{t\tau_j^{-1}d_j\Delta} u_j(\cdot, \tau) + \int_\tau^t e^{(t-s)\tau_j^{-1}d_j\Delta} f_j(u(\cdot, s)) \, ds, \quad 1 \leq j \leq N$$

$$\tag{4.97}$$

for any $0 < \tau \leq t < T$.

Furthermore, we have

$$\left[\int_\Omega u_j\varphi(\cdot,t)\right]_{t=t_1}^{t=t_2} = \iint_{\Omega\times(t_1,t_2)} \tau_j u_j\varphi_t - d_j\nabla u_j\cdot\nabla\varphi + f_j(u)\varphi\ dxdt$$

for any $1 \leq j \leq N$, $0 < t_1 \leq t_2 < T$, and $\varphi = \varphi(x,t) \in C^1(\overline{\Omega}\times[t_1,t_2])$.

Besides (4.95)–(4.96) we assume at most quadratic growth of the non-linearity $f(u) = (f_j(u))$,

$$|f(u)| \leq C_1(1+|u|^2), \quad u = (u_j) \geq 0, \tag{4.98}$$

and also its dissipativity indicated by

$$\sum_{j=1}^{N} f_j(u) \leq 0, \quad u = (u_j) \geq 0. \tag{4.99}$$

We also assume

$$\frac{\partial f_j}{\partial u_j}(u) \geq -C_2(1+|u|), \ 1 \leq j \leq N, \quad 0 \leq u = (u_j) \in \mathbb{R}^N. \tag{4.100}$$

For such a system, global-in-time existence of the weak solution is known as in Theorem 4.4 below, where $\|\ \|_p$, $1 \leq p \leq \infty$, stands for the standard L^p norm.

Theorem 4.4. *Assume (4.95), (4.96), (4.98), (4.99), and (4.100), and let*

$$0 \leq u_0 = (u_{j0}(x)) \in L^1(\Omega)^N$$

be given. Then there is a weak solution to (4.94) global-in-time, denoted by $0 \leq u = (u_j(\cdot,t)) \in C([0,+\infty), L^1(\Omega)^N)$, which satisfies

$$u \in L_{loc}^2(\overline{\Omega}\times(0,+\infty))^N,$$

$$\nabla u_j \in L_{loc}^p(\overline{\Omega}\times(0,+\infty))^N, \ 1 \leq p < \frac{4}{3}, \ 1 \leq j \leq N,$$

$$\|u(\cdot,t)\|_1 \leq C_3\|u_0\|_1 \quad for\ t \geq 0. \tag{4.101}$$

Inequality (4.99) is used to guarantee for the limit of approximate solutions to be a sub-solution to (4.94).

Generally, weak solution can include blowup time and may not be unique. The first result proven here is concerned with the orbit constructed in Theorem 4.4.

Theorem 4.5. *The orbit $\mathcal{O} = \{u(\cdot,t) \mid t \geq 0\}$ made by the solution $u = (u_j(\cdot,t))$ in Theorem 4.4 is relatively compact in $L^1(\Omega)^N$.*

The second result is the regularity of this solution.

Theorem 4.6. *Assume (4.100) in addition to (4.95), (4.96), (4.98), and (4.99), and let $n \leq 2$ and $0 \leq u_0 = (u_{j0}(x))$ be sufficiently smooth. Then the weak solution $u = (u_j(\cdot, t))$ to (4.94) obtained in Theorem 4.4 is classical, and takes relatively compact orbit $\mathcal{O} = \{u(\cdot, t) \mid t \geq 0\}$ in $C(\overline{\Omega})^N$.*

Since the classical solution is unique, Theorem 4.6 assures the existence of a unique classical solution to (4.94), which is global-in-time and uniformly bounded.

We recall that a fundamental property derived from (4.99) is the total mass control, indicated by

$$\frac{d}{dt} \int_{\Omega} \tau \cdot u \, dx \leq 0, \quad \tau = (\tau_j) > 0. \tag{4.102}$$

Here we use the inequality

$$\sum_{j=1}^{N} f_j(u) \log u_j \leq C_5(1 + |u|^2), \quad u = (u_j) \geq 0 \tag{4.103}$$

valid for $f = (f_j(u))$ satisfying (4.100) to establish additional a priori estimates of the solution.

Lemma 4.16. *If the nonlinearity $f = (f_j(u))$, $u = (u_j)$, satisfies (4.95), (4.96), (4.99), and (4.100), then inequality (4.103) holds true.*

Proof. Assuming (4.95), (4.96), (4.98), (4.99), and (4.100), we shall show (4.103). Put

$$\tilde{f}_j(u) = f_j(u_1, \cdots, u_{j-1}, 0, u_{j+1}, \cdots, u_N) \geq 0, \quad 0 \leq u = (u_j) \in \mathbb{R}^N.$$

If $|u| \leq 1$ is the case, we have $0 \leq u_j \leq 1$ for $1 \leq j \leq N$. Then, for $u_j > 0$ it holds that

$$f_j(u) \log u_j = (f_j(u) - \tilde{f}_j(u)) \log u_j + \tilde{f}_j(u) \log u_j$$
$$\leq (f_j(u) - \tilde{f}_j(u)) \log u_j \leq C_{42} u_j |\log u_j| \leq C_{43},$$

and hence

$$\sum_{j=1}^{N} f_j(u) \log u_j \leq N C_{36}, \quad |u| \leq 1. \tag{4.104}$$

Assume $|u| > 1$, and put $s_j = u_j / |u| \in (0, 1]$. It holds that

$$\sum_{j=1}^{N} s_j^2 = 1 \tag{4.105}$$

and

$$\sum_{j=1}^{N} f_j(u) \log u_j = \log |u| \cdot \sum_{j=1}^{N} f_j(u) + \sum_{j=1}^{N} f_j(u) \log s_j$$

$$\leq \sum_{j=1}^{N} f_j(u) \log s_j \qquad (4.106)$$

by (4.99). Here we have

$$f_j(u) \log s_j = (f_j(u) - \tilde{f}_j(u)) \log s_j + \tilde{f}_j(u) \log s_j$$
$$\leq (f_j(u) - \tilde{f}_j(u)) \log s_j \qquad (4.107)$$

and

$$f_j(u) - \tilde{f}_j(u)$$
$$= \int_0^1 \frac{d}{ds} f_j(s_1|u|, \cdots, s_{j-1}|u|, s \cdot s_j|u|, s_{j+1}|u|, \cdots, s_N|u|) \, ds$$
$$= \int_0^1 \frac{\partial f_j}{\partial u_j}(u(s)) \, ds \cdot s_j|u|,$$

where

$$u(s) = (s_1|u|, \cdots, s_{j-1}|u|, s \cdot s_j|u|, s_{j+1}|u|, \cdots, s_N|u|).$$

Since

$$|u(s)| \leq |u|, \quad 0 \leq s \leq 1$$

it follows from (4.100) that

$$(f_j(u) - \tilde{f}_j(u)) \log s_j \leq C_2(1 + |u|)|u| \cdot s_j |\log s_j|$$
$$\leq C_{44}|u|^2, \quad |u| \geq 1. \qquad (4.108)$$

Inequalities (4.106)–(4.108) imply

$$\sum_{j=1}^{N} f_j(u) \log u_j \leq N C_{44}|u|^2, \quad |u| \geq 1 \qquad (4.109)$$

and then we obtain (4.103) by (4.104) and (4.109). □

Now we use the point-wise inequality derived from (4.99),

$$\frac{\partial}{\partial t}(\tau \cdot u) - \Delta(d \cdot u) \leq 0 \text{ in } Q_T, \quad \frac{\partial}{\partial \nu}(d \cdot u)\Big|_{\partial \Omega} \leq 0, \qquad (4.110)$$

where $d = (d_j) > 0$. We actually have the equality for the boundary condition on $d \cdot u$ in (4.110). Obviously, (4.102) is a direct consequence of

(4.110), which, however, deduces several other important properties. The estimate below is obtained by the duality argument recently developed.

Theorem 4.7. *If $0 \leq u = (u_j(x,t))$ is smooth on $\overline{\Omega} \times [0,T]$ and satisfies (4.110), then it follows that*

$$\|u\|_{L^2(Q_T)} \leq C_6 T^{1/2} \|u_0\|_2, \quad u|_{t=0} = u_0. \tag{4.111}$$

Proof. Let $u_0 = u|_{t=0}$. By (4.110) we have

$$\tau \cdot u(\cdot,t) - \tau \cdot u_0 \leq \int_0^t \Delta(d \cdot u(\cdot,s)) \, ds,$$

and hence

$$(\tau \cdot u(\cdot,t), d \cdot u(\cdot,t)) - (\tau \cdot u_0, d \cdot u(\cdot,t))$$

$$\leq -(\nabla d \cdot u(\cdot,t), \nabla \int_0^t d \cdot u(\cdot,s) \, ds)$$

$$= -\frac{1}{2} \frac{d}{dt} \|\nabla \int_0^t d \cdot u(\cdot,s) \, ds\|_2^2, \tag{4.112}$$

where $(\ ,\ )$ denotes the L^2-inner product. Integration of (4.112) over $(0,T)$ implies

$$\int_0^T (\tau \cdot u(\cdot,t), d \cdot u(\cdot,t)) \, dt$$

$$\leq \|\tau \cdot u_0\|_2 \cdot \int_0^T \|d \cdot u(\cdot,t)\|_2 \, dt$$

$$\leq T^{1/2} \|\tau \cdot u_0\|_2 \cdot \left(\int_0^T \|d \cdot u(\cdot,t)\|_2^2 \, dt \right)^{1/2},$$

and hence (4.111) holds by $u = (u_j(\cdot,t)) \geq 0$. $\qquad\square$

We shall use the duality argument, relying on the study of the parabolic problem

$$\frac{\partial v}{\partial t} - \Delta(av) = f \text{ in } Q_T, \quad \frac{\partial}{\partial \nu}(av)\Big|_{\partial\Omega} = 0, \quad v|_{t=0} = v_0(x) \tag{4.113}$$

where

$$0 < C_9^{-1} \leq a = a(x,t) \leq C_9, \quad f \in L^2(Q_T), \quad v_0 \in L^2(\Omega). \tag{4.114}$$

This study takes an important role, because (4.110) implies

$$\frac{\partial v}{\partial t} - \Delta(av) \leq 0 \text{ in } Q_T, \quad \frac{\partial}{\partial \nu}(av)\bigg|_{\partial\Omega} \leq 0 \qquad (4.115)$$

for $v = \tau \cdot u + 1$ and $a = \frac{d \cdot u + 1}{\tau \cdot u + 1}$.
Here we use the following fact.

Lemma 4.17. *For (4.114), there is a unique solution* $v = v(x,t) \in L^2(Q_T)$ *to (4.113) such that* $\int_0^t av \in L^2(0,T;H^2(\Omega))$ *in the sense that*

$$v - \Delta\left(\int_0^t av(\cdot,s)\, ds\right) = v_0 + \int_0^t f(\cdot,s)\, ds$$

$$\frac{\partial}{\partial \nu}\int_0^t av(\cdot,s)\, ds\bigg|_{\partial\Omega} = 0. \qquad (4.116)$$

Similarly to (4.111), the estimate

$$\|v\|_{L^2(Q_T)} \leq C_{46}T^{1/2}(\|v_0\|_2 + \|f\|_{L^2(Q_T)}) \qquad (4.117)$$

is proven for the above $v = v(x,t)$, which ensures the following result by the dominated convergence theorem.

Lemma 4.18. *Let* $0 < C_9^{-1} \leq a_k = a_k(x,t) \leq C_9$, $v_{k0} \in L^2(\Omega)$, *and* $f_k \in L^2(Q_T)$, $k = 1,2,\cdots$, *be sequences of coefficients, initial values, and inhomogeneous terms, respectively, satisfying*

$$a_k \to a \quad a.e. \text{ in } Q_T = \Omega \times (0,T)$$

$$v_{k0} \to v_0 \text{ in } L^2(\Omega), \quad f_k \to f \quad \text{in } L^2(Q_T). \qquad (4.118)$$

Let $v_k = v_k(x,t) \in L^2(Q_T)$ *be the solution to*

$$\frac{\partial v_k}{\partial t} - \Delta(a_k v_k) = f_k, \quad \frac{\partial}{\partial \nu}(a_k v_k)\bigg|_{\partial\Omega} = 0, \quad v_k|_{t=0} = v_{k0}(x) \qquad (4.119)$$

in the sense of Lemma 4.17. Then it holds that

$$v_k \to v \quad \text{in } L^2(Q_T),$$

where $v = v(x,t)$ *is the solution to (4.113).*

Lemma 4.18 implies the following result.

Lemma 4.19. *The solution* $v = v(x,t)$ *to (4.113) in Lemma 4.17 satisfies*

$$\|v(\cdot,t)\|_1 \leq \|v_0\|_1 + \int_0^t \|f(\cdot,s)\|_1\, ds \quad \text{for a.e. } t \in (0,T). \qquad (4.120)$$

Proof. Letting $v_0^\pm = \max\{0, \pm v\}$, $f^\pm = \max\{0, \pm f\}$, we take smooth $C_9^{-1} \leq a_k = a_k(x,t) \leq C_9$, $f_{\pm k} = f_{\pm k}(x,t)$, and $v_{\pm 0k} = v_{\pm 0k}(x)$, $k = 1, 2, \cdots$, such that

$$a_k \to a, \text{ a.e.,} \quad v_{\pm k0} \to v_0^\pm \text{ in } L^2(\Omega), \quad f_{\pm k} \to f^\pm \text{ in } L^2(Q_T).$$

There is a unique classical solution $v_{\pm k} = v_{\pm k}(x,t) \geq 0$ to

$$\frac{\partial v_{\pm k}}{\partial t} - \Delta(a_k v_{\pm k}) = f_{\pm k} \text{ in } Q_T, \quad \frac{\partial}{\partial \nu}(a_k v_{\pm k})\Big|_{\partial \Omega} = 0, \quad v_{\pm k}|_{t=0} = v_{\pm k0}(x)$$

$$(4.121)$$

which satisfies

$$\|v_{\pm k}(\cdot, t)\|_1 = \|v_{\pm k0}\|_1 + \int_0^t \|f_{\pm k}(\cdot, s)\|_1 \, ds, \quad 0 \leq t \leq T. \qquad (4.122)$$

Here we have $v_{\pm k} \to v_\pm$ in $L^2(Q_T)$ by Proposition 4.18, which solves

$$\frac{\partial v_\pm}{\partial t} - \Delta(a v_\pm) = f^\pm \text{ in } Q_T, \quad \frac{\partial}{\partial \nu}(a v_\pm)\Big|_{\partial \Omega} = 0, \quad v_\pm|_{t=0} = v_0^\pm,$$

in the sense of (4.116). Hence it follows that $v = v_+ - v_-$ from the uniqueness of the solution and also

$$\|v_\pm(\cdot, t)\|_1 = \|v_0^\pm\|_1 + \int_0^t \|f^\pm(\cdot, s)\|_1 \, ds, \quad 0 \leq t \leq T.$$

Then we obtain (4.120) by

$$\|v(\cdot, t)\|_1 = \|v_+(\cdot, t) - v_-(\cdot, t)\|_1 \leq \|v_+(\cdot, t)\|_1 + \|v_-(\cdot, t)\|_1$$
$$\|v_0\|_1 = \|v_0^+\|_1 + \|v_0^-\|_1$$
$$\|f(\cdot, s)\|_1 = \|f^+(\cdot, s)\|_1 + \|f^-(\cdot, s)\|_1.$$

$\square$

Finally, the following lemma is derived similarly to Theorem 4.11 in §4.9.

Lemma 4.20. *Let* $0 < C_9^{-1} \leq a = a(x,t) \leq C_9$ *and let* $v = v(x,t) \geq 0$ *be a smooth function on* $\overline{\Omega} \times [0, T]$ *satisfying*

$$\frac{\partial v}{\partial t} - \Delta(a v) \leq 0 \text{ in } Q_T, \quad \frac{\partial}{\partial \nu}(a v)\Big|_{\partial \Omega} \leq 0.$$

Then it holds that

$$\|v\|_{L^2(\Omega \times (\eta, T))} + \left\|\int_\eta^T a v(\cdot, s) \, ds\right\|_\infty \leq C_{47}(\eta, T)\|v\|_{L^1(Q_T)}$$

for any $0 < \eta < T$.

We also note the regularity of the weak solution to the heat equation

$$\frac{\partial w}{\partial t} = \Delta w + H \text{ in } Q_T, \quad \frac{\partial w}{\partial \nu}\Big|_{\partial\Omega} = 0, \quad w|_{t=0} = w_0(x) \qquad (4.123)$$

for

$$w_0 \in L^1(\Omega), \quad H \in L^1(Q_T). \qquad (4.124)$$

Here, compactness of the mapping

$$(w_0, H) \in L^1(\Omega) \times L^1(Q_T) \mapsto w \in L^1(Q_T)$$

is particularly important for the proof of Theorem 3.3.

Theorem 4.8. *The mapping* $\mathcal{F} : (w_0, H) \in L^1(\Omega) \times L^1(Q_T) \mapsto w \in L^1(Q_T)$ *is compact, where* $w = w(x,t)$ *is the solution to (4.123) in Proposition 4.10. In other words, image of each bounded set in* $L^1(\Omega) \times L^1(Q_T)$ *by* $\mathcal{F}$ *is relatively compact in* $L^1(Q_T)$.

Proof. By (4.211), the dual operator

$$\mathcal{F}^* : L^\infty(Q_T) \to L^\infty(\Omega) \times L^\infty(Q_T)$$

is realized as $\mathcal{F}^*(h) = (\theta|_{t=0}, \theta)$, where $\theta = \theta(\cdot, t)$ is the solution to the backward heat equation

$$\frac{\partial \theta}{\partial t} + \Delta \theta = h \text{ in } Q_T, \quad \frac{\partial \theta}{\partial \nu}\Big|_{\partial\Omega} = 0, \quad \theta|_{t=T} = 0.$$

Then the assertion follows because $\mathcal{F}^*$ is compact by the parabolic regularity. $\square$

4.6 L^1-compactness of the Orbit

As is described in the previous section, global-in-time existence of the weak solution is known under the assumptions of Theorem 3.3. Here we show that this orbit is relatively compact in $L^1(\Omega)$. Given $t_k \uparrow +\infty$, we construct a compact family of functions in $L^1(Q_0)^N$ which dominates $u_k = u_k(x,t) = u(x, t + t_k) \geq 0$ above, where $Q_0 = \Omega \times (-\eta_0, 1)$ for $\eta_0 > 0$. We prove that this dominating sequence is bounded in $L^2(Q_{\eta_0})$ which implies that $\{f_j(u_k)\}$ is bounded in $L^1(Q_{\eta_0})$. This bound implies the compactness of $\{u_k\}$ in $L^1(Q_{\eta_0})$ due to the compactness of the mapping $(w_0, H) \in L^1(\Omega) \times L^1(Q_T) \mapsto w \in L^1(Q_T)$ in (4.123). Then, we even prove that the dominating sequence is relatively compact in $L^2(Q_\eta)$, $\eta \in (0, \eta_0)$. From dominating convergence, it follows that $\{u_k\}$ is itself relatively compact in $L^2(Q_\eta)$. Then a sub-sequence of $f_j(u_k)$ converges in $L^1(Q_\eta)$ so that

u_k converges in $C([-\eta, 1]; L^1(\Omega))$. In particular, $u(\cdot, t_k)$ converges in $L^1(\Omega)$ which is our main objective; the proof of Theorem 4.5.

To begin with, we use the following scheme to construct the global-in-time weak solution to (4.94). In fact, the initial value $0 \leq u_0 = (u_{0j}) \in L^1(\Omega)^N$ is approximated by smooth $\tilde{u}_0^\ell = (\tilde{u}_{j0}^\ell)$, $\ell = 1, 2, \cdots$, satisfying

$$\tilde{u}_{j0}^\ell = \tilde{u}_{j0}^\ell(x) \geq \max\{\frac{1}{\ell}, u_{j0}(x)\} \quad \text{a.e. in } \Omega$$

$$\tilde{u}_{j0}^\ell \to u_{j0} \text{ in } L^1(\Omega) \text{ and a.e. in } \Omega, \quad 1 \leq j \leq N. \qquad (4.125)$$

Second, the nonlinearity is modified by a smooth, non-decreasing truncation $T_\ell : [0, +\infty) \to [0, \ell + 1]$, such that $T_\ell(s) = s$ for $0 \leq s \leq \ell$. Then the nonlinearity $f^\ell = (f_j \circ T_\ell)$ satisfies (4.95), (4.96), and (4.99) for $f = (f_j^\ell)$. Then we take the unique global-in-time classical solution $\tilde{u}^\ell = (\tilde{u}_j^\ell(\cdot, t))$ to

$$\tau_j \frac{\partial \tilde{u}_j^\ell}{\partial t} - d_j \Delta \tilde{u}_j^\ell = f_j^\ell(\tilde{u}^\ell) \quad \text{in } \Omega \times (0, +\infty)$$

$$\left. \frac{\partial \tilde{u}_j^\ell}{\partial \nu} \right|_{\partial\Omega} = 0, \quad \tilde{u}_j^\ell\big|_{t=0} = \tilde{u}_{j0}^\ell(x) \qquad (4.126)$$

to obtain

$$\|\tau \cdot \tilde{u}^\ell(\cdot, t)\|_1 \leq \|\tau \cdot \tilde{u}^\ell(\cdot, s)\|_1, \quad 0 \leq s \leq t < +\infty \qquad (4.127)$$

and in particular,

$$\sup_{t \geq 0} \|\tilde{u}^\ell(\cdot, t)\|_1 \leq C_{10}. \qquad (4.128)$$

Third, we have

$$\|\tilde{u}_j^\ell\|_{L^2(Q(\eta, T))} + \|\nabla \tilde{u}_j^\ell\|_{L^p(Q(\eta, T))^N} \leq C_{11}(\eta, T, p, \|u_0\|_1), \quad 1 \leq j \leq N \qquad (4.129)$$

for $0 < \eta < T$ and $1 \leq p < \frac{4}{3}$, recalling $Q(\eta, T) = \Omega \times (\eta, T)$. Finally, up to a subsequence we have

$$\tilde{u}^\ell \to u \quad \text{in } L^1_{loc}(\overline{\Omega} \times [0, +\infty))^N \text{ and a.e. in } \Omega \times (0, +\infty). \qquad (4.130)$$

Summing up, we obtain

$$\|\tau \cdot u(\cdot, t)\|_1 \leq \|\tau \cdot u(\cdot, s)\|_1, \quad 0 \leq s \leq t < +\infty$$

$$\sup_{t \geq 0} \|u(\cdot, t)\|_1 \leq C_{10} \qquad (4.131)$$

by (4.127)–(4.128). It holds also that

$$\|u_j\|_{L^2(Q(\eta, T))} + \|\nabla u_j\|_{L^p(Q(\eta, T))^N} \leq C_{11}(\eta, T, p, \|u_0\|_1), \quad 1 \leq j \leq N \qquad (4.132)$$

by (4.129), and this $u = (u_j(\cdot, t))$ is a weak solution to (4.94) satisfying (4.101). In particular, we obtain $u = (u_j(\cdot, t)) \in C([0, +\infty), L^1(\Omega)^N)$ by (4.97).

Given $t_k \uparrow +\infty$, let

$$u_{jk}(\cdot, t) = u_j(\cdot, t + t_k), \quad u_k = (u_{jk}(\cdot, t)), \quad Q = \Omega \times (-2, 1). \quad (4.133)$$

It holds that

$$\|u_k\|_{L^2(Q)^N} \leq C_{12} \quad (4.134)$$

by (4.132) and hence

$$\|f(u_k)\|_{L^1(Q)^N} \leq C_{13}.$$

Since

$$\|u_k(\cdot, -2)\|_1 \leq C_{10}$$

holds by (4.131), passing to a subsequence, we have

$$u_k \to u_\infty \quad \text{in } L^1(Q)^N \text{ and a.e. in } Q \quad (4.135)$$

by Lemma 4.8. From (4.132), furthermore, this u_∞ is a weak solution to (4.94) (for a different initial value) satisfying (4.101). In particular, it holds that

$$u_k \rightharpoonup u_\infty \text{ weakly in } L^2(Q)^N, \quad \|u_\infty\|_{L^2(Q)^N} \leq C_{12} \quad (4.136)$$

by (4.134).

The coefficients

$$\underline{a} \leq a_k(x, t) \equiv \frac{d \cdot u_k + 1}{\tau \cdot u_k + 1} \leq \overline{a}, \quad \underline{a} \leq a_\infty(x, t) \equiv \frac{d \cdot u_\infty + 1}{\tau \cdot u_\infty + 1} \leq \overline{a} \quad (4.137)$$

are well-defined, provided with the property

$$a_k \to a_\infty \quad \text{a.e. in } Q \quad (4.138)$$

where

$$\underline{a} = \inf_{s>0} \frac{\underline{d}s + 1}{\overline{\tau}s + 1} > 0, \quad \overline{a} = \sup_{s>0} \frac{\overline{d}s + 1}{\underline{\tau}s + 1} < +\infty$$

for $\underline{d} = \min_j d_j$, $\overline{d} = \max_j d_j$, $\underline{\tau} = \min_j \tau_j$, and $\overline{\tau} = \max_j \tau_j$.

Since the first convergence in (4.135) means

$$\lim_{k \to \infty} \int_{-2}^{1} \|u(\cdot, t + t_k) - u_\infty(\cdot, t)\|_1 \, dt = 0, \quad (4.139)$$

we have

$$\lim_{k \to \infty} \|u_k(\cdot, t) - u_\infty(\cdot, t)\|_1 = 0 \quad \text{for a.e. } t \in (-2, 1),$$

passing to a subsequence. In particular, there is $\eta_0 \in (1,2)$ such that

$$u_k(\cdot, -\eta_0) \to u_\infty(\cdot, -\eta_0) \quad \text{in } L^1(\Omega) \tag{4.140}$$

as $k \to \infty$. The convergence (4.140), combined with (4.136), is not sufficient to apply Lemma for the proof of the strong convergence

$$u_k \to u_\infty \quad \text{in } L^2(Q_0), \quad Q_0 = \Omega \times (-\eta_0, 1).$$

We bound u_k from above by the solution w_k of an appropriate majorizing system, and prove that w_k is compact in $L^2(Q_0)$. For justification purposes, furthermore, we do it on regularized approximate systems by the introduction of w_k^ℓ below.

Similarly to (4.140), we may assume

$$\tilde{u}_k^\ell(\cdot, -\eta_0) \to u_k(\cdot, -\eta_0) \quad \text{in } L^1(\Omega), \; k = 1, 2, \cdots \tag{4.141}$$

as $\ell \to \infty$ by (4.130), where

$$\tilde{u}_k^\ell(\cdot, t) = \tilde{u}^\ell(\cdot, t + t_k).$$

Now we take smooth $w_k^\ell = w_k^\ell(x, t)$, satisfying

$$\frac{\partial w_k^\ell}{\partial t} - \Delta(a_k^\ell w_k^\ell) = 0 \quad \text{in } Q_0 = \Omega \times (-\eta_0, 1)$$

$$\left. \frac{\partial}{\partial \nu}(a_k^\ell w_k^\ell) \right|_{\partial \Omega} = 0, \quad w_k^\ell \big|_{t=-\eta_0} = \tau \cdot \tilde{u}_k^\ell(\cdot, -\eta_0), \tag{4.142}$$

where

$$a_k^\ell(x, t) = \frac{d \cdot \tilde{u}_k^\ell + 1}{\tau \cdot \tilde{u}_k^\ell + 1}.$$

Since $w_k^\ell(\cdot, t) \geq 0$ it follows that

$$\|w_k^\ell(\cdot, t)\|_1 \leq \|\tau \cdot \tilde{u}_k^\ell(\cdot, -\eta_0)\|_1 \leq C_{10}, \quad -\eta_0 \leq t \leq 1 \tag{4.143}$$

from (4.142). Therefore, by Lemma 4.20, each $\eta_1 \in (1, \eta_0)$ admits the estimate

$$\left\| \int_{-\eta_1}^1 a_k^\ell w_k^\ell \, dt \right\|_\infty + \|w_k^\ell\|_{L^2(Q_1)^N} \leq C_{14}(\eta_1), \quad Q_1 = \Omega \times (-\eta_1, 1). \tag{4.144}$$

Furthermore, inequality

$$\sum_{j=1}^N f_j^\ell(u) \leq 0, \quad 0 \leq u = (u_j) \in \mathbb{R}^N$$

implies

$$\frac{\partial}{\partial t}(\tau \cdot \tilde{u}_k^\ell + 1) - \Delta(a_k^\ell(\tau \cdot \tilde{u}_k^\ell + 1)) \leq 0, \quad \left. \frac{\partial}{\partial \nu}(\tau \cdot \tilde{u}_k^\ell + 1) \right|_{\partial \Omega} = 0,$$

and hence

$$\tau \cdot \tilde{u}_k^\ell + 1 \leq w_k^\ell \quad \text{in } Q_0 \tag{4.145}$$

by the maximum principle.

In the following, first, we shall show that $\{w_k^\ell\}_\ell$ is relatively compact in $L^2_{loc}(\overline{\Omega} \times (-\eta_0, 1])$ for each $k = 1, 2, \cdots$ (Lemma 4.21). Second, assuming $w_k^\ell \to w_k^\infty$ in $L^2_{loc}(\overline{\Omega} \times (-\eta_0, 1])$ up to a subsequence, we shall show that $\{w_k^\infty\}$ is relatively compact in $L^2_{loc}(\overline{\Omega} \times (-\eta_0, 1])$ (Lemma 4.22). Since

$$0 \leq \tau \cdot u_k + 1 \leq w_k^\infty \quad \text{a.e. in } Q_0 \tag{4.146}$$

this property implies the relatively compactness of $\{\tau \cdot u_k\}$ (and hence that of $\{u_k\}$) in $L^2_{loc}(\overline{\Omega} \times (\eta_0, 1])$, by $u_k = (u_{jk}) \geq 0$ and $\tau = (\tau_j) > 0$.

Lemma 4.21. *For each* $k = 1, 2, \cdots$, *the family* $\{w_k^\ell\}_\ell \subset L^2(Q_1)^N$ *is relatively compact.*

Proof. In the following proof, we fix k and let $\ell \to \infty$. By (4.130), we have

$$\underline{a} \leq a_k^\ell(x,t) \leq \overline{a}, \ a_k^\ell(x,t) \to a_k(x,t) \equiv a(x,t+t_k) \text{ for a.e. } (x,t) \in Q_1. \tag{4.147}$$

Since (4.144) holds, there is a subsequence satisfying

$$w_k^\ell \rightharpoonup w_k^\infty \quad \text{weakly in } L^2(Q_1).$$

From (4.147) and standard duality argument, it follows also that

$$\left\| \int_{-\eta_1}^1 a_k w_k^\infty \, dt \right\|_\infty + \|w_k^\infty\|_{L^2(Q_1)} \leq C_{14}(\eta_1). \tag{4.148}$$

Here we shall show

$$w_k^\ell(\cdot, t) \to w_k^\infty(\cdot, t) \quad \text{in } L^1(\Omega) \text{ and for a.e. } t \in (-\eta_0, 1). \tag{4.149}$$

For this purpose, we take smooth $r_0 = r_0(x)$ and define $z_k^\ell = z_k^\ell(x,t)$ by

$$\frac{\partial z_k^\ell}{\partial t} - \Delta(a_k^\ell z_k^\ell) = 0 \quad \text{in } Q_0$$

$$\left. \frac{\partial z_k^\ell}{\partial \nu} \right|_{\partial \Omega} = 0, \quad z_k^\ell \big|_{t=-\eta_0} = r_0. \tag{4.150}$$

By (4.142) and (4.150) we obtain

$$\sup_{-\eta_0 \leq t \leq 1} \|w_k^\ell(\cdot, t) - z_k^\ell(\cdot, t)\|_1 \leq \|\tau \cdot \tilde{u}_k^\ell(\cdot, -\eta_0) - r_0\|_1, \tag{4.151}$$

using Lemma 4.19.

Since (4.147), we have

$$z_k^\ell \to z_k^\infty \quad \text{in } L^2(Q_0) \tag{4.152}$$

by Lemma 4.18. In particular, it follows that

$$z_k^\ell(\cdot, t) \to z_k^\infty(\cdot, t) \quad \text{in } L^2(\Omega)^N \text{ and for a.e. } t \in (-\eta_0, 1). \tag{4.153}$$

Here, $z_k^\infty = z_k^\infty(x, t)$ is the L^2 solution to

$$\frac{\partial z_k^\infty}{\partial t} - \Delta(a_k z_k^\infty) = 0 \text{ in } Q_0, \quad \left.\frac{\partial z_k^\infty}{\partial \nu}\right|_{\partial\Omega} = 0, \quad z_k^\infty|_{t=-\eta_0} = r_0.$$

Using

$$\|w_k^\ell(\cdot, t) - w_k^{\ell'}(\cdot, t)\|_1$$
$$\leq \|w_k^\ell(\cdot, t) - z_k^\ell(\cdot, t)\|_1 + \|z_k^\ell(\cdot, t) - z_k^{\ell'}(\cdot, t)\|_1 + \|z_k^{\ell'}(\cdot, t) - w_k^{\ell'}(\cdot, t)\|_1$$
$$\leq \|z_k^\ell(\cdot, t) - z_k^{\ell'}(\cdot, t)\|_1 + 2\|\tau \cdot \tilde{u}_k^\ell(\cdot, -\eta_0) - r_0\|_1, \quad -\eta_0 \leq t \leq 1, \tag{4.154}$$

we obtain

$$\limsup_{\ell, \ell' \to \infty} \|w_k^\ell(\cdot, t) - w_k^{\ell'}(\cdot, t)\|_1 \leq 2\|\tau \cdot u_k(\cdot, -\eta_0) - r_0\|_1 \quad \text{for a.e. } t \in (-\eta_0, 1)$$

by (4.141) and (4.153). Since r_0 is an arbitrary smooth function, there holds that

$$\limsup_{\ell, \ell' \to \infty} \|w_k^\ell(\cdot, t) - w_k^{\ell'}(\cdot, t)\|_1 \leq 0 \quad \text{for a.e. } t \in (-\eta_0, 1)$$

and hence (4.149). In particular, we may assume

$$\lim_{\ell \to \infty} \|w_k^\ell(\cdot, -\eta_1) - w_k^\infty(\cdot, -\eta_1)\|_1 = 0. \tag{4.155}$$

Reducing (4.142) to

$$[w_k^\ell(\cdot, t)]_{t=t_1}^{t=t_2} = \Delta \int_{t_1}^{t_2} a_k^\ell w_k^\ell(\cdot, t)\, dt$$

$$\left.\frac{\partial}{\partial \nu} \int_{t_1}^{t_2} a_k^\ell w_k^\ell(\cdot, t)\, dt\right|_{\partial\Omega} = 0, \quad -\eta_1 < t_1, t_2 < 1,$$

we obtain

$$[w_k^\infty(\cdot, t)]_{t=t_1}^{t=t_2} = \Delta \int_{t_1}^{t_2} a_k w_k^\infty(\cdot, t)\, dt$$

$$\left.\frac{\partial}{\partial \nu} \int_{t_1}^{t_2} a_k w_k^\infty(\cdot, t)\, dt\right|_{\partial\Omega} = 0 \quad \text{for a.e. } t_1, t_2 \in (-\eta_1, 1),$$

in the sense of distributions on $\overline{\Omega}$, recalling (4.148). It thus follows that

$$
\left[w_k^\ell(\cdot, t) - w_k^\infty(\cdot, t) \right] - \Delta \int_{-\eta_1}^t \left[a_k^\ell w_k^\ell - a_k w_k^\infty \right] (\cdot, t')\, dt'
$$
$$
= \left[w_k^\ell(\cdot, -\eta_1) - w_k^\infty(\cdot, -\eta_1) \right]
$$
$$
\frac{\partial}{\partial \nu} \int_{-\eta_1}^t \left[a_k^\ell w_k^\ell - a_k w_k^\infty \right] (\cdot, t')\, dt' \bigg|_{\partial\Omega} = 0 \quad \text{for a.e. } t \in (-\eta_1, 1) \quad (4.156)
$$

in the same sense. From the elliptic regularity, (4.144), and (4.148), we get

$$
\int_{-\eta_1}^t \left[a_k^\ell w_k^\ell - a_k w_k^\infty \right] (\cdot, t')\, dt' \in H^2(\Omega) \quad \text{for a.e. } t \in (-\eta_1, 1).
$$

Then, taking $L^2(Q)$ inner product of the first equation of (4.156) with $a_k^\ell w_k^\ell - a_k w_k^\infty$ leads to

$$
\iint_{Q_1} (w_k^\ell - w_k^\infty)(a_k^\ell w_k^\ell - a_k w_k^\infty)\, dxdt
$$
$$
\leq \int_\Omega (w_k^\ell(\cdot, -\eta_1) - w_k^\infty(\cdot, -\eta_1))\, dx \cdot \int_{-\eta_1}^1 [a_k^\ell w_k^\ell - a_k w_k^\infty](\cdot, t)\, dt.
$$

Then it follows that

$$
\iint_{Q_1} (w_k^\ell - w_k^\infty)(a_k^\ell w_k^\ell - a_k w_k^\infty)\, dxdt
$$
$$
\leq 2C_{14}(\eta_1) \| w_k^\ell(\cdot, -\eta_1) - w_k^\infty(\cdot, -\eta_1) \|_1
$$

from (4.144) and (4.148). We thus end up with

$$
\limsup_{\ell \to \infty} \iint_{Q_1} (w_k^\ell - w_k^\infty)(a_k^\ell w_k^\ell - a_k w_k^\infty)\, dxdt \leq 0 \quad (4.157)
$$

by (4.155).

Here, we use

$$
\underline{d} \| w_k^\ell - w_k^\infty \|_{L^2(Q_1)^N}^2 \leq \iint_{Q_1} a_k^\ell (w_k^\ell - w_k^\infty)^2\, dxdt
$$
$$
= \iint_{Q_1} (w_k^\ell - w_k^\infty)(a_k^\ell w_k^\ell - a_k w_k^\infty) + (w_k^\ell - w_k^\infty) w_k^\infty (a_k - a_k^\ell)\, dxdt
$$
$$
\leq \iint_{Q_1} (w_k^\ell - w_k^\infty)(a_k^\ell w_k^\ell - a_k w_k^\infty) + \frac{d}{2}(w_k^\ell - w_k^\infty)^2
$$
$$
+ \frac{1}{2\underline{d}}(w_k^\infty)^2 (a_k - a_k^\ell)^2\, dxdt
$$

to deduce

$$\underline{d}\|w_k^\ell - w_k^\infty\|_{L^2(Q_1)^N}^2 \le \iint 2(w_k^\ell - w_k^\infty)(a_k^\ell w_k^\ell - a_k w_k^\infty)$$
$$+\frac{1}{\underline{d}}(w_k^\infty)^2(a_k - a_k^\ell)^2 \, dxdt.$$

Then it follows that

$$w_k^\ell \to w_k^\infty \quad \text{in } L^2(Q_1)^N$$

from (4.147), (4.157), and the dominated convergence theorem. □

By Lemma 4.21, passing to a subsequence, we have

$$w_k^\ell \to w_k^\infty \quad \text{in } L^2_{loc}(\overline{\Omega} \times (-\eta_0, 1]) \text{ and a.e. in } \Omega \times (-\eta_0, 1) \qquad (4.158)$$

as $\ell \to \infty$, where $k = 1, 2, \cdots$.

Lemma 4.22. *The family $\{w_k^\infty\}$ is relatively compact in $L^2_{loc}(\overline{\Omega} \times (-\eta_0, 1])^N$.*

Proof. We repeat the proof of the previous lemma, replacing w_k^ℓ by w_k^∞. First, we have (4.148) for any $\eta_1 \in (1, \eta_0)$. Second, it follows that

$$\frac{\partial w_k^\infty}{\partial t} - \Delta(a_k w_k^\infty) = 0 \quad \text{in } Q_0 = \Omega \times (-\eta_0, 1)$$
$$\frac{\partial}{\partial \nu}(a_k w_k^\infty)\Big|_{\partial\Omega} = 0, \quad w_k^\infty|_{t=-\eta_0} = \tau \cdot u_k(\cdot, -\eta_0) \qquad (4.159)$$

from (4.142). We define $z_k^\ell = z_k^\ell(x,t)$ by (4.150) for smooth $r_0 = r_0(x)$. Passing to a subsequence, we obtain (4.152), where $z_k^\infty = z_k^\infty(x,t)$ is the L^2 solution to

$$\frac{\partial z_k^\infty}{\partial t} - \Delta(a_k z_k^\infty) = 0 \quad \text{in } Q_0, \quad \frac{\partial z_k^\infty}{\partial \nu}\Big|_{\partial\Omega} = 0, \quad z_k^\infty|_{t=-\eta_0} = r_0$$

defined by Lemma 4.17. Then, Lemma 4.18 guarantees

$$z_k^\infty \to z_\infty \quad \text{in } L^2(Q_0) \qquad (4.160)$$

by (4.137)–(4.138). Here, $z_\infty = z_\infty(x,t)$ is the L^2 solution to

$$\frac{\partial z_\infty}{\partial t} - \Delta(a_\infty z_\infty) = 0 \quad \text{in } Q_0, \quad \frac{\partial z_\infty}{\partial \nu}\Big|_{\partial\Omega} = 0, \quad z_\infty|_{t=-\eta_0} = r_0.$$

We modify (4.154) as

$$\|w_k^\ell(\cdot,t) - w_{k'}^\ell(\cdot,t)\|_1$$
$$\le \|w_k^\ell(\cdot,t) - z_k^\ell(\cdot,t)\|_1 + \|z_k^\ell(\cdot,t) - z_{k'}^\ell(\cdot,t)\|_1 + \|z_{k'}^\ell(\cdot,t) - w_{k'}^\ell(\cdot,t)\|_1$$
$$\le \|z_k^\ell(\cdot,t) - z_{k'}^\ell(\cdot,t)\|_1 + \|\tau \cdot \tilde{u}_k^\ell(\cdot, -\eta_0) - r_0\|_1 + \|\tau \cdot \tilde{u}_{k'}^\ell(\cdot, -\eta_0) - r_0\|_1,$$

so that letting $\ell \to \infty$ leads to

$$\|w_k^\infty(\cdot,t) - w_{k'}^\infty(\cdot,t)\|_1 \leq \|z_k^\infty(\cdot,t) - z_{k'}^\infty(\cdot,t)\|_1 + \|\tau \cdot u_k(\cdot,-\eta_0) - r_0\|_1$$
$$+\|\tau \cdot u_{k'}(\cdot,-\eta_0) - r_0\|_1 \quad \text{for a.e. } t \in (-\eta_0,1). \tag{4.161}$$

From (4.140), and (4.160), (4.161), it follows that

$$\lim_{k,k'\to\infty} \|w_k^\infty - w_{k'}^\infty\|_1 = 0 \quad \text{for a.e. } t \in (-\eta,1) \tag{4.162}$$

because r_0 is arbitrary. Inequality (4.148), and equations of (4.159) and (4.162) imply the result as in the proof of Lemma 4.21. □

Proof of Theorem 4.5. Since (4.146) follows from (4.130), (4.145), and (4.158), we obtain

$$0 \leq u_{jk} + 1 \leq \underline{\tau}^{-1} w_k^\infty \quad \text{a.e. in } Q_0, \quad 1 \leq j \leq N \tag{4.163}$$

where $\underline{\tau} = \min_j \tau_j > 0$. It also holds that

$$w_k^\infty \to w_\infty \quad \text{in } L^2_{loc}(\overline{\Omega} \times (-\eta_0,1])^N \text{ and a.e. in } \Omega \times (-\eta_0,1), \tag{4.164}$$

passing to a subsequence. From (4.135), (4.163)–(4.164), and the dominated convergence theorem it follows that

$$\iint_{\Omega\times(-\eta_1,1)} (u_{jk})^2 \, dxdt \to \iint_{\Omega\times(-\eta_1,1)} (u_{j\infty})^2 \, dxdt \tag{4.165}$$

for any $\eta_1 \in (\eta_0,2)$ where $u_\infty = (u_{j\infty})$.

Therefore, it holds that

$$u_k \to u_\infty \quad \text{in } L^2_{loc}(\overline{\Omega} \times (-\eta_0,1])^N \text{ and a.e. in } \Omega \times (-\eta_0,1) \tag{4.166}$$

by (4.136), and hence

$$f(u_k) \to f(u_\infty) \quad \text{in } L^1_{loc}(\overline{\Omega} \times (-\eta_0,1])^N \tag{4.167}$$

by (4.98) and the dominated convergence theorem.

From (4.135), on the other hand, there is $\eta \in (1,\eta_0)$ such that

$$u_k(\cdot,-\eta) \to u_\infty(\cdot,-\eta) \quad \text{in } L^1(\Omega)^N. \tag{4.168}$$

Proposition 4.25, combined with (4.167) and (4.168), now implies

$$u_k \to u_\infty \quad \text{in } C([-\eta,1], L^1(\Omega)^N),$$

and hence

$$u_k(\cdot,0) = u(\cdot,t_k) \to u_\infty(\cdot,0) \quad \text{in } L^1(\Omega)^N.$$

Thus, any $t_k \uparrow +\infty$ admits a subsequence such that $\{u(\cdot,t_k)\}$ converges in $L^1(\Omega)^N$, and the proof is complete. □

4.7　Classical Solutions for $2D$ Case

The argument for the proof of $n = 1$ is included in that for $n = 2$ and is easier. Hence we assume $n = 2$ in this section. As is noted in §4.5, $n = 2$ is the critical dimension for the uniform boundedness of the classical solution $u = (u_j(\cdot, t))$ to (4.94) with (4.98)–(4.99). We have, therefore, $T = +\infty$ and $\sup_{t \geq 0} \|u(\cdot, t)\|_\infty < +\infty$, provided that $\|u_0\|_1$ is sufficiently small. This property is called ε-regularity in §1.2, and furthermore, monotonicity formula is valid as for (1.9) with (1.12). Then we have the formation of finitely many delta-functions to $u = (u_j(\cdot, t))$ as the blowup time approaches. To show Theorem 4.6, first, we derive a bound on $\sup_{0 \leq t < T} \|u(\cdot, t)\|_{L \log L}$, using (4.103) and (4.111). This bound is improved to the one on $\sup_{0 \leq t < T} \|u(\cdot, t)\|_2$ by the Gagliardo-Nirenberg inequality. Once this estimate is achieved, we get a bound of $\sup_{0 \leq t < T} \|u(\cdot, t)\|_\infty$ by the semi-group estimate and bootstrap argument, which implies $T = +\infty$. Since these bounds are not uniform in T, we exclude the possibility of blowup in infinite time in the second step. For this purpose we assume the contrary, and derive the above described blowup mechanism for the solution sequence, obtained by the translation in time of the original global-in-time and classical solution. Then this property, formation of finitely many delta functions, contradicts Theorem 3.3, the relative compactness of the orbit in $L^1(\Omega)$ made by this classical solution.

Thus, assuming the smooth initial value $0 \leq u_0 = (u_{j0}(x))$, we have the unique local-in-time classical solution denoted by $u = (u_j(\cdot, t))$, $0 \leq t < T$. We may assume $u_{j0} = u_{j0}(x) > 0$, $1 \leq j \leq N$, on $\overline{\Omega}$ by the strong maximum principle, which implies $u_j(\cdot, t) > 0$ on $\overline{\Omega}$ for any $1 \leq j \leq N$. Below we shall take the case $n = 2$.

Proof of Theorem 4.6. The fundamental estimate is (4.131), particularly,

$$\sup_{0 \leq t < T} \|u(\cdot, t)\|_1 \leq C_{10}. \tag{4.169}$$

First, we show the a priori estimate

$$\sup_{0 \leq t < T} \|u(\cdot, t)\|_\infty \leq C_{15}(T), \tag{4.170}$$

which guarantees for this $u = u(\cdot, t)$ to be global-in-time. To this end, we multiply (4.94) by $\log u_j$. Then (4.103) implies

$$\frac{d}{dt}\sum_{j=1}^{N}\tau_j\int_\Omega \Phi(u_j)\,dx + \underline{d}\sum_{j=1}^{N}\int_\Omega u_j^{-1}|\nabla u_j|^2\,dx$$

$$\leq C_{16}\left(\int_\Omega |u|^2\,dx + 1\right) \quad \text{with } \underline{d}=\min_j d_j > 0, \qquad (4.171)$$

where

$$\Phi(s) = s(\log s - 1) + 1, \quad s > 0.$$

This inequality, combined with Proposition 4.16, implies

$$\sup_{0\leq t<T}\|\Phi(u_j(\cdot,t))\|_1 \leq C_{17}(T), \quad 1\leq j\leq N. \qquad (4.172)$$

Here we use

$$\|w\|_3^3 \leq \varepsilon\|w\|_{H^1}^2\|w\log w\|_1 + C_{18}(\varepsilon), \quad 0\leq w\in L^3(\Omega) \qquad (4.173)$$

derived from (4.54), where $\varepsilon > 0$ is arbitrary. In fact, inequality (4.98) implies

$$\frac{\tau_j}{2}\frac{d}{dt}\|u_j\|_2^2 + d_j\|\nabla u_j\|_2^2 \leq C_{19}(\|u\|_3^3 + 1).$$

Then we obtain

$$\tau_j\frac{d}{dt}\|u_j\|_2^2 + d_j\|\nabla u_j\|_2^2 \leq C_{20}(T), \quad 1\leq j\leq N$$

by (4.169), (4.172)–(4.173), and Poincaré-Wirtinger's inequality, and hence

$$\sup_{0\leq t<T}\|u(\cdot,t)\|_2 \leq C_{21}(T). \qquad (4.174)$$

Once (4.174) is proven, the semigroup estimate (4.12) applied to (4.97) implies (4.170) by the quadratic growth (4.98). More precisely, we put

$$g_j = \mu u_j + C_1(1+|u|^2)$$

for $\mu \gg 1$, and define $\tilde{u}_j = \tilde{u}_j(\cdot,t)$ by

$$\tau_j\frac{\partial \tilde{u}_j}{\partial t} - d_j\Delta\tilde{u}_j + \mu\tilde{u}_j = g_j(\cdot,t), \quad \left.\frac{\partial\tilde{u}_j}{\partial\nu}\right|_{\partial\Omega}=0, \quad \tilde{u}_j|_{t=0}=u_{j0}(x).$$

Then the comparison principle guarantees $0\leq u_j\leq \tilde{u}_j$, and it holds also that

$$\tilde{u}_j(\cdot,t) = e^{tL_j}u_{j0} + \int_0^t e^{(t-s)L_j}\tau_j^{-1}g_j(\cdot,s)\,ds,$$

where $L_j = \tau_j^{-1}[-d_j\Delta+\mu]$ provided with the Neumann boundary condition. Then inequality (4.170) follows from the iteration scheme as in §4.4.

Second, we show that (4.170) is improved as

$$\sup_{t\geq 0} \|u(\cdot,t)\|_\infty \leq C_{25}. \qquad (4.175)$$

If this is not the case, we have the non-empty blowup set

$$\mathcal{S} = \{x_0 \in \overline{\Omega} \mid 1 \leq \exists j \leq N, \; \exists x_k \to x_0, \; \exists t_k \uparrow +\infty, \; \lim_{k\to\infty} u_j(x_k,t_k) = +\infty\}.$$

Given $x_0 \in \mathcal{S}$, we have $t_k \uparrow +\infty$ and $x_k \to x_0$ such that

$$\lim_{k\to\infty} |u|(x_k,t_k) = +\infty, \qquad (4.176)$$

where $|u| = \sqrt{\sum_{j=1}^N u_j^2}$. By Theorem 3.3 and its proof, we have a subsequence denoted by the same symbol, satisfying (4.166) and

$$u_k \to u_\infty \quad \text{in } C([-1,1], L^1(\Omega)^N) \qquad (4.177)$$

for $u_k = u_k(\cdot,t)$ defined by (4.133).

Given $x_0 \in \overline{\Omega}$ and $0 < R \ll 1$, let $0 \leq \varphi = \varphi_{x_0,R}(x) \in C^\infty(\overline{\Omega})$ be the cut-off function introduced by [Senba and Suzuki (2001)], that is,

$$\varphi_{x_0,R}(x) = \begin{cases} 1, & x \in \Omega \cap B(x_0, R/2) \\ 0, & x \in \Omega \setminus B(x_0, R), \end{cases} \qquad \left.\frac{\partial \varphi}{\partial \nu}\right|_{\partial\Omega} = 0, \qquad (4.178)$$

and

$$|\nabla\varphi| \leq C_{26} R^{-1}\varphi^{5/6}, \quad |\Delta\varphi| \leq C_{26} R^{-2}\varphi^{2/3}. \qquad (4.179)$$

Given $\varepsilon > 0$, we take sufficiently small $R > 0$ such that

$$\|u_\infty(\cdot,0)\|_{L^1(\Omega\cap B(x_0,4R))} < \frac{\varepsilon}{4}.$$

Then we obtain

$$\int_\Omega u_\infty^j(\cdot,0)\varphi_{x_0,4R} \, dx < \frac{\varepsilon}{4} \quad \text{for } 1 \leq j \leq N.$$

Since the mapping

$$t \mapsto \int_\Omega u_\infty^j(\cdot,t)\varphi_{x_0,4R} \, dx$$

is continuous by $u_\infty \in C([-1,1], L^1(\Omega)^N)$, there exists $\delta \in (0,1)$ such that

$$\int_\Omega u_\infty^j(\cdot,t)\varphi_{x_0,4R} \, dx < \frac{\varepsilon}{2}, \quad |t| < \delta$$

which implies

$$\sup_{|t|\le\delta} \|u_\infty(\cdot,t)\|_{L^1(\Omega\cap B(x_0,2R))} < \frac{\varepsilon}{2}. \tag{4.180}$$

By (4.177), inequality (4.180) implies

$$\sup_{|t|\le\delta} \|u_k(\cdot,t)\|_{L^1(\Omega\cap B(x_0,R))} < \varepsilon \tag{4.181}$$

for $k \gg 1$, similarly. Henceforth, we assume (4.181) for $k = 1, 2, \cdots$.
This inequality guarantees

$$\|u(\cdot,t_k)\|_{L^\infty(\Omega\cap B(x_0,R/8))} \le C_{27}, \quad k = 1, 2, \cdots, \tag{4.182}$$

which contradicts (4.176). Thus the uniform boundedness (4.175) has been
shown. We complete the proof of Theorem 4.6 with this inequality, because
it implies relative compactness of the orbit $\mathcal{O} = \{u(\cdot,t) \mid t \ge 0\}$ in $C(\overline{\Omega})^N$.

For the sake of completeness, we describe how to derive (4.182). This
process is actually called the local ε regularity. In our setting, we can take
$s_k \in (0,\delta)$ satisfying

$$\|u_k(\cdot,-s_k)\|_2 \le C_{28} \tag{4.183}$$

by (4.166). Here we use

$$\int_\Omega u_j^3 \varphi_{x_0,R} \, dx \le C_{29}\|u_j\|_{L^1(\Omega\cap B(x_0,R))} \cdot \int_\Omega |\nabla u_j|^2 \varphi_{x_0,R} \, dx + C_{29}\|u_j\|_1 \tag{4.184}$$

derived from the Gagliardo-Nirenberg inequality, valid to any smooth $u = (u_j(\cdot,t)) \ge 0$. Furthermore, the inequality

$$\frac{\tau_j}{2}\frac{d}{dt} \int_\Omega u_j^2 \varphi_{x_0,R} \, dx + d_j \int_\Omega |\nabla u_j|^2 \, \varphi_{x_0,R} \, dx$$

$$\le C_{30}(R) \left(\int_\Omega |u|^3 \varphi_{x_0,R} \, dx + 1 \right), \tag{4.185}$$

follows from (4.98). We thus end up with

$$\sup_{t\in[-s_k,\delta]} \|u_k(\cdot,t)\|_{L^2(\Omega\cap B(x_0,R/2))}^2$$

$$+ \int_{-s_k}^\delta \|\nabla u_k(\cdot,t)\|_{L^2(\Omega\cap B(x_0,R/2))}^2 \, dt \le C_{31} \tag{4.186}$$

by (4.183)–(4.185), recalling $u_k = (u_{jk}(\cdot,t)) = (u_j(\cdot,t+t_k))$. Then we take
$0 < s_k' < s_k$ such that

$$\|\nabla u_k(\cdot,s_k')\|_{L^2(\Omega\cap B(x_0,R/2))} \le C_{32},$$

using (4.186), which implies

$$\|u_k(\cdot, s'_k)\|_p \leq C_{33}(p), \quad 1 \leq p < \infty \qquad (4.187)$$

by (4.169) and Sobolev's embedding theorem. Using an analogous inequality to (4.185), with u_j replaced by $u_j^{3/2}$, we obtain

$$\sup_{t \in [-s'_k, \delta]} \|u_k(\cdot, t)\|_{L^3(\Omega \cap B(x_0, R/4))} \leq C_{34}.$$

This inequality is improved as

$$\sup_{t \in [-s'_k, \delta]} \|u_k(\cdot, t)\|_{L^4(\Omega \cap B(x_0, R/4))} \leq C_{35} \qquad (4.188)$$

by repeating the same argument.

Here we use

$$\tau_j \frac{\partial \tilde{u}_{jk}}{\partial t} - d_j \Delta \tilde{u}_{jk} = \tilde{g}_{jk}, \quad \left. \frac{\partial \tilde{u}_j^k}{\partial \nu} \right|_{\partial \Omega} = 0$$

with $\tilde{u}_{jk} = u_{jk}\varphi$ and $\varphi = \varphi_{x_0, R/4}$, where

$$\tilde{g}_{jk} = -d_j(u_{jk}\Delta\varphi + 2\nabla u_{jk} \cdot \nabla\varphi) + f_j(u_k)\varphi.$$

We have

$$\int_{-s'_k}^{\delta} \|\tilde{g}_{jk}(\cdot, t)\|_2^2 \, dt \leq C_{36}$$

by (4.186) and (4.188). Then, using

$$\tilde{u}_{jk}(\cdot, t) = e^{(t+s_k)\tau_j^{-1}d_j\Delta}\tilde{u}_{jk}(\cdot, -s'_k) + \int_{-s'_k}^{t} e^{(t-s)\tau_j^{-1}d_j\Delta}\tau_j^{-1}\tilde{g}_{jk}(\cdot, s) \, ds$$

for $t \in (-s_k, \delta)$, and the semi-group estimate (4.92), that is,

$$\|\nabla e^{t\Delta}w\|_r \leq C_{37}(q, r) \max\{1, t^{-\frac{n}{2}(\frac{1}{q}-\frac{1}{r})-\frac{1}{2}}\}\|w\|_q, \quad 1 \leq q \leq r \leq \infty$$

with $n = 2$, we obtain

$$\sup_{t \in [-s''_k, \delta]} \|\nabla u_{jk}(\cdot, t)\|_r \leq C_{38}$$

for $0 < s''_k < s_k$ and $1 \leq r < \infty$, and hence (4.182) by (4.169). $\qquad \square$

In the above proof, inequality (4.100) is used to exclude blowup in finite time. This condition can be replaced by (4.217) as is described in §4.5.

4.8 Spatial Homogenization

Theorem 4.9 below says that the solution becomes spatially homogeneous under the presence of an entropy functional. This assertion follows from the LaSalle principle and the relatively compactness of the orbit. Since we are concerned with the weak solution, we use the approximate solution.

Theorem 4.9. *Assume (4.95), (4.96), (4.98), and (4.99), and let*

$$0 \le u = (u_j(\cdot,t)) \in C([0,+\infty),L^1(\Omega)^N) \qquad (4.189)$$

be the global-in-time weak solution to (4.94) in Theorem 4.4. Define its ω-limit set by

$$\omega(u_0) = \{u_* \in L^1(\Omega)^N \mid \exists t_k \uparrow +\infty, \ \lim_{k\to\infty} \|u(\cdot,t_k) - u_*\|_1 = 0\}.$$

Then we have the following properties:

(1) Assume $f_j(u) = u_j g_j(u)$, $1 \le j \le N_1$, with

$$|g_j(u)| \le C_8(1+|u|), \quad \sum_{j=1}^{N_1} b_j \tau_j^{-1} g_j(u) \ge 0, \quad 0 \le u = (u_j) \in \mathbb{R}^N,$$

$$(4.190)$$

where $0 < b = (b_j) \in \mathbb{R}^{N_1}$ and $1 \le N_1 \le N$. Assume, furthermore,

$$\log u_{j0} \in L^1(\Omega), \quad 1 \le j \le N_1. \qquad (4.191)$$

Then it holds that

$$P_1\omega(u_0) \subset \mathbb{R}_+^{N_1} = \{u = (u_1,\cdots,u_{N_1}) \in \mathbb{R}^N \mid u_1,\cdots,u_{N_1} > 0\}$$

where $P_1 : (u_1,\cdots,u_N) \mapsto (u_1,\cdots,u_{N_1})$.
(2) Assume that inequality (4.99) is improved as

$$\sum_{j=1}^{N} f_j(u) \le -e \cdot u, \quad 0 \le u = (u_j) \in \mathbb{R}^N \qquad (4.192)$$

with $0 \le e = (e_j) \in \mathbb{R}^N$ satisfying $e_{N_2+1},\cdots,e_N > 0$ for $N_2 \ge N_1$. Then it holds that $P_2\omega(u_0) = \{0\}$, where $P_2 : (u_1,\cdots,u_N) \mapsto (u_{N_2+1},\cdots,u_N)$.

The second inequality of (4.190) provides with a Lyapunov function to (4.94). Instead of (4.191), on the other hand, we may assume $u_{j0} \in L^\infty(\Omega)$ with $u_{j0} \not\equiv 0$, $1 \le j \le N_1$, by the strong maximum principle and the parabolic regularity.

Lemma 4.23. *Under the assumptions of the first case of Theorem 4.9, it holds that*

$$\log u_j \in L^1_{loc}(\overline{\Omega} \times [0, +\infty)), \ \nabla \log u_j \in L^2(\Omega \times (0, +\infty))^N, \quad 1 \le j \le N_1$$

and

$$\frac{d}{dt} H(u) \ge \sum_{j=1}^{N_1} b_j \tau_j^{-1} d_j \int_\Omega |\nabla \log u_j|^2 \ dx \ge 0 \qquad (4.193)$$

in the sense of distributions with respect to t, where

$$H(u) = \sum_{j=1}^{N_1} \int_\Omega b_j \log u_j \ dx.$$

Proof. Let $\tilde{u}^\ell = (\tilde{u}^\ell_j(\cdot, t))$ be the approximate solution of $u = (u_j(\cdot, t))$ defined by (4.126). It satisfies (4.130), and also $\tilde{u}^\ell_j(\cdot, t) > 0$ on $\overline{\Omega}$ for $1 \le j \le N$. Letting $g^\ell_j = g_j \circ T_\ell$, we have

$$\frac{d}{dt} H(\tilde{u}^\ell) = \sum_{j=1}^{N_1} b_j \tau_j^{-1} \int_\Omega |\nabla \log \tilde{u}^\ell_j|^2 + g^\ell_j(\tilde{u}^\ell) \ dx$$

$$\ge \sum_{j=1}^{N_1} \int_\Omega b_j \tau_j^{-1} |\nabla \log \tilde{u}^\ell_j|^2 \ dx \ge 0$$

and hence

$$H(\tilde{u}^\ell(\cdot, t)) \ge H(\tilde{u}^\ell_0) \ge H(u_0) > -\infty \qquad (4.194)$$

by (4.191) and (4.125). Therefore, using

$$\log_+ \tilde{u}^\ell_j \le \tilde{u}^\ell_j \to u_j \text{ in } L^1_{loc}(\overline{\Omega} \times [0, +\infty)) \text{ and a.e. in } \Omega \times (0, +\infty) \quad (4.195)$$

valid to $1 \le j \le N$ and Fatou's lemma, we have

$$\log u_j \in L^1_{loc}(\overline{\Omega} \times [0, +\infty)), \quad 1 \le j \le N_1$$

$$H(u(\cdot, t)) \ge H(u_0) \quad \text{for a.e. } t, \qquad (4.196)$$

where $\log_+ s = \max\{\log s, 0\}$. Furthermore, (4.128) implies

$$H(\tilde{u}^\ell(\cdot, t)) \le C_{40},$$

and, therefore,

$$\iint_{\Omega \times (0, +\infty)} |\nabla \log \tilde{u}^\ell_j|^2 \ dx dt \le C_{41}, \quad 1 \le j \le N_1. \qquad (4.197)$$

Thus $\{\nabla \log \tilde{u}^\ell_j\}$, $1 \le j \le N_1$, is weakly relatively compact in $L^2(\Omega \times (0, +\infty))^N$. Consequently, it holds that

$$\nabla \log u_j \in L^2(\Omega \times (0, +\infty))^N, \quad 1 \le j \le N_1 \qquad (4.198)$$

and (4.193) in the sense of distributions with respect to t. $\qquad \square$

We have already shown (4.166) for $u_k = (u_{jk}(\cdot, t))$, $u_{jk}(\cdot, t) = u_j(\cdot, t + t_k)$, and $\eta_0 \in (1,2)$. Let $u_\infty = (u_{j\infty}(\cdot, t))$. We take $\eta_1 \in (1, \eta_0)$ and put $Q_1 = \Omega \times (-\eta_1, 1)$.

Lemma 4.24. *Under the assumptions of the first case of Theorem 4.9, it holds that*

$$\log u_{j\infty} \in L^1(Q_1), \quad \log u_{jk} \to \log u_{j\infty} \text{ in } L^1(Q_1)$$

as $k \to \infty$ for $1 \le j \le N_1$.

Proof. We take $\eta_2 \in (\eta_1, \eta_0)$ and put $Q_2 = \Omega \times (-\eta_2, 1)$. By (4.193) we have

$$\iint_{Q_2} \sum_{j=1}^{N_1} b_j \log u_{jk} \, dxdt \ge (1 + \eta_2) \cdot H(u_0) > -\infty, \tag{4.199}$$

recalling (4.191). Then $\log u_{j\infty} \in L^1(Q_2)$, $1 \le j \le N_1$, follow from (4.166), (4.195), and Fatou's lemma. In particular, we obtain

$$\log u_{jk} \to \log u_{j\infty} \quad \text{a.e. in } Q_2, \ 1 \le j \le N_1. \tag{4.200}$$

By (4.126) we obtain

$$\tau_j \frac{\partial}{\partial t} \log \tilde{u}_j^\ell - d_j \Delta \log \tilde{u}_j^\ell \ge g_j^\ell(\tilde{u}^\ell), \quad \left. \frac{\partial}{\partial \nu} \log \tilde{u}_j \right|_{\partial \Omega} = 0,$$

which implies

$$\tau_j \frac{\partial}{\partial t} \log u_{jk} - d_j \Delta \log u_{jk} \ge g_j(u_k), \quad \left. \frac{\partial}{\partial \nu} \log u_{jk} \right|_{\partial \Omega} = 0, \quad 1 \le j \le N_1$$

in the sense of distributions in Q_1, recalling (4.190), (4.130), and (4.197)–(4.198).

By (4.199) there is $\eta \in (\eta_1, \eta_2)$ such that $\{\log u_{jk}(\cdot, -\eta)\}$, $1 \le j \le N_1$, is bounded in $L^1(\Omega)$. Then we take the solution

$$w_j^k = w_j^k(\cdot, t) \in L^\infty(-\eta, 1; L^1(\Omega)) \cap L^1_{loc}(-\eta, 1; W^{1,1}(\Omega)) \tag{4.201}$$

to

$$\tau_j \frac{\partial w_j^k}{\partial t} - d_j \Delta w_j^k = g_j(u_k) \quad \text{in } \Omega \times (-\eta, 1) \equiv Q_\eta$$

$$\left. \frac{\partial w_j^k}{\partial \nu} \right|_{\partial \Omega} = 0, \quad w_j^k \big|_{t=-\eta} = \log u_{jk}(\cdot, -\eta). \tag{4.202}$$

Then we obtain

$$w_j^k \le \log u_{jk} (\le u_{jk}) \text{ in } Q_\eta, \quad 1 \le j \le N_1 \tag{4.203}$$

from the comparison principle. By (4.190) and (4.166) we have

$$g_j(u_k) \to g_j(u_\infty) \quad \text{in } L^1(Q_\eta)$$

by the dominated convergence theorem which implies

$$w_j^k \to w_j \quad \text{in } L^1(Q_\eta) \tag{4.204}$$

with some w_j by Proposition 4.8. The result follows from (4.200)–(4.204) and the dominated convergence theorem. $\qquad \square$

Proof of Theorem 4.9. Since $\{u(\cdot, t) \mid t \geq 0\}$ is relatively compact in $L^1(\Omega)^N$, the ω-limit set $\omega(u_0)$ is non-empty. Let $t_k \uparrow +\infty$ and $u(\cdot, t_k) \to u_*$ in $L^1(\Omega)^N$. Passing to a subsequence, we obtain (4.177) for $u_k(\cdot, t) = (u_j(\cdot, t + t_k))$.

Under the assumptions of the first case, we have the existence of

$$\lim_{t \uparrow +\infty} H(u(\cdot, t))$$

by (4.131) and (4.193), which implies the LaSalle principle,

$$\lim_{k \to \infty} \int_{t_k-1}^{t_k+1} dt \cdot \sum_{j=1}^{N_1} b_j \tau_j^{-1} d_j \int_\Omega |\nabla \log u_j|^2 \, dx = 0$$

again by (4.193). Then we obtain

$$\nabla \log u_{j\infty} = 0 \quad \text{in } \Omega \times (-1, 1), \quad 1 \leq j \leq N_1$$

in the sense of distributions, recalling Lemma 4.24. Then it follows that $0 < u_{j\infty} \in \mathbb{R}$ for $1 \leq j \leq N_1$.

In the second case we use (4.94) in the form of

$$\tau_j \frac{\partial u_j}{\partial t} + e_j u_j = d_j \Delta u_j + f_j(u) + e_j u_j, \qquad \left. \frac{\partial u_j}{\partial \nu} \right|_{\partial \Omega} = 0.$$

It holds that

$$\frac{d}{dt} \int_\Omega \tau \cdot u \, dx + \int_\Omega e \cdot u \, dx \leq 0$$

in the sense of distributions with respect to t, and hence there exists

$$\lim_{t \uparrow +\infty} \int_\Omega \tau \cdot u \, dx.$$

Then we obtain

$$\iint_{\Omega \times (-1,1)} e \cdot u_\infty(x, t) \, dx dt = 0$$

from the LaSalle principle, and hence

$$u_{j*} = 0, \quad N_2 + 1 \leq j \leq N$$

for $u_* = (u_{j*})$. The proof is complete. $\qquad \square$

4.9 Notes

Transport theory in Section 4.1 is provided in Section 2.1.4 of [Suzuki and Senba (2011)]. The fact that quasi-positivity (4.3) guarantees the positivity of the solution is proven by the method of invariant regions, valid to reaction-diffusion systems. See Chapter 14 of [Smoller (1986)] for this notion.

For actual proof of the positivity of the solution, we put $G_i(u) = -u_i$ ($i = 1, 2, \ldots, N$) for $u = (u_1, u_2, \ldots, u_N)$, and take the first quadrant with boundary,

$$K = \bigcap_{i=1}^{N} \{u \mid G_i(u) \leqq 0\}.$$

Since

$$dG_i(u) = \sum_{j=1}^{N} \frac{\partial G_i}{\partial u_j} f_j(u) = -f_i(u),$$

if $f_i(u) \leq 0$ for $u \in K$ satisfying $\{u \mid G_i(u) = 0\}$, then it holds that $dG_i(u) \leqq 0$. Hence, if $f_i(u_1, \ldots, u_{i-1}, 0, u_{i+1}, \ldots, u_N) \geqq 0$ for $i = 1, 2, \ldots, N$, the first quadrant K is an invariant region in the sense of Theorem 14.11 of [Smoller (1986)]. The strong maximum principle now guarantees the positivity of the solution.

For the semi-group estimate (4.12), we refer to Lemma 3 in Part I of [Rothe (1984)]. The other estimate (4.92) follows from

$$\|\nabla e^{t\Delta} w\|_p \leq C \max\{1, t^{-1/2}\} \|w\|_p.$$

For the proof we use the fractional powers of $-\Delta$. We recall that the domain of the fractional power $(-\Delta)^\alpha$ is the Sobolev space $W^{2\alpha, p}(\Omega)$ if $0 \leq \alpha < (p+1)/(2p)$, and $W_N^{2\alpha, p}$ if $(p+1)/(2p) < \alpha \leq 1$, where $W_N^{2\alpha, p}(\Omega)$ is the Sobolev space with the Neumann boundary condition. See Theorem 16.11 of [Yagi (2010)]. Moreover, it holds that

$$\|(-\Delta)^\alpha e^{t\Delta}\|_{\mathcal{B}(L^p(\Omega), L^p(\Omega))} \leq C t^{-\alpha} \quad (t > 0),$$

where $\|\cdot\|_{\mathcal{B}(L^p(\Omega), L^p(\Omega))}$ is the norm for the bounded linear operators on $L^p(\Omega)$ into $L^p(\Omega)$. See Theorem 3.3 in Chapter 3 of [Tanabe (1979)]. Since

$$(-\Delta)^\alpha e^{t\Delta} u = (-\Delta)^\alpha e^{(t/2)\Delta} e^{(t/2)\Delta} u,$$

we obtain

$$\|(-\Delta)^\alpha e^{t\Delta} u\|_p \leq C \max \left\{1, t^{-\frac{n}{2}\left(\frac{1}{q} - \frac{1}{p}\right) - \alpha}\right\} \|u\|_q \quad (t > 0) \tag{4.205}$$

by (4.12), where $1 \leq q \leq p \leq \infty$ and $\alpha \geq 0$. Since

$$\|(-\Delta)^{1/2}v\|_p \approx \|v\|_{W^{1,p}(\Omega)}$$
$$\|\Delta v\|_p \approx \|v\|_{W^{2,p}(\Omega)} \tag{4.206}$$

for $1 < p < \infty$, the left-hand side on (4.205) can be replaced by $\|\nabla e^{t\Delta}u$ $Vert_p$. See Theorem 16.11 of [Yagi (2010)] and Section 5.6 of [Evans (1998)] for the Poincaré's inequality.

Local-in-time well-posedness and the blowup of the solution to equation (4.10) in $L^p(\Omega)$ is studied by [Weissler (1980, 1985)] for $f(u) = |u|^{p-1}u$, $1 < p < \infty$. If the nonlinearity satisfies

$$\|f(u) - f(v)\|_2 \leq C(\|A^\alpha u\|_2 + \|A^\alpha v\|_2)^{p-1}\|A^\alpha u - A^\alpha v\|_2$$

for $A = -\Delta$, a unique solution local-in-time is constructed in

$$E_{\alpha,T} = \{u : [0,T] \to L^2(\Omega) \mid t^{\alpha-1/2}u \in BC(0,T; D(A^\alpha))$$

for $0 < T \ll 1$, where $A = -\Delta$. See [Ikehata and Suzuki (2000)].

Theorem 4.1 is due to [Alikakos (1979); Latos, Suzuki, and Ya-mada (2012)]. A standard monograph on the parabolic regularity is [Ladyženskaya, Solonikov, and Ural'ceva (1968)]. Theory of dynamical systems applied to partial differential equations is described in [Henry (1981)]. See [Hale (1988)] for the general theory. Classification of ODE orbits to (4.24) is done by [Ni, Suzuki, and Takagi (2006)]. See [Karali, Suzuki, and Yamada (2013)] for a similar result to Theorem 4.1 concerning Gierer-Meinhard equation. System (4.32) is studied by [Suzuki and Yamada (2013)]. The method developed in this work is applied to the general dissipative Lotka-Volterra system by [Suzuki and Yamada (2015)] in accordance with the thermodynamical structure. Then this result is refined in the context of weak solutions by [Pierre, Suzuki, and Yamada (2018)]. Skew symmetric Lotka-Volterra ODE systems with entropy, on the other hand, are classified in [Kobayashi, Suzuki, and Yamada (2017)]. Proof of $T = +\infty$ for $n = 1$ is done by [Masuda and Takahashi (1994)] for several cases of $A = (a_{jk})$ and $e = (e_j)$ in (4.33).

Models with diffusion for virus dynamics are used in many references. Here we refer to [Prüss, Zacher, and Schnaubelt (2008); Wang, Yang, and Kuniya (2016)]. Model (4.68) is due to [Nowak and Bangham (1996)]. Reproduction number R_0 in (4.69) is formulated by [Bonhoeffer, May, Shaw, and Nowak (1997)]. The Lyapunov functions for $R_0 > 1$ and $R_0 \leq 1$ are introduced by [Korobeinikov (2004)]. The proof of Lemma 4.11 is based on the method of [Latos, Suzuki, and Yamada (2012)]. See Corollary

7.11 of [Gilbarg and Trudinger (1983)] and its notes for Morrey's theorem used in the proof of Lemma 4.14. See Theorem 4.3.3 of [Henry (1981)] for dynamical theory on ω-limit set, and Section 4.4 is due to [Sasaki and Suzuki (2018)].

The function $\Phi(s)$ in (4.70) is different from the entropy density effective to the thermodynamical model, $\Phi(s) = s(\log s - 1) + 1$, which guarantees the Csizár-Kullback inequality

$$\|g - G\|_1 \leq 4M \int_\Omega G\Phi\left(\frac{g}{G}\right)\,dx \qquad (4.207)$$

valid to

$$0 < g, G \in L^1(\Omega), \quad \|g\|_1 = \|G\|_1 = M. \qquad (4.208)$$

It is now well-known that this inequality can be a basis to infer the exponential convergence of the spatially inhomogeneous solution to the spatially homogeneous equilibrium in L^1 norm. See Theorem 31 of [Carrillo, Jüngel, Markowich, Toscani, Unterriter (2001)].

Without (4.207), however, the convergence (4.76) is exponential, thanks to the theory of linearization. In fact, in the case of $R_0 > 1$, for example, the ODE theory has revealed that the stationary solution $u^* = (u_1^*, u_2^*, u_3^*)$ is linearly asymptotically stable for the ODE part. This means that the eigenvalues of the linearized matrix

$$A = \begin{pmatrix} -m - \beta u_3^* & 0 & -\beta u_1^* \\ \beta u_3^* & -a & \beta u_1^* \\ 0 & ar & -b \end{pmatrix}$$

lie in the left-half space on the complex plane,

$$\mathrm{Re}\,\sigma(A) < -\delta, \qquad (4.209)$$

where $\sigma(A)$ denotes the spectrum of A and $\delta > 0$. If $u = (u_i(x,t))$ is the solution to (4.65), (4.66), and (4.67), we obtain

$$\frac{\partial \tilde{u}}{\partial t} = (L + A)(\tilde{u}) + F(\tilde{u}) \qquad (4.210)$$

for $\tilde{u} = (\tilde{u}_i(x,t))$ with $\tilde{u}_i = u_i - u_i^*$, where

$$L = \begin{pmatrix} d_1\Delta & 0 & 0 \\ 0 & d_2\Delta & 0 \\ 0 & 0 & d_3\Delta \end{pmatrix}, \quad F(\tilde{u}) = \beta\begin{pmatrix} -\tilde{u}_1\tilde{u}_3 \\ \tilde{u}_1\tilde{u}_3 \\ 0 \end{pmatrix}.$$

Here we recall that Δ is provided with the Neumann boundary condition.

Regarding that the solution u approaches to the equilibrium u^*, we take the linear part of (4.210),

$$\frac{\partial v}{\partial t} = Lv + Av, \quad v|_{t=0} = v_0.$$

Since $(Lv, v) \leqq 0$ for $v \in D(L)$, it holds that

$$\frac{1}{2}\frac{d}{dt}\|v\|_2^2 \leq -\delta\|v\|_2^2$$

and hence $\|v(\cdot, t)\|_2 \leq \|v_0\|_2 e^{-\delta t}$. Since $LA = AL$, we obtain

$$\|L^\gamma v(\cdot, t)\|_2 \leq \|L^\gamma v_0\|_2 e^{-\delta t}$$

for any $\gamma > 0$, in particular, $\|v(\cdot, t)\|_\infty \leq Ce^{-\delta t}$. The exponential decay of $\|\tilde{u}(\cdot, t)\|_\infty$ as $t \to \infty$ now follows. See Theorem 6.10 of [Yagi (2010)] for this linear theory.

The description of Sections 4.5–4.8 is due to [Pierre, Suzuki, and Yamada (2018)]. Equality (4.97) for the weak solution $u = (u_j(\cdot, t))$ is shown in [Baras and Pierre (1984)] for the Dirichlet boundary condition. The proof is similar to this case of (4.94). See also Lemma 5.1 of [Pierre (2003)]. Thus, equality (4.97) is a consequence of the following theorem proven by the comparison principle. See Lemma 3.4 of [Baras and Pierre (1984)].

Theorem 4.10. *Given $w_0 \in L^1(\Omega)$ and $H \in L^1(Q_T)$, let*

$$w = w(\cdot, t) \in L^\infty(0, T; L^1(\Omega)) \cap L^1_{loc}(0, T; W^{1,1}(\Omega))$$

be the solution to (4.123). More precisely, for any $\varphi \in W^{1,\infty}(\Omega)$ it holds that

$$\frac{d}{dt}\int_\Omega w\varphi \, dx + \int_\Omega \nabla w \cdot \nabla\varphi \, dx = \int_\Omega H\varphi \, dx$$

in the sense of distributions with respect to t and

$$\lim_{t\downarrow 0} w(\cdot, t) = w_0$$

in the sense of measures on $\overline{\Omega}$. Then it follows that

$$w(\cdot, t) = e^{t\Delta}w_0 + \int_0^t e^{(t-s)\Delta}H(\cdot, s) \, ds, \quad 0 \leq t \leq T. \tag{4.211}$$

In particular, $w \in C([0, T], L^1(\Omega))$ and this solution exists uniquely.

The existence of the solution in the above theorem can be proven by the duality argument as in Lemma 3.3 of [Baras and Pierre (1984)]. By (4.211), a result comparable to Lemma 4.18 below is obtained.

Lemma 4.25. *The mapping $\mathcal{F} : (w_0, H) \in L^1(\Omega) \times L^1(Q_T) \mapsto w \in C([0, T], L^1(\Omega))$ is continuous, where $w = w(x, t)$ is the solution to (4.123) in Proposition 4.10.*

The compactness result Theorem 4.8 is known even to the nonlinear contraction semigroup. See [Baras (1978)] and Lemma 3.3 of [Baras and Pierre (1984)].

Theorem 4.4 is due to [Pierre and Rolland (2016)]. Provided with (4.95), (4.96), (4.98), and (4.99), global-in-time existence of the weak solution to (4.94) is proven for $u_0 = (u_{j0}) \in L^2(\Omega)^N$ in [Pierre (2003)]. Theorem 4.4 is an extension of this result, in the sense that it admits general $0 \le u_0 \in L^1(\Omega)^N$. Inequality (4.99) is used to guarantee for the limit of approximate solutions to be a sub-solution to (4.94). This inequality may be relaxed as

$$\sum_{j=1}^{N} f_j(u) \le C_4(b \cdot u + 1), \quad 0 \le u = (u_j) \in \mathbb{R}^N$$

for Theorem 4.4 to hold, where $0 \le b = (b_j) \in \mathbb{R}^N$. See Theorem 5.14 of [Pierre (2010)]). Inequality (4.100) may be so relaxed as (H6) in [Pierre and Rolland (2016)] in Theorem 4.4. This inequality, however, is used also in the proof of Theorem 4.6. Analogous results to Theorem 4.6 are valid under the assumption

$$|f_j(u)| \le C\left(1 + |u|^{1+\frac{2}{n}}\right), \quad \frac{\partial f_j}{\partial u_j} \ge -C\left(1 + |u|^{\frac{2}{n}}\right)$$

if n stands for the space dimension.

The first example covered by Theorems 4.4–4.6 is the four-component system describing chemical reaction $A_1 + A_3 \leftrightarrow A_2 + A_4$:

$$N = 4, \quad f_j(u) = (-1)^j(u_1 u_3 - u_2 u_4), \ 1 \le j \le 4. \tag{4.212}$$

There is a weak solution global-in-time (4.212) which converges exponentially to a unique spatially homogeneous stationary state in L^1 norm. Similar results hold for the renormalized solution involving higher growth rate. These cases are treated in the following chapter in this monograph. This solution is classical even in higher space dimensions, which is proven by [Fellner, Latos, and Suzuki (2016)] when the diffusion coeffcients are quasi-uniform and [Caputo, Goudon, and Vasseur (2017)] for general case. The second example is the Lotka-Volterra system, where

$$f_j(u) = \left(-e_j + \sum_k a_{jk}u_k\right)u_j, \quad 1 \le j \le N, \tag{4.213}$$

in (4.94). For (4.213) the assumptions of Theorem 4.4 are fulfilled if

$$0 \le (e_j) \in \mathbb{R}^N \tag{4.214}$$

and

$$(Au, u) \leq 0, \quad 0 \leq u = (u_j) \in \mathbb{R}^N \tag{4.215}$$

where $A = (a_{jk})$. The nonlinearities (4.212) and (4.213) with

$$(e_j) = 0, \quad {}^t A + A = 0, \ A = (a_{jk}) \tag{4.216}$$

satisfy the equality in (4.99):

$$\sum_{j=1}^{N} f_j(u) = 0, \quad 0 \leq u = (u_j) \in \mathbb{R}^N. \tag{4.217}$$

Under this condition, blowup in finite time is excluded if $n \leq 2$. See [Goudon and Vasseur (2010)] and also Proposition 3.2 of [Cãnizo, Desvilletes, and Fellner (2014)]. Blowup in infinite time is also excluded by the proof of Proposition 5.1 of [Suzuki and Yamada (2015)]. Hence Theorem 4.6 is still valid without (4.100) if (4.217) is assumed for (4.99). This result holds even if $-e_j u_j$ is added to $f_j(u)$ satisfying (4.217) for each $1 \leq j \leq N$, where $e_j \geq 0$ is a constant.

Besides (4.102), blowup analysis is used in [Suzuki and Yamada (2015)] for the study of (4.213)–(4.214), based on the scaling

$$u_\mu(x, t) = \mu^2 u(\mu x, \mu^2 t), \quad \mu > 0. \tag{4.218}$$

At this process, the inequality (4.103) is confirmed, and plays a key role in establishing a priori estimates of the solution in [Suzuki and Yamada (2015)]. Theorem 4.7 is due to [Pierre (2010)].

By the argument developed in [Suzuki and Yamada (2015)], inequality (4.111) guarantees global-in-time existence of the classical solution, indicated by $T = +\infty$, under the assumptions of Theorem 4.6. Here we show the following proof.

Alternative Proof of Theorem 4.7. Let $a = \dfrac{d \cdot u + 1}{\tau \cdot u + 1}$,

$$0 < a_0 = \min_{u \geq 0} \frac{(\min d_i) u + 1}{(\max d_i) u + 1} \leq a = a(x, t) \leq a_1 = \frac{(\max d_i) u + 1}{(\min \tau_i) u + 1},$$

and $z = \tau \cdot u + 1$. Then it holds that

$$z_t - \Delta(az) \leq 0, \quad z \geq 0, \quad \left.\frac{\partial z}{\partial \nu}\right|_{\partial \Omega} = 0. \tag{4.219}$$

Given $\theta - \theta(x, t) \geq 0$, let $w = w(x, t)$ be the solution to

$$w_t + a\Delta w = -\theta, \quad w|_{t=T} = 0, \quad \left.\frac{\partial w}{\partial \nu}\right|_{\partial \Omega} = 0. \tag{4.220}$$

Here we recall Lemma 4.17. By (4.220) it holds that $w \geq 0$. Multiplying Δw, we obtain

$$-\frac{1}{2}\frac{d}{dt}\|\nabla w\|_2^2 + a_0\|\Delta w\|_2^2 \leq -(\theta, \Delta w) \leq \|\theta\|_2 \|\Delta w\|_2$$

$$\leq \frac{a_0}{2}\|\Delta w\|_2^2 + \frac{8}{a_0}\|\theta\|_2^2$$

which implies

$$\|\nabla w(0)\|_2^2 + \int_0^T \|\Delta w\|_2^2 \, dt \leq C\|\theta\|_{L^2(Q_T)}^2.$$

Multiplying w to (4.220), on the other hand, we get

$$-\frac{1}{2}\frac{d}{dt}\|w\|_2^2 = (a\Delta w, w) + (w, \theta) \leq a_1\|\Delta w\|_2\|w\|_2 + \|\theta\|_2\|w\|_2$$

$$\leq \frac{1}{4}\|w\|_2^2 + 16a_1^2\|\Delta w\|_2^2 + \frac{1}{4}\|w\|_2^2 + 16\|\theta\|_2^2$$

and hence

$$-\frac{1}{2}\frac{d}{dt}(e^t\|w\|_2^2) \leq 16e^T(a_1^2\|\Delta w\|_2^2 + \|\theta\|_2^2), \quad 0 < t < T.$$

It follows that

$$\|w(0)\|_2^2 \leq 32e^T\left(a_1^2 \int_0^T \|\Delta w\|_2^2 + \|\theta\|_2^2 \, dt\right) \leq C_T\|\theta\|_{L^2(Q_T)}^2.$$

We thus end up with

$$\|w(0)\|_{H^1}^2 + \int_0^T \|\Delta w\|_2^2 \, dt \leq C_T\|\theta\|_{L^2(Q_T)}^2. \tag{4.221}$$

From (4.219) and (4.220) it follows that

$$\frac{d}{dt}(z, w) = (z_t, w) + (z, w_t)$$

$$\leq (\Delta(az), w) + (z, -\theta - a\Delta w) = -(z, \theta)$$

and hence

$$\int_0^T (z, \theta) \, dt \leq (z_0, w(0)) \leq \|z_0\|_2 \cdot \|w(0)\|_2$$

$$\leq C_T\|z_0\|_2 \cdot \|\theta\|_{L^2(Q_T)}.$$

Thus we obtain

$$\|z\|_{L^2(Q_T)} \leq C_T\|z_0\|_2$$

by $z = z(x, t) \geq 0$ and hence

$$\|u\|_{L^2(Q_T)} \leq C_T\|u_0\|_2.$$

$\square$

The other inequality derived from (4.110) for $u = (u_j(x,t)) \geq 0$ is

$$\int_0^T \|u(\cdot,t)\|_\infty \, dt \leq C. \tag{4.222}$$

In fact, we have

$$-\Delta \int_0^T d \cdot u \, dt \leq \tau \cdot u_0, \quad \frac{\partial}{\partial \nu} \int_0^T d \cdot u \, dt \bigg|_{\partial \Omega} \leq 0,$$

which implies

$$\int_0^T d \cdot u \, dt \leq C$$

and hence (4.222).

The four-component system (4.212) is provided with

$$\sum_{i=1}^4 (\log u_i) f_i(u) = -(u_1 u_3 - u_2 u_4 [\log(u_1 u_3) - \log(u_2 u_4)] \leq 0.$$

Let $f = (f_j(u)) : \mathbf{R}_k^N \to \mathbf{R}^N$ be locally Lipschitz and quasi-positive (4.96), and assume

$$\sum_{j=1}^N f_j(u) \log u_j \leq 0$$

for (4.103). Writing $z_j = \Phi(u_j) \geq 0$ with $\Phi(s) = s(\log s - 1) + 1$, we obtain

$$\tau_j \frac{\partial z_j}{\partial t} - d_j \Delta z_j$$

$$= (\log u_j) \left(\frac{\partial u_j}{\partial t} - d_j \Delta u_j \right) - d_j u_j^{-1} |\nabla u_j|^2$$

$$= (\log u_j) f_j(u) - 4 d_j |\nabla u_j^{1/2}|^2$$

and hence

$$\frac{\partial}{\partial t} (\tau \cdot z) - \Delta(d \cdot z) \leq -4 \sum_j d_j |\nabla u_j^{1/2}|^2 \leq 0.$$

Inequality (4.111) is thus improved as

$$\|u \cdot \log u\|_{L^2(Q_T)} \leq C(1 + T^{1/2})$$

in this case, where $\log u = (\log u_j)$ for $u = (u_j) \geq 0$. This estimate is a basis of the existence of the renormalized solution $u = u(\cdot,t) \in L_{loc}^\infty([0,+\infty), L \log L(\Omega))$. See [Cãnizo, Desvilletes, and Fellner (2014)].

There is a case where component-wise estimate is effective. An example is

$$\frac{\partial u_1}{\partial t} - d_1 \Delta u_1 = -u_1^\alpha u_2^\beta, \quad \frac{\partial u_2}{\partial t} - d_2 = u_1^\alpha u_2^\beta \quad \text{in } \Omega \times (0, T)$$

$$\frac{\partial}{\partial \nu}(u_1, u_2)\Big|_{\partial\Omega} = 0, \quad (u_1, u_2)|_{t=0} = (u_{10}(x), u_{20}(x)) > 0, \quad (4.223)$$

where α, β are positive integers. Given smooth initial value $(u_{10}(x), u_{20}(x))$, we have a unique classical solution $(u_1, u_2) = (u_1(x, t), u_2(x, t))$ local-in-time, satisfying

$$u = (u_i(\cdot, t)) > 0, \quad 0 < u_1(\cdot, t) \leq \|u_{10}\|_\infty \quad (4.224)$$

by the maximum principle. To estimate $u_2 = u_2(\cdot, t)$ we take $0 \leq \theta = \theta(x, t) \in C_0^\infty(Q_T)$, to define $\psi = \psi(x, t) \geq 0$ by

$$\frac{\partial \psi}{\partial t} + d_2 \Delta \psi = -\theta, \quad \psi|_{t=T} = 0, \quad \frac{\partial \psi}{\partial \nu}\Big|_{\partial\Omega} = 0. \quad (4.225)$$

Then it holds that

$$\frac{d}{dt} \int_\Omega u_2 \psi \, dx = \int_\Omega \frac{\partial u_2}{\partial t} \psi + u_1 \frac{\partial \psi}{\partial t} \, dx$$

$$= \int_\Omega (d_2 \Delta u_2 - \frac{\partial u_1}{\partial t} + d_1 \Delta u_1) \psi + u_2(-d_2 \Delta \psi - \theta) \, dx$$

$$= \int_\Omega -u_2 \theta + (-\frac{\partial u_1}{\partial t} + d_2 \Delta u_1) \psi \, dx$$

$$= -\frac{d}{dt} \int_\Omega u_1 \psi \, dx + \int_\Omega -u_2 \theta + u_1(\psi_t + d_1 \Delta \psi) \, dx.$$

Integrating in t, therefore, we obtain

$$\iint_{Q_T} u_2 \theta \, dxdt \leq (u_{10}, \psi(0)) + (u_{20}, \psi(0))$$

$$+ \left(\|\psi_t\|_{L^{p'}(Q_T)} + d_1 \|\Delta \psi\|_{L^{p'}(Q_T)} \right) \|u_1\|_{L^p(Q_T)}$$

for $1 < p < \infty$ and $\frac{1}{p'} + \frac{1}{p} = 1$. Here we use the maximal regularity to (4.225),

$$\|\psi_t\|_{L^{p'}(Q_T)} + \|\Delta \psi\|_{L^{p'}(Q_T)} \leq C_{T,p} \|\theta\|_{L^{p'}(Q_T)}.$$

See Section 8.5 of [Prüss (1993)]. Since

$$\psi(0) = -\int_0^T \psi_t \, dt$$

it holds that

$$\iint_{Q_T} u_2 \theta \ dx dt \le C_{T,p}(\|u_0\|_p + \|u_1\|_{L^p(Q_T)})\|\theta\|_{L^{p'}(Q_T)}$$

for $u_0 = (u_{i0})$, and therefore,

$$\|u_2\|_{L^p(Q_T)} \le C_{T,p}(\|u_0\|_p + \|u_1\|_{L^p(Q_T)}) \tag{4.226}$$

for any $1 < p < \infty$. We obtain

$$\|u_2\|_{L^p(Q_T)} \le C_{T,p}, \quad 1 < p < \infty \tag{4.227}$$

by (4.224) and (4.226). These estimates guarantee the global-in-time existence of the solution by the semi-group estimate. The same argument is valid to a three component system associated with the chemical reaction

$$\alpha U + \beta V \leftrightarrow \gamma W.$$

See [Laamri (2011)].

The next fact due to [Lepoutre, Pierre, and Rolland (2012)] is a refinement of Lemma 4.7.

Theorem 4.11. *Under the assumptions of Lemma 4.7, it holds that*

$$\|u\|_{L^2(Q(\eta,T))} \le C_7(\eta, T)\|u_0\|_1^{1/2}\|u\|_{L^1(Q_T)}^{1/2} \tag{4.228}$$

for any $0 < \eta < T$ where $Q(\eta, T) = \Omega \times (\eta, T)$.

Proof. It follows from (4.110) that

$$\tau \cdot u(\cdot, T) - \tau \cdot u(\cdot, t) \le \int_t^T \Delta(d \cdot u(\cdot, s)) \ ds, \quad 0 \le t \le T. \tag{4.229}$$

It holds that

$$V_t = -d \cdot u$$

for

$$V(\cdot, t) = \int_t^T d \cdot u(\cdot, s) \ ds, \tag{4.230}$$

and hence (4.229) implies

$$\Delta V \ge -\tau \cdot u(\cdot, t) \ge \tilde{\tau} V_t \text{ in } Q_T, \quad \left.\frac{\partial V}{\partial \nu}\right|_{\partial \Omega} \le 0 \quad \text{for } \tilde{\tau} = \max_j \tau_j d_j^{-1} \tag{4.231}$$

by $u = (u_j(\cdot, t)) \ge 0$. It follows also that

$$\|V(\cdot, 0)\|_1 \le \int_0^T \|d \cdot u(\cdot, s)\|_1 \ ds \le \tilde{d}\|u\|_{L^1(Q_T)} \quad \text{for } \tilde{d} = \max_j d_j$$

from (4.230). Therefore, the parabolic regularity to (4.231) implies

$$\sup_{\eta \le t \le T} \|V(\cdot, t)\|_\infty \le C_{45}(\eta, \tilde\tau) \|V(\cdot, 0)\|_1$$

$$\le C_{45}(\eta, \tilde\tau) \cdot \tilde d \cdot \|u\|_{L^1(Q_T)} \qquad (4.232)$$

by $u = (u_j(\cdot, t)) \ge 0$.

Taking $0 \le t_0 \le t \le T$, we apply (4.110) again, to obtain

$$\tau \cdot u(\cdot, t) \le \tau \cdot u(\cdot, t_0) + \int_{t_0}^t \Delta(d \cdot u)(\cdot, s)) \, ds.$$

Then it follows that

$$(\tau \cdot u(\cdot, t), d \cdot u(\cdot, t)) \le (\tau \cdot u(\cdot, t_0), d \cdot u(\cdot, t)) - \frac{1}{2}\frac{d}{dt}\|\nabla \int_{t_0}^t d \cdot u(\cdot, s) \, ds\|_2^2$$

where $(\ ,\ )$ denotes the L^2-inner product. Integrating the above inequality with respect to $t \in [0, T]$ leads to

$$\iint_{\Omega \times (t_0, T)} (\tau \cdot u)(d \cdot u) \, dxdt + \frac{1}{2}\|\nabla \int_{t_0}^T d \cdot u(\cdot, s) \, ds\|_2^2$$

$$\le \int_{t_0}^T (\tau \cdot u(\cdot, t_0), d \cdot u(\cdot, t)) \, dt = (\tau \cdot u(\cdot, t_0), \int_{t_0}^T d \cdot u(\cdot, t) \, dt)$$

$$\le \|\tau \cdot u(\cdot, t_0)\|_1 \cdot \|V(\cdot, t_0)\|_\infty. \qquad (4.233)$$

Inequality (4.228) is a direct consequence of (4.232)–(4.233) and (4.102).
$\square$

Inequality (4.181) can be shown alternatively by the relative compactness of $\{u(\cdot, t_k)\} \subset L^1(\Omega)$ and an inequality derived from (4.98), (4.228), and (4.169), that is,

$$\int_{-1}^1 \left| \frac{d}{dt} \int_\Omega u_j(\cdot, t + t_k)\varphi \, dx \right| \, dt \le C\|\varphi\|_{W^{2,\infty}}, \quad k \gg 1 \qquad (4.234)$$

valid to $\varphi \in C^2(\overline\Omega)$ with $\left.\frac{\partial\varphi}{\partial\nu}\right|_{\partial\Omega} = 0$. Inequality (4.234) is called the monotonicity formula by [Suzuki (2005); Suzuki (2015)]. See Section 1.2.

Lemma 4.17 is shown in the proof of Lemma 2.3 of [Lepoutre, Pierre, and Rolland (2012)]. In accordance with this lemma, we note the following fact.

Theorem 4.12. *It holds that*

$$\|v\|_{L^2(Q_T)} \le C_T\|v_0\|_2 \qquad (4.235)$$

for $v = v(x, t) \ge 0$ satisfying (4.115) with (4.114).

Proof. Given $\theta = \theta(x, t) \geq 0$, we take $w = w(x, t) \geq 0$ satisfying

$$w_t + a\Delta w = -\theta, \quad w|_{t=T} = 0, \quad \left.\frac{\partial w}{\partial \nu}\right|_{\partial\Omega} = 0. \qquad (4.236)$$

Multiplying Δw to (4.236), we have

$$-\frac{1}{2}\frac{d}{dt}\|\nabla w\|_2^2 + a_0\|\Delta w\|_2^2 \leq \|\theta\|_2\|\Delta w\|_2$$

$$\leq \frac{a_0}{2}\|\Delta w\|_2^2 + \frac{8}{a_0}\|\theta\|_2^2$$

and hence

$$\|\nabla w(0)\|_2^2 + \int_0^T \|\Delta w\|_2^2 \, dt \leq C\|\theta\|_{L^2(Q_T)}^2.$$

Multiplying $-w$ to (4.236), on the other hand, ensures

$$-\frac{1}{2}\frac{d}{dt}\|w\|_2^2 = (a\Delta w, w) + (w, \theta) \leq \|w\|_2(a_1\|\Delta w\|_2 + \|\theta\|_2)$$

$$\leq \frac{1}{2}\|w\|_2^2 + C(\|\Delta w\|_2^2 + \|\theta\|_2^2),$$

which implies

$$-\frac{d}{dt}(e^t\|w\|_2^2) \leq 2Ce^t(\|\Delta w\|_2^2 + \|\theta\|_2^2).$$

Thus we obtain

$$\|w(0)\|_2^2 \leq Ce^T \int_0^T \|\Delta w\|_2^2 + \|\theta\|_2^2 \, dt \leq C_T\|\theta\|_{L^2(Q_T)}^2$$

and hence

$$\|w(0)\|_{H^1}^2 + \int_0^T \|\Delta w\|_2^2 \, dt \leq C_T\|\theta\|_{L^2(Q_T)}^2. \qquad (4.237)$$

Since $\theta \geq 0$ we obtain

$$\frac{d}{dt}(v, w) = (v_t, w) + (v, w_t) \leq (\Delta(av), w) + (v, -\theta - a\Delta w)$$

$$= -(v, \theta) \qquad (4.238)$$

and then it holds that

$$\int_0^T (v, \theta) \, dt \leq (v_0, w(0) \leq \|v_0\|_2 \cdot \|w(0)\|_2 \leq C_T\|v_0\|_2 \cdot \|\theta\|_{L^2(Q_T)}$$

by (4.237). Since $v \geq 0$, we obtain

$$\|v\|_{L^2(Q_T)} \leq C_T\|v_0\|_2.$$

$\square$

The following fact is applicable to (4.116) from the proof.

Theorem 4.13. *It holds that*

$$\|v\|_{L^2(Q_T)} \leq C_T \left(\|v_0\|_2 + \min\{ \int_0^T \|f\|_{\frac{2n}{n+2}} \, dt, \, (\int_0^T \|f\|_{\frac{2n}{n+4}}^2 \, dt\})^{1/2} \right)$$
(4.239)

if (4.115) is replaced by

$$\frac{\partial v}{\partial t} - \Delta(av) \leq f, \qquad \left. \frac{\partial}{\partial \nu}(av) \right|_{\partial\Omega} \leq 0$$

in the previous lemma.

Proof. Given $\theta = \theta(x,t) \geq 0$, we take $w = w(x,t)$ satisfying (4.236), to obtain (4.237). Then (4.238) is replaced by

$$\frac{d}{dt}(v, w) \leq -(v, \theta) + (f, w)$$

and then it holds that

$$\int_0^T (z, \theta) \, dt \leq C_T \|v_0\|_2 \|\theta\|_{L^2(Q_T)} + \int_0^T (f, w) \, dt.$$

The second term is estimated as

$$\int_0^T (f, w) \, dt \leq \sup_{0 < t \leq T} \|(-\Delta + 1)^{1/2} w(t)\|_2 \cdot \int_0^T \|(-\Delta + 1)^{1/2} f\|_2 \, dt$$

$$\leq C_T \|\theta\|_{L^2(Q_T)} \int_0^T \|f\|_{\frac{2n}{n+2}} \, dt$$

and also

$$\int_0^T (f, w) \, dt \leq \left(\int_0^T \|(-\Delta + 1)^{-1} f\|_2^2 \, dt \right)^{1/2} \cdot \left(\int_0^T \|(-\Delta + 1) w\|_2^2 dt \right)^{1/2}$$

$$\leq C_T \|\theta\|_{L^2(Q_T)} \left(\int_0^T \|f\|_{\frac{2n}{n+4}} \, dt \right)^{1/2}.$$

Then we obtain (4.239). □

As an application, if

$$\sum_i (\log u_i) f_i(u) \leq C \left(1 + |u|^{\frac{n+2}{2}} \right), \qquad u_0 \log |u_0| \in L^2(\Omega)$$

then it holds that

$$\|u \log |u|\|_{L^2(Q_T)} \leq C_T \left(\|u_0 \log |u_0|\|_2 + 1 \right).$$

To see this property, we use $z_i = \Phi(u_i)$ for $\Phi(s) = s(\log s - 1) + 1$ which satisfies

$$\frac{\partial z_i}{\partial t} - d_i \Delta z_i = (\log u_i) f_i(u) - 4 d_i |\nabla u_i^{1/2}|^2.$$

Then we obtain

$$\frac{\partial}{\partial t} \tau \cdot z - \Delta d \cdot z \le C \left(1 + |u|^{\frac{n+2}{n}} \right) \equiv f, \quad \left. \frac{\partial}{\partial \nu} (\tau \cdot z) \right|_{\partial \Omega} \le 0$$

with

$$\int_0^T \|f\|_{\frac{2n}{n+2}} \, dt \le C_T \int_0^T \left(1 + \|u\|_2^2 \right) \, dt \le C_T$$

where $\tau = {}^t(1,1,\cdots,1)$. Then the result follows from Theorem 4.13.

The scheme in Section 4.6 to construct the global-in-time weak solution to (4.94) is due to [Pierre and Rolland (2016)]. See the proof of Theorem 1 of [Pierre and Rolland (2016)] for (4.129)–(4.130). By Lemma 2 of [Pierre and Rolland (2016)] the family $\{u_k\}$ in (4.133) is relatively compact in $L^p(Q_0)$ for $1 \le p < 2$. Therefore, we could replace the convergence in (4.140) by a convergence in $L^p(\Omega)$ for all $p < 2$, but it is not clear how to obtain the conclusion of Lemma 4.18 directly with this better convergence. See Theorem 4 in p.21 of [Evans and Gariepy (1992)] for the proof of (4.165).

Inequality (4.171) coincides with (3.18) in [Suzuki and Yamada (2015)] for $\varphi \equiv 1$. Inequality (4.173) is nothing but (22) of [Biler, Hilhorst, and Nadzieja (1994)]. Its local version is presented as in Lemma 11.1 of [Suzuki (2005)]. Iteration scheme used for inequality (4.170) is the same as in pp. 10–11 of [Suzuki and Yamada (2015)]. More precisely, assuming

$$\sup_{t \in [0,T)} \|u(\cdot,t)\|_q \le C_1(T), \quad q \ge 2,$$

we obtain

$$\sup_{t \in [0,T)} \|\tilde{u}_j(\cdot,t)\|_r \le C_2(T), \quad q \le r \le \infty$$

satisfying $\frac{2}{q} - \frac{1}{r} < 1$, by $n = 2$. Repeating this argument twice, we reach (4.170). We use Lemma 5.2 of [Suzuki and Yamada (2015)] applied to $u_k(\cdot,t) = u(\cdot, t + t_k)$ to derive (4.182) from (4.181). Inequality (4.183) makes the argument simpler. Hence we apply the argument in pp. 14–15 of [Suzuki and Yamada (2015)] in the later. Inequality (4.184) is nothing but inequality (3.19) in [Suzuki and Yamada (2015)], or Lemma 11.1 of [Suzuki (2005)]. The proof of (4.185) is automatic as in (3.8) of [Suzuki and

Yamada (2015)]. The analogous inequality to (4.185), with u_j replaced by $u_j^{3/2}$, is given in (3.12) of [Suzuki and Yamada (2015)].

Spatially asymptotic homogenization is observed for (4.94) with (4.213)–(4.214) under the presence of entropy [Masuda and Takahashi (1994); Suzuki and Yamada (2015)]. Theorem 4.9 is applicable to the Lotka-Volterra system. Thus we have a wide class of (4.99) with (4.216) provided with $(N - 2)$ entropies, where any non-stationary spatially homogeneous solutions are periodic-in-time [Kobayashi, Suzuki, and Yamada (2017)]. For such a system, the ω-limit set $\omega(u_0)$ forms a spatially homogeneous periodic solution or a unique spatially homogeneous stationary state. In particular, the ω-limit set $\omega(u_0)$ in Theorem 4.9 is not always contained in the set of stationary solutions. In [Suzuki and Yamada (2015)] the argument using ω-limit set is executed in the framework of the classical solution. See Theorem 4.10 for the construction of the solution to (4.202) in (4.201). For the comparison principle to guarantee (4.203), we refer to Lemma 3.4 of [Baras and Pierre (1984)].

Chapter 5

Network

This chapter is devoted to the mathematical study of reaction network. First, we introduce the notion of chemical reaction network. Then we study the asymptotic behavior of the solution to reaction-diffusion systems modeling multi-components reversible chemistry with diffusion. This is the weak solution realized as a limit of adequate approximate solutions. It is proved in any space dimension that, as time tends to infinity, the solution converges exponentially to the unique homogeneous stationary solution. Having viewed this result, we turn to several striking phenomena emerging from the actual reaction network observed in the study of mathematical oncology, that is, grouping of reactions, foliation composed of nested periodic orbits, break down of dynamical equilibrium, and transient limit cycle.

5.1 Chemical Reaction Network

Chemical reaction network is composed of three factors, substances $S = \{S_i\}_{i=1}^N$, complexes $C = \{y_r\}_{r=1}^\ell \subset \mathbf{N}_*^N$, and reactions $R = \{y_r \to y_r' \ (k_r) \mid r = 1, \cdots, m\}$, where $\mathbf{N}_* = \{0, 1, 2, \cdots\}$. Hence each $y_r \in C$ takes the form $y_r = (y_{r1}, \cdots, y_{rN})$.

In the fundamental process $S_1 + S_2 \to S_3$ with the reaction rate k, for example, it holds that

$$S = \{S_1, S_2, S_3\}, \ C = \{(1,1,0), (0,0,1)\}, \ R = \{(1,1,0) \to (0,0,1) \ (k)\}. \tag{5.1}$$

In (5.1), the elements $(1,1,0)$ and $(0,0,1)$ stand for $S_1 + S_2$ and S_3, respectively. We obtain $N = 3$ and $C = \{y_r\}_{r=1}^2$, where

$$y_1 = (1,1,0) \in C, \quad y_{1,1} = 1, \ y_{1,2} = 1, \ y_{1,3} = 0$$
$$y_2 = (0,0,1) \in C, \quad y_{2,1} = 0, \ y_{2,2} = 0, \ y_{2,3} = 1.$$

We say that $\{S, C, R\}$ is a chemical reaction network if the following requirements hold:

(1) Each $S_i \in S$ takes $[y_r \to y'_r] \in R$ such that $y_{ri} > 0$ or $y'_{ri} > 0$.
(2) Any $y \in C$ does not admit $[y \to y] \in R$.
(3) Each $y \in C$ takes $y' \in C$ such that $[y \to y'] \in R$ or $[y' \to y] \in R$.

Henceforth, we put

$$c^y = \prod_{i=1}^{N} c_i^{y_i}$$

for $0 \le y = (y_i) \in \mathbf{N}_*^N$ and $0 \le c = (c_i) \in \mathbf{R}^N$. If $c_i = c_i(t)$ stands for the concentration of S_i, the mass action law for this chemical network is formulated by

$$\frac{dc}{dt} = R(c), \ C(0) = c_0 \ge 0, \quad 0 \le c = (c_i(t)) \in C^1([0, T], \mathbf{R}^N), \qquad (5.2)$$

where

$$R(c) = \prod_{r=1}^{m} k_r c^{y_r} (y'_r - y_r).$$

The Wegscheider matrix is defined by

$$W = {}^t\left[(y'_r - y_r)\right]_{r=1, \cdots, m} .$$

It is an $m \times N$ matrix, and if $s = \dim \operatorname{Ker} W > 0$, there is $Q \in \mathbf{R}^{s \times N}$ with rank $Q = s$ such that

$$QR(c) = 0, \quad 0 \le c \in \mathbf{R}^N.$$

Then, mass conservation of (5.2) is realized as

$$Qc(t) = Qc_0 \equiv M \in \mathbf{R}^s.$$

Given $0 \le c_\infty \in \mathbf{R}^N$ with $Qc_\infty = M$, we say the following.

(1) c_∞ is an equilibrium if $R(c_\infty) = 0$.
(2) c_∞ is a detailed balance equilibrium if any $[y \to y' \ (k_f)] \in R$ admits $[y' \to y \ (k_b)] \in R$ such that $k_f c_\infty^y = k_b c_\infty^{y'}]$.
(3) c_∞ is a complex balance equilibrium if any $y \in C$ admits

$$\sum_{r|y_r=y} k_r c_\infty^{y_r} = \sum_{\{r'|y'_{r'}=y\}} k_{r'} c_\infty^{y_{r'}},$$

where $R = \{y_r \to y'_r \ (k_r)\}_{r=1}^{m}$.

First, a detailed balance equilibrium is a complex balance equilibrium. Second, a complex balance equilibrium is an equilibrium. We say that the above $c_\infty > 0$ is positive. Generally, a non-positive equilibrium $0 \leq c_\infty \in \mathbf{R}^N$ is called a boundary equilibrium. The following fact, proven immediately, indicates the entropy dissipation toward a positive complex balance equilibrium. Here we note $\Psi(x,y) = 0$ if and only if $x = y$ for $x, y > 0$.

Theorem 5.1. *Let* $0 < c_\infty \in \mathbf{R}^N$ *be a positive complex balance equilibrium, and define the relative entropy by*

$$E(c \mid c_\infty) = \sum_{i=1}^{N} c_{i,\infty} \Phi\left(\frac{c_i}{c_{i\infty}}\right),$$

where $\Phi(s) = s(\log s - 1) + 1$. *Then it holds that*

$$\frac{d}{dt} E(c \mid c_\infty) = - \sum_{r=1}^{m} k_r c_\infty^{y_r} \Psi\left(\frac{c^{y_r}}{c_\infty^{y_r}}, \frac{c^{y_r'}}{c_\infty^{y_r'}}\right) \leq 0,$$

where $c = c(t)$ *is a solution to (5.2) and* $\Psi(x,y) = y\Phi(\frac{x}{y})$ *for* $x, y > 0$.

We also say that a chemical reaction network $\{S, C, R\}$ is complex balanced if any $0 < M \in \mathbf{R}^s$ admits a complex balance equilibrium $0 < c_\infty \in \mathbf{R}^N$ such that $Qc_\infty = M$. The following fact is known.

Theorem 5.2. *Let* $\{S, C, R\}$ *be a complex balanced network.*

(1) Any equilibrium $c_\infty \geq 0$ *is complex balanced.*
(2) Given $0 < M \in \mathbf{R}^s$, *we have a unique complex balance equilibrium* $0 < c_\infty \in \mathbf{R}^N$ *such that* $Qc_\infty = M$.

A boundary equilibrium may exist in the second case.

5.2 Multi-component Reaction

Here we treat the simplest network composed of one reversible chemical reaction with multi-components $\{A_i\}_{1 \leq i \leq N}$,

$$\alpha_1 A_1 + \cdots + \alpha_m A_m \rightleftharpoons \alpha_{m+1} A_{m+1} + \cdots + \alpha_N A_N, \qquad (5.3)$$

provided with the diffusion, where $m, N, \alpha_k, k = 1, ..., N$ are positive integers with $1 \leq m < N$.

Let $u_k = u_k(x,t)$ be the concentration of A_k at position $x \in \Omega \subset \mathbf{R}^n$ and time $t \in [0, T), T > 0$, where Ω is an open, bounded domain with smooth boundary $\partial\Omega$. According to the mass action law, with reaction rates c_1 from left to right and c_2 from right to left, and according to Fick's law for the diffusion, the evolution of $u = (u_1, ..., u_N)$ is described by the reaction-diffusion system

$$\tau_k \frac{\partial u_k}{\partial t} - d_k \Delta u_k = \chi_k f(u) \qquad \text{in } Q_T = \Omega \times (0, T)$$

$$\left.\frac{\partial u_k}{\partial \nu}\right|_{\partial\Omega} = 0, \quad u_k|_{t=0} = u_{k0}(x) \geq 0 \qquad (5.4)$$

for $1 \leq k \leq N$, where $d_k > 0$, $1 \leq k \leq N$, ν is the outer unit normal vector and

$$f(u) = c_1 \prod_{j=1}^{m} u_j^{\alpha_j} - c_2 \prod_{j=m+1}^{N} u_j^{\alpha_j} \qquad (5.5)$$

for

$$\chi_k = \begin{cases} -\alpha_k, & 1 \leq k \leq m \\ \alpha_k, & m+1 \leq k \leq N. \end{cases}$$

For this system we can define a renormalized solution to (5.4) global-in-time. Then we show that it converges exponentially in $L^1(\Omega)$ as $t \to +\infty$ to a unique homogeneous stationary solution.

The nonlinearity of this model is quadratic if $m = 2$, $N = 4$, and $\alpha_k = 1$, that is $f(u) = c_1 u_1 u_2 - c_2 u_3 u_4$. Global-in-time classical solutions exist for this f in space dimension $n = 1, 2$ as is described in §4.7 and §4.9.

Even if the weak solution is not obtained global-in-time, weaker solutions are defined in the spirit of the renormalized solutions by DiPerna-Lions for the Boltzmann equation. Here we do not need the definition of such solutions; only use the fact that they are obtained as limit of solutions of a standard approximate regularized system.

This approximate solution $u^\varepsilon = (u_k^\varepsilon(x, t))$ satisfies

$$\tau_k \frac{\partial u_k^\varepsilon}{\partial t} - d_k \Delta u_k^\varepsilon = \chi_k f_\varepsilon(u^\varepsilon) \quad \text{in } Q_T = \Omega \times (0, T)$$

$$\left.\frac{\partial u_k^\varepsilon}{\partial \nu}\right|_{\partial\Omega} = 0, \quad u_k^\varepsilon|_{t=0} = u_{k0}^\varepsilon(x) \geq 0 \qquad (5.6)$$

for $1 \leq k \leq N$, where $\tau_k \in (0, \infty)$ and

$$f_\varepsilon(u) = \frac{f(u)}{1 + \varepsilon|f(u)|}, \quad u_{k0}^\varepsilon = \inf\{u_{k0}, \varepsilon^{-1}\}, \quad u_{k0} \geq 0. \qquad (5.7)$$

Given $(u_{k0}) \in L^1(\Omega)^N$, there exists a unique classical solution to (5.6)–(5.7) global-in-time by $|f_\epsilon(u)| \leq 1/\epsilon$. Since the nonlinearity is quasipositive,

$$\chi_k f_\epsilon(u) \geq 0, \quad u \in [0, \infty)^N, \quad u_k = 0, \quad 1 \leq k \leq N,$$

this solution u^ϵ is nonnegative. Then, the following convergence result holds.

Theorem 5.3. *Assume $u_{k0} \log u_{k0} \in L^1(\Omega)$ for $1 \le k \le N$. Then each $\{u^{\epsilon_\ell}\}$ with $\epsilon_\ell \downarrow 0$ admits a subsequence converging in $L^1_{loc}([0, \infty); L^1(\Omega)^N)$ and a.e. to some $u \in L^\infty([0, \infty); L^1(\Omega))^N$ such that*

$$u_k \log u_k \in L^\infty_{loc}([0, \infty); L^1(\Omega))$$

for $1 \le k \le N$.

The conservation properties

$$\frac{1}{|\Omega|} \int_\Omega \tau_i u_i^\epsilon + \tau_j u_j^\epsilon \, dx = \frac{1}{|\Omega|} \int_\Omega \tau_i u_{i0}^\epsilon + \tau_j u_{j0}^\epsilon \, dx \qquad (5.8)$$

hold for $1 \le i \le m < j \le N$, thanks to the homogeneous Neumann boundary conditions and they are preserved at the limit for u. For $w : \Omega \to \mathbb{R}$, we will throughout denote

$$\overline{w} = \frac{1}{|\Omega|} \int_\Omega w \, dx.$$

Then we can show the following theorem.

Theorem 5.4. *Let u be as in Theorem 5.3. Assume, moreover,*

$$\overline{u}_{i0} + \overline{u}_{j0} > 0, \quad 1 \le i \le m < j \le N. \qquad (5.9)$$

Then, there exists $C, a > 0$ depending on $\|u_0\|_{L^1(\Omega)^N}$ such that

$$\|u(\cdot, t) - z\|_{L^1(\Omega)^N} \le Ce^{-at}, \quad t \ge 0 \qquad (5.10)$$

where $z = (z_j)_{1 \le j \le N} \in (0, \infty)^N$ is the unique nonnegative solution of

$$f(z) = 0, \ \tau_i z_i + \tau_j z_j = \tau_i \overline{u}_{i0} + \tau_j \overline{u}_{j0}, \quad 1 \le i \le m < j \le N. \qquad (5.11)$$

The same conclusion would actually hold for any limit u of adequate approximate solutions of system (5.4), and not only for the solutions of the specific system (5.6), (5.7).

The positivity condition (5.9) in Theorem 5.4 is not a restriction. Indeed, if one has

$$\frac{1}{|\Omega|} \int_\Omega u_{i0} + u_{j0} \, dx = 0$$

for some $1 \le i \le m < j \le N$, in other words if $u_{i0} \equiv 0 \equiv u_{j0}$, then by uniqueness, $u_i^\epsilon(t) \equiv 0 \equiv u_j^\epsilon(t)$, $f(u^\epsilon) \equiv 0$, and therefore, system (5.4) is reduced to the heat equation for each u_k. It is well known in this case that $u_k = u_k(\cdot t)$ converges exponentially as $t \uparrow +\infty$ to the average

$$\frac{1}{|\Omega|} \int_\Omega u_{k0} \, dx.$$

The proof of the asymptotic result of Theorem 5.4 is based on the use of the relative entropy. Let

$$E(w \mid v) = \frac{1}{|\Omega|} \int_\Omega v \, \Phi\left(\frac{w}{v}\right) \, dx, \quad \Phi(s) = s(\log s - 1) + 1 \geq 0, \, \forall s > 0,$$
(5.12)

where w, v are measurable nonnegative functions (with $v(x)^2 + w^2(x) > 0$ a.e. $x \in \Omega$). This entropy is extended to the vector valued functions $u = (u_k)_{1 \leq k \leq N}, z = (z_k)_{1 \leq k \leq N}$ as

$$\mathbf{E}(u \mid z) = \sum_{k=1}^N \tau_k E(u_k \mid z_k).$$
(5.13)

We will more simply write

$$E(w \mid 1) = E(w), \quad \mathbf{E}(u) = \sum_{k=1}^N \tau_k E(u_k), \quad \mathbf{E}(z) = \sum_{k=1}^N \tau_k E(z_k).$$
(5.14)

The main point is to show the following lemma.

Lemma 5.1. *With the notation and assumptions of Theorem 5.4*

$$\frac{d}{dt} \mathbf{E}(u(t) \mid z) \leq -2a \, \mathbf{E}(u(t) \mid z),$$
(5.15)

in the sense of distributions in $(0, \infty)$.

By Theorem 5.3, $\mathbf{E}(u(t) \mid z)$ is bounded for t near 0, say by C_0. Therefore, (5.15) implies

$$\mathbf{E}(u(t) \mid z) \leq C_0 \, e^{-2a \, t}, \quad t \geq 0.$$
(5.16)

We then apply a Cziszár-Kullback type inequality, namely (see (4.207)–(4.208))

$$\|u(t) - z\|_{L^1(\Omega)^N} \leq C \, \mathbf{E}(u(t) \mid z),$$

which implies (5.10).

Now we show the strategy to prove the main inequality (5.15). Assume for simplicity that, in the definition (5.5) of f and χ_k, we have

$$c_1 = c_2 = 1 = \alpha_k, \quad 1 \leq k \leq N.$$
(5.17)

Actually, we will see later that there is no loss of generality when considering this specific case. This is achieved by extending the system, see §5.9. Henceforth, we always assume that

$$c_1 = c_2 = 1, \quad \alpha_k = 1, \ 1 \le k \le N. \tag{5.18}$$

Then, if u is a solution of (5.4), we have, at least formally

$$\frac{d}{dt}E(u_k(t)) = \frac{1}{|\Omega|} \int_\Omega (\log u_k)\partial_t u_k \, dx$$

$$= \frac{1}{|\Omega|} \int_\Omega -d_k \frac{|\nabla u_k|^2}{u_k} + \chi_k(\log u_k)f(u_k) \, dx.$$

This implies that for $\mathbf{E}(u) = \sum_{k=1}^N E(u_k)$, since here $\tau_k = 1$ for all k,

$$\frac{d}{dt}\mathbf{E}(u(t)) = -D(u(t)), \tag{5.19}$$

where

$$D(u) = \frac{1}{|\Omega|} \int_\Omega 4 \sum_{k=1}^N d_k |\nabla \sqrt{u_k}|^2$$

$$+ \left(\log \prod_{k=1}^m u_k - \log \prod_{k=m+1}^N u_k \right) \left(\prod_{k=1}^m u_k - \prod_{k=m+1}^N u_k \right) \, dx. \tag{5.20}$$

Thanks to the definition of z, as proved in Lemma 5.4 below,

$$\mathbf{E}(u(t) \mid z) = \mathbf{E}(u(t)) - \mathbf{E}(z)$$

so that

$$\frac{d}{dt}\mathbf{E}(u(t) \mid z) = \frac{d}{dt}\mathbf{E}(u(t)). \tag{5.21}$$

Now, Lemma 5.1 will be a consequence of the following lemma.

Lemma 5.2. *Assume (5.17). With the notation and assumptions of Theorem 5.4 it holds that*

$$D(u(t)) \ge 2a \, \mathbf{E}(u(t)|z) \tag{5.22}$$

in the sense of distribution on $(0, \infty)$.

It is now clear that combining (5.19), (5.21) and (5.22) yields Lemma 5.1.

The derivation in (5.19) is indeed very formal since here u is only obtained as the limit of regular solutions but may not be regular itself. In fact, we will only prove the inequality

$$\frac{d}{dt}\mathbf{E}(u(t)) \le -D(u(t))$$

which, obviously, is sufficient to deduce inequality (5.15) in Lemma 5.1.

The proof of Lemma 5.2 is completely algebraic. It only uses from the solution $u(t)$ that satisfies the conservation properties

$$\bar{u}_i(t) + \bar{u}_j(t) = \bar{u}_{i0} + \bar{u}_{j0} \equiv U_{ij}, \quad 1 \leq i \leq m < j \leq N. \tag{5.23}$$

We conclude this section with the uniqueness of z as defined in Theorem 5.4.

Lemma 5.3. *Under the assumptions of Theorem 5.4, there exists a unique $z = (z_k) \in [0, \infty)^N$ such that*

$$f(z) = 0, \quad \tau_i z_i + \tau_j z_j = \tau_i \bar{u}_{i0} + \tau_j \bar{u}_{j0}, \ 1 \leq i \leq m < j \leq N. \tag{5.24}$$

Moreover, $z_k > 0$ for any $1 \leq k \leq N$.

Proof. Let $U_{ij} = \tau_i \bar{u}_{i0} + \tau_j \bar{u}_{j0}$. By (5.17), $U_{ij} > 0$ for $1 \leq i \leq m < j \leq N$. The relations (5.24) are equivalent to

$$z_j = [U_{1j} - \tau_1 z_1]/\tau_j \geq 0, \quad m+1 \leq j \leq N$$
$$z_i = [\tau_1 z_1 + U_{iN} - U_{1N}]/\tau_i \geq 0, \quad 2 \leq i \leq m$$
$$g(z_1) = 0, \tag{5.25}$$

where

$$g(z_1) = c_1 z_1 \prod_{i=2}^{m} \frac{[\tau_1 z_1 + U_{iN} - U_{1N}]^{\alpha_i}}{\tau_i^{\alpha_i}} - c_2 \prod_{j=m+1}^{N} \frac{[U_{1j} - \tau_1 z_1]^{\alpha_j}}{\tau_j^{\alpha_j}}.$$

Let us define

$$M_0 \equiv \min_{m+1 \leq j \leq N} U_{1j}/\tau_1, \quad m_0 \equiv \max_{2 \leq i \leq m} [U_{1N} - U_{iN}]^+/\tau_1.$$

We note that

$$U_{1N} - U_{iN} = U_{1j} - U_{ij} = \tau_1 \bar{u}_{10} - \tau_i \bar{u}_{i0}$$

is independent of $j = m+1, ..., N$. It follows that $m_0 < M_0$. The function $g : [m_0, M_0] \to \mathbf{R}$ is continuous, strictly increasing and satisfies $g(m_0) < 0$, $g(M_0) > 0$. Therefore there exists a unique $z_1 \in (m_0, M_0)$ such that $g(z_1) = 0$. For this z_1, the z_i, z_j defined by (5.25) are nonnegative and do satisfy the expected relations (5.24). They are all strictly positive: indeed, if one had $z_i = 0$ for some $1 \leq i \leq m$, then $f(z) = 0$ would imply that $z_j = 0$ also for some $m + 1 \leq j \leq N$ which is a contradiction with $\tau_i z_i + \tau_j z_j = U_{ij} > 0$. $\qquad \square$

5.3 Entropy Dissipation

We show that Lemma 5.2 implies Theorem 5.4. We begin with the following identity.

Lemma 5.4. *Under the assumptions of Lemma 5.2 it holds that*

$$\mathbf{E}(u(t)|z) = \mathbf{E}(u(t)) - \mathbf{E}(z), \quad t \geq 0. \tag{5.26}$$

Proof. The functions $E(\cdot \mid \cdot)$, $\mathbf{E}(\cdot \mid \cdot)$, $E(\cdot)$, and $\mathbf{E}(\cdot)$ are defined in (5.12), (5.13), and (5.14). The following property is valid for any $0 \leq w \in L^1(\Omega)$ and $w_* \in (0, \infty)$:

$$E(w \mid w_*) = E(w) - E(w_*) - (\overline{w} - w_*)\log w_*. \tag{5.27}$$

We apply this equality to $w = u_k(t)$, $w_* = z_k$ for $1 \leq k \leq N$ and sum over k. Then (5.26) is reduced to checking

$$\sum_{k=1}^{N} \tau_k(\overline{u}_k(t) - z_k)\log z_k = 0. \tag{5.28}$$

In fact, we have

$$\tau_i \overline{u}_i^\epsilon(t) + \tau_j \overline{u}_j^\epsilon(t) = \tau_i \overline{u}_{i0}^\epsilon + \tau_j \overline{u}_{j0}^\epsilon, \quad 1 \leq i \leq m < j \leq N$$

by (5.8) for $\epsilon > 0$. This equality is preserved at the limit and gives

$$\tau_i \overline{u}_i(t) + \tau_j \overline{u}_j(t) = \tau_i \overline{u}_{i0} + \tau_j \overline{u}_{j0}, \quad 1 \leq i \leq m < j \leq N. \tag{5.29}$$

Since $\tau_i z_i + \tau_j z_j = \tau_i \overline{u}_{i0} + \tau_j \overline{u}_{j0}$, this equality may be rewritten as

$$\tau_k(\overline{u}_k(t) - z_k) = \begin{cases} \tau_1(\overline{u}_1(t) - z_1), & 1 \leq k \leq m \\ -\tau_1(\overline{u}_1(t) - z_1), & m+1 \leq k \leq N. \end{cases} \tag{5.30}$$

Then we write (5.28) as

$$\sum_{k=1}^{N} \tau_k(\overline{u}_k(t) - z_k)\log z_k = \tau_1(\overline{u}_1(t) - z_1)\left\{ \sum_{k=1}^{m}\log z_k - \sum_{k=m+1}^{N}\log z_k \right\} = 0,$$

using $f(z) = 0$, because (5.18) holds so that

$$f(z) = \prod_{i=1}^{m} z_i - \prod_{j=m+1}^{N} z_j.$$

$\square$

We show a key lemma.

Lemma 5.5. *With the notation and assumptions of Theorem 5.4, together with (5.18), we have*

$$\frac{d}{dt}\mathbf{E}(u) \le -D(u) \tag{5.31}$$

in the sense of distributions on $(0,\infty)$.

Proof. For the classical solution $u^\varepsilon = (u_k^\varepsilon(\cdot,t))$ to approximate scheme (5.6)–(5.7), it holds that

$$\frac{d}{dt}\mathbf{E}(u^\varepsilon) + D_\epsilon(u^\varepsilon) = 0 \tag{5.32}$$

by (5.18), where

$$D_\epsilon(u) = 4\sum_{k=1}^{N} d_k \|\nabla\sqrt{u_k}\|_2^2$$
$$+ \frac{1}{|\Omega|} \int_\Omega \frac{f(u)}{1+\epsilon|f(u)|} \log \frac{\prod_{k=1}^{m} u_k}{\prod_{k=m+1}^{N} u_k} \ge 0. \tag{5.33}$$

Inequality (5.32) implies after integration in time

$$\mathbf{E}(u^\varepsilon(\cdot,t)) \le \mathbf{E}(u_0^\varepsilon), \quad \iint_{Q_T} |\nabla\sqrt{u_k^\varepsilon}|^2 \, dxdt \le C, \ 1 \le k \le N. \tag{5.34}$$

From the first inequality in (5.34), using Proposition 5.3 and Fatou's lemma, we deduce

$$\mathbf{E}(u(\cdot,t)) \le \mathbf{E}(u_0) \quad \text{a.e. } t. \tag{5.35}$$

Let us prove that, up to a subsequence,

$$\lim_{\ell\to\infty} \mathbf{E}(u^{\varepsilon_\ell}(\cdot,t)) = \mathbf{E}(u(\cdot,t)) \quad \text{a.e. } t \in (0,\infty). \tag{5.36}$$

In fact we have

$$\frac{\partial}{\partial t}(\tau_i u_i^\varepsilon + \tau_j u_j^\varepsilon) - \Delta(d_i u_i^\varepsilon + d_j u_j^\varepsilon) = 0 \quad \text{in } Q_T$$

$$\frac{\partial}{\partial \nu}(d_i u_i^\varepsilon + d_j u_j^\varepsilon)\Big|_{\partial\Omega} = 0, \quad u_k^\varepsilon|_{t=0} = u_{k0}^\varepsilon$$

for $1 \le i \le m < j \le N$ and $1 \le k \le N$. Then Theorem 4.7 in §4.5 implies

$$\|u^\varepsilon\|_{L^2(Q_{\tau,T})} \le C_{\tau,T} \tag{5.37}$$

for any $\tau \in (0,T)$ with $C_{\tau,T} > 0$ independent of ε, where $Q_{\tau,T} = \Omega \times (\tau,T)$. Since u^{ε_ℓ} tends to u a.e. (see Proposition 5.3), we classically deduce (5.36) from Egorov's theorem and the estimate (5.37).

Indeed, given $\alpha > 0$, there exists a compact set $K_\alpha \subset Q_{\tau,T}$ such that $u^{\varepsilon_\ell} \to u$ uniformly on K_α and $|Q_{\tau,T} \setminus K_\alpha| < \alpha$. With $\Phi(s) = s[\log s - 1] + 1$ as in (5.12), since for some $C \in (0, \infty)$

$$0 \le \Phi(s)^{3/2} \le C(s^2 + 1), \quad s > 0,$$

it holds by (5.37) that

$$\iint_{Q_{\tau,T} \setminus K_\alpha} |\Phi(u^\varepsilon) - \Phi(u)| \, dx dt \le |Q_{\tau,T} \setminus K_\alpha|^{1/3}$$

$$\cdot \left(\iint_{Q_{\tau,T} \setminus K_\alpha} |\Phi(u^\varepsilon) - \Phi(u)|^{3/2} dx \right)^{2/3} \le C\alpha.$$

Hence

$$\limsup_{\ell \to \infty} \iint_{Q_{\tau,T}} |\Phi(u^{\varepsilon_\ell}) - \Phi(u)| \, dx dt \le C\alpha.$$

Letting $\alpha \downarrow 0$, we obtain (recall the definition of $\mathbf{E}$ in (5.13), (5.14))

$$\lim_{\ell \to \infty} \int_\tau^T |\mathbf{E}(u^{\varepsilon_\ell}(\cdot, t)) - \mathbf{E}(u(\cdot, t))| \, dt = 0,$$

and therefore (5.36) passing to a subsequence.

Let $\phi \in C_0^\infty[0, T)^+$. It holds that

$$\phi(0)\mathbf{E}(u_0^\varepsilon) + \int_0^\infty \phi'(t)\mathbf{E}(u^\varepsilon(\cdot, t)) \, dt = \int_0^\infty \phi(t) D_\varepsilon(u^\varepsilon(\cdot, t)) \, dt \qquad (5.38)$$

by (5.32). As $\varepsilon = \varepsilon^\ell \downarrow 0$, the left-hand side of (5.38) converges to

$$\phi(0)\mathbf{E}(u_0) + \int_0^\infty \phi'(t)\mathbf{E}(u(\cdot, t)) \, dt.$$

Here, we used the dominated convergence theorem, recalling (5.36) with (5.35) and $(u_{k0} \log u_{k0}) \in L^1(\Omega)^N$.

To treat the right-hand side of (5.38), we recall the expression of $D_\varepsilon(u^\varepsilon)$ in (5.33). For its first term, we use (5.34) to deduce the weak convergence,

$$\nabla \sqrt{u_k^{\varepsilon_\ell}} \rightharpoonup \nabla \sqrt{u_k} \quad \text{in } L^2(Q_T)^N \text{ for } 1 \le k \le N,$$

passing to a subsequence. Fatou's lemma is applicable to the second term and it follows that

$$\liminf_{\ell \to \infty} \int_0^\infty \phi(t) D_{\varepsilon_\ell}(u^{\varepsilon_\ell}(\cdot, t)) \, dt \ge \int_0^\infty \phi(t) D(u(\cdot, t)) \, dt.$$

We thus end up with

$$\phi(0)\mathbf{E}(u_0) + \int_0^\infty \phi'(t)\mathbf{E}(u(\cdot, t)) \, dt \ge \int_0^\infty \phi(t) D(u(\cdot, t)) \, dt$$

which means (5.31) on $[0, \infty)$ in the sense of distributions, because $T > 0$ and $\phi \in C_0^\infty[0, \infty)^+$ are arbitrary. $\qquad \square$

The following lemma is an adaptation of the classical Cziszár-Kullback inequality.

Lemma 5.6. *With the notation and assumptions of Theorem 5.4, it holds that*

$$\|u(t) - z\|_{L^1(\Omega)^N} \le C \, \mathbf{E}(u(t) \mid z), \quad t \ge 0,$$

for some $C > 0$ depending on u_0, z.

Proof. For $\Phi(s) = s(\log s - 1) + s$ as defined by (5.12), we have

$$|s - 1|^2 \le C(M) \, \Phi(s), \quad s \in [0, M].$$

We deduce

$$|\overline{u}_k(t) - z_k|^2 = z_k^2 \left| \frac{\overline{u}_k(t)}{z_k} - 1 \right|^2 \le C \, z_k \Phi\left(\frac{\overline{u}_k(t)}{z_k} \right), \quad 1 \le k \le N,$$

where C depends only on $\|u_0\|_{L^1(\Omega)^N}$, $\|z\|$. It follows that, for some $C_1 > 0$

$$C_1 \|\overline{u}(t) - z\|_{L^1(\Omega)^N}^2 \le \sum_{k=1}^{N} \tau_k |\overline{u}_k(t) - z_k|^2 \le C \, \mathbf{E}(\overline{u}(t) \mid z). \tag{5.39}$$

Then the classical Cziszár-Kullback inequality says

$$\left[\frac{1}{|\Omega|} \int_{\Omega} |u_k(x, t) - \overline{u}_k(t)| \, dx \right]^2 \le 4 \overline{u}_k(t) E(u_k(t) \mid \overline{u}_k(t)). \tag{5.40}$$

This inequality implies, for some other constant C

$$\|u(\cdot, t) - \overline{u}(t)\|_{L^1(\Omega)^N}^2 \le C \, \mathbf{E}(u(t) \mid \overline{u}(t)). \tag{5.41}$$

Using the obvious relation

$$\mathbf{E}(u(t) \mid z) = \mathbf{E}(u(t) \mid \overline{u}(t)) + \mathbf{E}(\overline{u}(t) \mid z)$$

together with (5.39) and (5.41), we obtain with another constant C that

$$\|u(t) - z\|_{L^1(\Omega)^N}^2 \le C \, \mathbf{E}(u(t) \mid z),$$

which is the estimate of Lemma 5.6. $\qquad\square$

Proof of Theorem 5.4. We continue to assume (5.18). By Lemmas 5.5, 4.21, and 5.2, it holds that

$$\frac{d}{dt} \mathbf{E}(u \mid z) \le -2a \, \mathbf{E}(u \mid z)$$

in the sense of distributions on $(0, \infty)$. This is the statement of Lemma 5.1 and it implies

$$\mathbf{E}(u(\cdot, t) \mid z) \le C e^{-2at}, \quad t \ge 0. \tag{5.42}$$

Together with Lemma 5.6, this inequality implies Theorem 5.4. $\qquad\square$

5.4 Logarithmic Sobolev Inequality

Here we show Lemma 5.6. We denote by u_k, u any of the functions $u_k(t), u(t)$ without indicating the t dependence which is actually not used in this section. Only the conservation laws, see (5.29),

$$\tau_i \overline{u}_i^k + \tau_j \overline{u}_j^k = U_{ij} \equiv \tau_i \overline{u}_{i0} + \tau_j \overline{u}_{j0}, \quad 1 \le i \le m < j \le N,$$

will be used together with the simplified assumption (5.18) and the following properties

$$0 < U_0 \equiv \min_{i,j} U_{ij}, \quad \max_{i,j} U_{i;j} \le \left(\sum_{k=1}^{N} \tau_k \right) \max_k u_k. \qquad (5.43)$$

All constants below denoted by C in short will depend only on U_0, $\|u_0\|_{L^1(\Omega)^N}$, τ_k for $1 \le k \le N$.

Lemma 5.7. *It holds that*

$$\mathbf{E}(\overline{u} \mid z) \le C \sum_{k=1}^{N} \left(\sqrt{\overline{u}_k} - \sqrt{z_k} \right)^2. \qquad (5.44)$$

Proof. It is easily seen that

$$B(s) = \Phi(s)/(\sqrt{s} - 1)^2$$

is continuous on $[0, \infty)$. Thus $B(\overline{u}_k/z_k)$ is bounded above by the constants in (5.43). It holds also that

$$\mathbf{E}(\overline{u} \mid z) = \sum_{k=1}^{N} \tau_k z_k \Phi \left(\frac{\overline{u}_k}{z_k} \right) = \sum_{k=1}^{N} \tau_k z_k \left(\frac{\sqrt{\overline{u}_k}}{\sqrt{z_k}} - 1 \right)^2 B \left(\frac{\overline{u}_k}{z_k} \right)$$

$$\le C \sum_{k=1}^{N} (\sqrt{\overline{u}_k} - \sqrt{z_k})^2,$$

and hence Lemma 5.7. $\qquad \square$

Lemma 5.8. *It holds that*

$$\sum_{k=1}^{N} (\sqrt{\overline{u}_k} - \sqrt{z_k})^2 \le C \left[f(\sqrt{\overline{u}}) \right]^2, \quad \sqrt{\overline{u}} = (\sqrt{\overline{u}_k})_{1 \le k \le N}. \qquad (5.45)$$

Proof. Recall that, under the assumption (5.18),

$$f(u) = \prod_{i=1}^{m} u_i - \prod_{j=m+1}^{N} u_j.$$

According to (5.30), we have

$$\bar{u} - z = \theta e, \quad \theta = \bar{u}_1 - z_1, \quad e = (e_k)_{1 \le k \le N}$$
$$e_i = \tau_1/\tau_i, \; e_j = -\tau_1/\tau_j, \quad 1 \le i \le m < j \le N. \tag{5.46}$$

Therefore

$$f(\bar{u}) = f(\bar{u}) - f(z) = \left[\int_0^1 \nabla f((1-s)z + s\bar{u}) \, ds \right] \cdot (\bar{u} - z)$$
$$= L(\bar{u})(\bar{u}_1 - z_1), \tag{5.47}$$

where

$$L(\zeta) = \int_0^1 \nabla f((1-s)z + s\zeta) \cdot e \, ds, \quad 0 \le \zeta \in \mathbf{R}^N. \tag{5.48}$$

We have

$$\bar{u} = z + (\bar{u}_1 - z_1) e, \quad \bar{u}_1 \in I = [0, \min_{m<j\le N} U_{1j}].$$

The mapping $\sigma \in I \mapsto L(z + (\sigma - z_1) e)$ is continuous and does not vanish. Indeed, if one had $L(\zeta) = 0$ for some $\zeta = z + (\sigma - z_1)e, \sigma \in I$, then, by the same computation as in (5.47) with $\bar{u}$ replaced by ζ, we would also have $f(\zeta) = 0$. But the uniqueness property of Proposition 5.3 would imply $\zeta = z$. This is impossible since then $L(z) = 0$ and by (5.48),

$$L(z) = \nabla f(z) \cdot e = \tau_1 \left[\sum_{i=1}^{m} (\tau_i z_i)^{-1} \prod_{k=1}^{m} z_k + \sum_{j=m+1}^{N} (\tau_j z_j)^{-1} \prod_{k=m+1}^{N} z_k \right],$$

a contradiction.

Thus, for

$$\delta = \min_{\sigma \in I} L(z + (\sigma - z_1)e) > 0,$$

it holds that $L(\bar{u}) \ge \delta$, which implies by (5.47) and (5.46)

$$f(\bar{u})^2 = (L(\bar{u}))^2 (\bar{u}_1 - z_1)^2 \ge \delta^2 \|\bar{u} - z\|^2 / \|e\|^2,$$

where $\|\cdot\|$ denotes here the euclidean norm in $\mathbf{R}^N$. We combine this inequality with the identities

$$(\bar{u}_k - z_k)^2 = (\sqrt{\bar{u}_k} - \sqrt{z_k})^2 (\sqrt{\bar{u}_k} + \sqrt{z_k})^2$$
$$\ge \left(\min_{1 \le k \le N} z_k \right) \cdot (\sqrt{\bar{u}_k} - \sqrt{z_k})^2, \quad 1 \le k \le N$$

and with

$$f(\overline{u})^2 = \left(\prod_{i=1}^{m} \overline{u}_i - \prod_{j=m+1}^{N} \overline{u}_j \right)^2 = f(\sqrt{\overline{u}})^2 \cdot \left(\prod_{i=1}^{m} \sqrt{\overline{u}_i} + \prod_{j=m+1}^{N} \sqrt{\overline{u}_j} \right)^2$$
$$\leq Cf(\sqrt{\overline{u}})^2$$

to deduce (5.45). $\qquad\square$

Lemma 5.9. *It holds that*

$$\left[f(\sqrt{\overline{u}}) \right]^2 \leq \frac{C}{|\Omega|} \int_{\Omega} f(\sqrt{u})^2 + \sum_{k=1}^{N} |\nabla \sqrt{u_k}|^2 \, dx \qquad (5.49)$$

for $\sqrt{u} = (\sqrt{u_k})_{1 \leq k \leq N}$.

Proof. All constant C in this proof may again differ from each other but will depend only on the value in (5.43). Define $\sigma = \sigma(x) \in \mathbf{R}^N$ for $x \in \Omega$ by $\sqrt{u} = \sqrt{\overline{u}} + \sigma$.

First, we have

$$f(\sqrt{u})^2 = f(\sqrt{\overline{u}} + \sigma)^2 = \left(f(\sqrt{\overline{u}}) + \nabla f(\sqrt{\overline{u}}) \cdot \sigma + M \right)^2,$$

where

$$M = \int_0^1 (1-s) D^2 f(\sqrt{\overline{u}} + s\sigma)[\sigma, \sigma] \, ds.$$

Using $(\nabla f(\sqrt{\overline{u}}) \cdot \sigma + M)^2 \geq 0$, this inequality implies

$$f(\sqrt{u})^2 \geq f(\sqrt{\overline{u}})^2 + 2f(\sqrt{\overline{u}})\nabla f(\sqrt{\overline{u}}) \cdot \sigma + 2f(\sqrt{\overline{u}})M.$$

By Young's inequality and the estimate $|\nabla f(\sqrt{\overline{u}}) \cdot \sigma| \leq C\|\sigma\|$, we have

$$2f(\sqrt{\overline{u}})\nabla f(\sqrt{\overline{u}}) \cdot \sigma \geq -\frac{1}{2} f(\sqrt{\overline{u}})^2 - 2(\nabla f(\sqrt{\overline{u}}) \cdot \sigma)^2$$
$$\geq -\frac{1}{2} f(\sqrt{\overline{u}})^2 - C\|\sigma\|^2.$$

It follows from the two previous inequalities and $|f(\sqrt{\overline{u}})| \leq C$ that

$$f(\sqrt{u})^2 \geq \frac{1}{2} f(\sqrt{\overline{u}})^2 - C(\|\sigma\|^2 + |M|). \qquad (5.50)$$

Next, since $\sqrt{u} \geq 0$ implies $\sigma \geq -\sqrt{\overline{u}}$ in $\mathbf{R}^N$, we have the partition $\Omega = \Omega_1 \cup \Omega_2$ where

$$\Omega_1 = \{ x \in \Omega \mid -\sqrt{\overline{u}_k} \leq \sigma_k(x) \leq 1, \quad 1 \leq k \leq N \}$$
$$\Omega_2 = \cup_{1 \leq k \leq N} \{ x \in \Omega \mid \sigma_k(x) > 1 \}.$$

For $x \in \Omega_1, s \in [0, 1]$, one has $0 \leq \sqrt{\overline{u}_k} + s\sigma_k \leq 1 + \sqrt{\overline{u}_k}$, so that

$$|M| \leq \int_0^1 (1 - s)\|D^2 f(\sqrt{\overline{u}} + s\sigma)\| \, ds \cdot \|\sigma\|^2 \leq C\|\sigma\|^2, \quad x \in \Omega_1.$$

Together with (5.50), we deduce

$$\int_{\Omega_1} f(\sqrt{\overline{u}})^2 \, dx \geq \int_{\Omega_1} \left[\frac{1}{2} f(\sqrt{\overline{u}})^2 - C\|\sigma\|^2 \right] dx. \tag{5.51}$$

We also have

$$\int_{\Omega_2} f(\sqrt{\overline{u}})^2 \, dx = |\Omega_2| f(\sqrt{\overline{u}})^2 \leq f(\sqrt{\overline{u}})^2 \sum_{k=1}^N |\{\sigma_k^2 > 1\}|$$

with

$$|\{\sigma_k^2 > 1\}| = \int_{\{\sigma_k^2 > 1\}} dx \leq \int_{\{\sigma_k^2 > 1\}} \sigma_k^2 \, dx \leq \int_\Omega \sigma_k^2 \, dx,$$

which implies

$$\int_{\Omega_2} f(\sqrt{\overline{u}})^2 \, dx \leq f(\sqrt{\overline{u}})^2 \int_\Omega \|\sigma\|^2 \, dx \leq C \int_\Omega \|\sigma\|^2 \, dx. \tag{5.52}$$

By (5.51)–(5.52), we obtain

$$f(\sqrt{\overline{u}})^2 = \frac{1}{|\Omega|} \int_\Omega f(\sqrt{\overline{u}})^2 \, dx \leq C \frac{1}{|\Omega|} \int_\Omega [f(\sqrt{u})^2 + \|\sigma\|^2] \, dx. \tag{5.53}$$

Then, using in particular the Schwarz inequality, $\sqrt{\overline{u}_k} \geq \frac{1}{|\Omega|} \int_\Omega \sqrt{u_k}$, we have

$$\frac{1}{|\Omega|} \int_\Omega \sigma_k^2 = \frac{1}{|\Omega|} \int_\Omega u_k \, fc - 2\sqrt{\overline{u}_k}\sqrt{u_k} + \overline{u}_k \, dx$$

$$\leq 2 \left\{ \frac{1}{|\Omega|} \int_\Omega u_k - \left(\frac{1}{|\Omega|} \int_\Omega \sqrt{u_k} \right)^2 \right\}$$

$$= 2\frac{1}{|\Omega|} \int_\Omega \left(\sqrt{u_k} - \frac{1}{|\Omega|} \int_\Omega \sqrt{u_k} \right)^2 \, dx.$$

Using now Poincaré-Wirtinger's inequality implies

$$\frac{1}{|\Omega|} \int_\Omega \sigma_k^2 \, dx = 2\frac{1}{|\Omega|} \int_\Omega \left(\sqrt{u_k} - \frac{1}{|\Omega|} \int_\Omega \sqrt{u_k} \right)^2 \, dx$$

$$\leq C\frac{1}{|\Omega|} \int_\Omega |\nabla \sqrt{u_k}|^2 dx$$

and hence (5.49) by plugging this inequality for $k = 1, \cdots, N$ into (5.53). $\qquad\square$

Proof of Lemma 5.2. Combining Lemmas 5.7, 5.8, and 5.9, we obtain

$$\mathbf{E}(\overline{u} \mid z) \leq C \frac{1}{|\Omega|} \int_\Omega f(\sqrt{u})^2 + \sum_{k=1}^N |\nabla \sqrt{u_k}|^2 \, dx. \qquad (5.54)$$

Here, the elementary inequality

$$\left(\sqrt{Y} - \sqrt{X}\right)^2 \leq (Y - X) \log \frac{Y}{X}, \quad X, Y \geq 0,$$

applied to $Y = \prod_{i=1}^m u_i$, $X = \prod_{j=m+1}^N u_j$, implies that

$$f(\sqrt{u})^2 \leq f(u) \left(\log \prod_{i=1}^m u_i - \log \prod_{j=m+1}^N u_j \right)$$

and hence

$$\frac{1}{|\Omega|} \int_\Omega f(\sqrt{u})^2 \, dx$$

$$\leq \frac{1}{|\Omega|} \int_\Omega \left(\log \prod_{i=1}^m u_i - \log \prod_{j=m+1}^N u_j \right) \left(\prod_{i=1}^m u_i - \prod_{j=m+1}^N u_j \right) \, dx.$$

From this inequality and (5.54), we obtain

$$\mathbf{E}(\overline{u} \mid z) \leq CD(u). \qquad (5.55)$$

Finally, we use the additivity property $\mathbf{E}(u \mid z) = \mathbf{E}(u \mid \overline{u}) + \mathbf{E}(\overline{u} \mid z)$ and the logarithmic Sobolev inequality

$$E(u_k \mid \overline{u}_k) \leq \frac{C}{|\Omega|} \int_\Omega |\nabla \sqrt{u_k}|^2, \quad 1 \leq k \leq N, \qquad (5.56)$$

to deduce the statement of Lemma 5.2. $\qquad \square$

5.5 Pathway Grouping

This topic is emerged from a mathematical modeling which arose from a biological scenario for activation of basal membrane degradation. MT1-MMP (Membrane type-1 matrix metalloproteinase) is a protease working as an invasion apparatus of cancer cells. It is observed at the invasion front of cancer cells in the form of so-called invadopodia. MT1-MMP first activates MMP2, a secreted basal membrane enzyme which degrades collagen IV in the basal membrane. After the basal membrane is degraded, MT1-MMP degrades collagen I, II, II and laminin 1 and 5 of the ECM.

The process of MMP2 activation is illustrated based on biological evidence. First, the extracellular pro-MMP2 is recruited. To attach MT1-MMP to the plasma membrane it binds another molecule called TIMP2. MT1-MMP itself forms dimers. Once the MT1-MMP, TIMP2, pro-MMP2, MT1-MMP is formed, the latter MT1-MMP cuts the connection pro-MMP2 and TIMP2. This process is called shedding activates MMP2, forming secretive type basement membrane degrading protease.

Henceforth we write a, b, and c for MMP2, TIMP2, and MT1-MMP, respectively. The above scenario indicates that the molecule $abcc$ is the origin to activate the secretive type basal membrane. Since this molecule is created only in the presence of b, it must be provided in sufficient amount at the beginning. If the amount is too much, however, another molecule $abccba$ will be produced mainly. Hence $b_0 - abcc_\infty$ curve will exhibit one peak, where b_0 and $abcc_\infty$ denote the initial and final concentrations of b and $abcc$, respectively.

This biological model is concerned with the attachment and detachment of three kinds of molecule, a, b, and c. Let us explain rule of their polymerization. First, a has a hand which can attach b. Next, b has a hand that can attach a. This b has also a hand that can attach c. Finally, c has a hand that can attach b and also has another hand that can attach c. Excluding the other possibilities, we have 9 compounds made by a, b, and c. We assume that these chemical reactions are subject to the law of mass action described in §4.1.

Thus we define the reaction rates k_1, k_2, k_3, ℓ_1, ℓ_2, and ℓ_3 by

$$a + b \to ab \ (k_1), \qquad ab \to a + b \ (\ell_1)$$
$$b + c \to bc \ (k_2), \qquad bc \to b + c \ (\ell_2)$$
$$c + c \to cc \ (k_3/2), \qquad cc \to c + c \ (\ell_3). \tag{5.57}$$

We have experimental data of these values, but we do not have any experimental data for the other reaction rates. Here we assume the rates in (5.57) for the reaction of modificated molecules, for example,

$$ab + c \to abc \ (k_2), \qquad abc \to ab + c \ (\ell_2).$$

Hence we take

$$\frac{dX_3}{dt} = -k_3 X_3 X_7 + \ell_3 X_9,$$

$$\frac{dX_7}{dt} = -k_3 X_3 X_7 + \ell_3 X_9,$$

$$\frac{dX_9}{dt} = k_3 X_3 X_7 - \ell_3 X_9$$

for $abc + c \to abcc$ (k_3), $abcc \to abc + c$ (ℓ_3), where $X_3 = [c]$, $X_7 = [abc]$, and $X_9 = [abcc]$. Similarly, we use

$$\frac{dX_3}{dt} = -k_3 X_3 X_5 + \ell_3 X_8,$$

$$\frac{dX_5}{dt} = -k_3 X_3 X_5 + \ell_3 X_8,$$

$$\frac{dX_8}{dt} = k_3 X_3 X_5 - \ell_3 X_8$$

for $bc + c \to bcc$ (k_3), $bcc \to bc + c$ (ℓ_3), where $X_5 = [bc]$, $X_8 = [bcc]$, and also

$$\frac{dX_5}{dt} = -k_3 X_5 X_7 + \ell_3 X_{11},$$

$$\frac{dX_7}{dt} = -k_3 X_5 X_7 + \ell_3 X_{11},$$

$$\frac{dX_{11}}{dt} = k_3 X_5 X_7 - \ell_3 X_{11}$$

for $abc + bc \to abccb$ (k_3), $abccb \to abc + bc$ (ℓ_3), where $X_{11} = [abccb]$, and so forth.

Based on the diagram of reaction network, we end up with

$$\frac{dX_1}{dt} = -k_1 X_1 X_2 - k_1 X_1 X_5 - k_1 X_1 X_8 - 2k_1 X_1 X_{10} - k_1 X_1 X_{11},$$

$$\frac{dX_2}{dt} = -k_1 X_1 X_2 - k_2 X_2 X_3 - 2k_2 X_2 X_6 - k_2 X_2 X_8 - k_2 X_2 X_9$$
$$+\ell_2 X_5 + \ell_2 X_8 + 2\ell_2 X_{10} + \ell_2 X_{11},$$

$$\frac{dX_3}{dt} = -k_2 X_2 X_3 - k_3 X_3 X_3 - k_2 X_3 X_4 - k_3 X_3 X_5 - k_3 X_3 X_7$$
$$+\ell_2 X_5 + 2\ell_3 X_6 + \ell_2 X_7 + \ell_3 X_8 + \ell_3 X_9,$$

$$\frac{dX_4}{dt} = k_1 X_1 X_2 - k_2 X_3 X_4 - 2k_2 X_4 X_6 - k_2 X_4 X_8 - k_2 X_4 X_9$$
$$+\ell_2 X_7 + \ell_2 X_9 + \ell_2 X_{11} + 2\ell_2 X_{12},$$

$$\frac{dX_5}{dt} = -k_1 X_1 X_5 + k_2 X_2 X_3 - k_3 X_3 X_5 - k_3 X_5 X_5 - k_3 X_5 X_7$$
$$-\ell_2 X_5 + \ell_3 X_8 + 2\ell_3 X_{10} + \ell_3 X_{11},$$

$$\frac{dX_6}{dt} = 2k_3 X_3 X_3 - 2k_2 X_2 X_6 - 2k_2 X_4 X_6 - \ell_3 X_6 + \ell_2 X_8 + \ell_2 X_9,$$

$$\frac{dX_7}{dt} = k_1 X_1 X_5 + k_2 X_3 X_4 - k_3 X_3 X_7 - k_3 X_5 X_7 - k_3 X_7 X_7$$
$$-\ell_2 X_7 + \ell_3 X_9 + \ell_3 X_{11} + 2\ell_3 X_{12},$$

$$\frac{dX_8}{dt} = -k_1 X_1 X_8 + k_2 X_2 X_6 - k_2 X_2 X_8 + k_3 X_3 X_5 - k_2 X_4 X_8$$
$$-\ell_2 X_8 + 2\ell_2 X_{10} + \ell_2 X_{11} - \ell_3 X_8,$$

$$\frac{dX_9}{dt} = k_1 X_1 X_8 - k_2 X_2 X_9 + k_3 X_3 X_7 + 2k_2 X_4 X_6 - k_2 X_4 X_9$$
$$-\ell_2 X_9 + \ell_2 X_{11} + 2\ell_2 X_{12} - \ell_3 X_9,$$

$$\frac{dX_{10}}{dt} = -2k_1 X_1 X_{10} + k_2 X_2 X_8 + 2k_3 X_5 X_5 - 2\ell_2 X_{10} - \ell_3 X_{10},$$

$$\frac{dX_{11}}{dt} = -k_1 X_1 X_{11} + 2k_1 X_1 X_{10} + k_2 X_2 X_9 + k_2 X_4 X_8 + k_3 X_5 X_7$$
$$-2\ell_2 X_{11} - \ell_3 X_{11},$$

$$\frac{dX_{12}}{dt} = k_1 X_1 X_{11} + k_2 X_4 X_9 + 2k_3 X_7 X_7 - 2\ell_2 X_{12} - \ell_3 X_{12}, \quad (5.58)$$

where $X_1 = [a]$, $X_2 = [b]$, $X_3 = [c]$, $X_4 = [ab]$, $X_5 = [bc]$, $X_6 = [cc]$, $X_7 = [abc]$, $X_8 = [bcc]$, $X_9 = [abcc]$, $X_{10} = [bccb]$, $X_{11} = [abccb]$, and $X_{12} = [abccba]$. Finally, we use the experimental data of reaction rates and initial concentrations of a, b, c for numerical simulations.

Mass action laws adapted above guarantee the total mass conservations of a, b, and c, which is actually confirmed by the model (5.58). The next observation is that the reaction rates (5.57) are used for all other processes. According to this assumption, all the paths are classified in three categories: the attachment and detachment of a-b, b-c, and c-c. Since $\ell_1 = 0$, the first case is reduced to

$$a + bB \to abB \quad (k_1),$$

where bB stands for all the compounds of $b's$. This process is listed as

$$X_1 + X_2 \to X_4, \quad X_1 + X_5 \to X_7, \quad X_1 + X_8 \to X_9,$$
$$X_1 + X_{10} \to X_{11}, \quad X_1 + X_{11} \to X_{12},$$

or overall,

$$X_1 + (X_2 + X_5 + X_8 + X_{10} + X_{11})$$
$$\to X_4 + X_7 + X_9 + X_{11} + X_{12} \quad (k_1). \quad (5.59)$$

The reactions

$$bB + c\,C \to BbcC \quad (k_2), \quad BbcC \to bB + c\,C \quad (\ell_2),$$

next, are listed as

$$X_2 + X_3 \leftrightarrow X_5, \quad X_2 + X_6 \leftrightarrow X_8,$$
$$X_2 + X_8 \leftrightarrow X_{10}, \quad X_2 + X_9 \leftrightarrow X_{11},$$

and

$$X_4 + X_3 \leftrightarrow X_7, \quad X_4 + X_6 \leftrightarrow X_9,$$
$$X_4 + X_8 \leftrightarrow X_{11}, \quad X_4 + X_9 \leftrightarrow X_{12},$$

or

$$(X_2 + X_4) + (X_3 + X_6 + X_8 + X_9)$$
$$\to X_5 + X_7 + X_8 + X_9 + X_{10} + X_{11} + X_{12} \quad (k_2),$$
$$X_5 + X_7 + X_8 + X_9 + X_{10} + X_{11} + X_{12}$$
$$\to (X_2 + X_4) + (X_3 + X_6 + X_8 + X_9) \qquad (\ell_2). \qquad (5.60)$$

Finally, the reactions

$$cC_1 + cC_2 \to C_1 cc C_2 \quad (k_3), \quad C_1 cc C_2 \to C_1 c + C_2 c \quad (\ell_3),$$

finally, are listed as

$$X_3 + X_3 \leftrightarrow X_6, \quad X_3 + X_5 \leftrightarrow X_8, \quad X_3 + X_7 \leftrightarrow X_9,$$

$$X_5 + X_3 \leftrightarrow X_8, \quad X_5 + X_5 \leftrightarrow X_{10}, \quad X_5 + X_7 \leftrightarrow X_{11},$$

and

$$X_7 + X_3 \leftrightarrow X_9, \quad X_7 + X_5 \leftrightarrow X_{11}, \quad X_7 + X_7 \leftrightarrow X_{12}.$$

Using this grouping of chemical reaction laws, we can decompose the system into several modules, in which the solution is represented explicitly.

This integrability is already observed in the fundamental process

$$A + B \to P \ (k), \quad P \to A + B \ (\ell) \qquad (5.61)$$

regarded as a basic module of chemical reaction. By the mass action law, in fact, model (5.61) is realized as the system of ordinary differential equations

$$\frac{d[A]}{dt} = -k[A][B] + \ell[P]$$
$$\frac{d[B]}{dt} = -k[A][B] + \ell[P]$$
$$\frac{d[P]}{dt} = k[A][B] - \ell[P]. \qquad (5.62)$$

There arises the total mass conservation of A, B-particles,

$$\frac{d}{dt}([A] + [P]) = 0, \quad \frac{d}{dt}([B] + [P]) = 0,$$

and hence

$$[A] + [P] = \alpha, \quad [B] + [P] = \beta$$

where α, β are constants. Then (5.62) is reduced to the single equation

$$\frac{d[P]}{dt} = k(\alpha - [P])(\beta - [P]) - \ell[P]. \tag{5.63}$$

The solution to this (5.63) is explicitly given through the roots of the characteristic equation

$$k(\alpha - \lambda)(\beta - \lambda) - \ell\lambda = 0.$$

We describe this situation for a simpler model composed of the two kind of particles a, b, and two reactions

$$b + c \to bc \ (k_2), \quad bc \to b + c \ (\ell_2)$$
$$c + c \to cc \ (k_3/2), \quad cc \to c + c \ (\ell_3). \tag{5.64}$$

Then we obtain six complices except for b, c, that is, bc, cc, bcc, $bccb$ and six pathways

$$b + c \leftrightarrow bc, \quad c + c \leftrightarrow cc, \quad bc + c \leftrightarrow bcc$$
$$b + cc \leftrightarrow bcc, \quad bc + bc \leftrightarrow bccb, \quad b + bcc \leftrightarrow bccb.$$

First, the dimerization is listed by $c + c \leftrightarrow cc$ and $bc + bc \leftrightarrow bccb$. Second, there is a double attachment in $b + cc \to bcc$. Third, $bccb \to b + bcc$ is a double detachment. Thus the model is constructed as

$$\frac{dX_2}{dt} = -k_2 X_2 X_3 + \ell_2 X_5 - 2k_2 X_2 X_6 + \ell_2 2 X_8 - k_2 X_2 X_8 + 2\ell_2 X_{10}$$

$$\frac{dX_3}{dt} = -k_2 X_2 X_3 + \ell_2 X_5 - k_3 X_3^2 + 2\ell_3 X_6 - k_3 X_3 X_5 + \ell_3 X_8 \tag{5.65}$$

$$\frac{dX_5}{dt} = k_2 X_2 X_3 - \ell_2 X_5 - k_3 X_3 X_5 + \ell_3 X_8 - k_3 X_5^2 + 2\ell_3 X_{10} \tag{5.66}$$

$$\frac{dX_6}{dt} = \frac{k_3}{2} X_3^2 - \ell_3 X_6 - 2k_2 X_2 X_6 + \ell_2 X_8 \tag{5.67}$$

$$\frac{dX_8}{dt} = 2k_2 X_2 X_6 - \ell_2 X_8 + k_3 X_3 X_5 - \ell_3 X_8 - k_2 X_2 X_8 + 2\ell_2 X_{10} \tag{5.68}$$

$$\frac{dX_{10}}{dt} = k_2 X_2 X_8 - 2\ell_2 X_{10} + \frac{k_3}{2} X_5^2 - \ell_3 X_{10}. \tag{5.69}$$

System (5.69) is classified into two modules by (5.64). The first one is on $b + c \leftrightarrow bc$, that is,

$$X_2 + (X_3 + 2X_6 + X_8) \to X_5 + X_8 + 2X_{10} \ (k_2)$$
$$X_5 + X_8 + 2X_{10} \to X_2 + (X_3 + 2X_6 + X_8) \ (\ell_2). \tag{5.70}$$

In (5.70), the double attachment and detachment are realized as $2X_6$ and $2X_{10}$, respectively. This grouping (5.70) is actually justified as

$$\frac{dX_2}{dt} = -k_2 X_2 (X_3 + 2X_6 + X_8) + \ell_2 (X_5 + X_8 + 2X_{10})$$

$$\frac{d}{dt}(X_3 + 2X_6 + X_8) = -k_2 X_2 (X_3 + 2X_6 + X_8) + \ell_2 (X_5 + X_8 + 2X_{10})$$

$$\frac{d}{dt}(X_5 + X_8 + 2X_{10}) = k_2 X_2 (X_3 + 2X_6 + X_8)$$
$$-\ell_2 (X_5 + X_8 + 2X_{10}), \tag{5.71}$$

which is derived from the original model, (5.69). The second one is the dimerization $c + c \leftrightarrow cc$, that is,

$$(X_3 + X_5) + (X_3 + X_5) \to X_6 + X_8 + X_{10} \ (k_3)$$
$$X_6 + X_8 + X_{10} \to (X_3 + X_5) + (X_3 + X_5) \ (\ell_3). \tag{5.72}$$

In fact, we obtain

$$\frac{d}{dt}(X_3 + X_5) = -k_3 (X_3 + X_5)^2 + 2\ell_3 (X_6 + X_8 + X_{10})$$

$$\frac{d}{dt}(X_6 + X_8 + X_{10}) = \frac{k_3}{2}(X_3 + X_5)^2 - \ell_3 (X_6 + X_8 + X_{10}) \tag{5.73}$$

by (5.69).

By (5.71) and (5.73), the five quantities $y_1 = X_2$, $y_2 = X_3 + 2X_6 + X_8$, $y_3 = X_5 + X_8 + 2X_{10}$, $y_4 = X_3 + X_5$, and $y_5 = X_6 + X_8 + X_{10}$ are explicitly given. Then we use the first equation of (5.69),

$$\frac{dX_2}{dt} = -k_2 X_2 (X_3 + 2X_6 + X_8) + \ell_2 (X_5 + X_{10})$$

to deduce an explicit form of $y_6 = X_5 + X_1 0$. These values y_i, $1 \leq i \leq 6$ determine X_i, $1 \leq i \leq 6$.

5.6   Nested Periodic Orbits

Here we pick up a class of ODE models provided with conserved quantities, which makes all the solutions to be periodic-in-time except for the equilibrium. This class is actually realized in the system of quadratic nonlinearities, of which principal part is associated with skew-symmetric matrices.

We recall that Lotka-Volterra system is described by

$$\tau_j \frac{dv_j}{dt} = \left(-e_j + \sum_{k=1}^{N} a_{jk} v_k \right) v_j, \quad v_j|_{t=0} = v_{j0} > 0, \quad 1 \leq j \leq N \quad (5.74)$$

where $v_j = v_j(t) > 0$ stands for the population of the j-th species, $A = (a_{jk})$ is a square matrix of order N, and $\tau_j > 0$, $e_j \in \mathbf{R}$ are constants. Any solution $v = (v_j(t))$ to (5.74) remains in the positive cone $\mathbf{R}_+^N \equiv \{(v_j) \in \mathbf{R}^N \mid v_j > 0, 1 \leq \forall j \leq N\}$ as far as it exists, because each coordinate plane in $\mathbf{R}^N$ denoted by $v_j = 0$ for some $j = 1, \cdots, N$ is an invariant set of (5.74). One of major questions in population dynamics, then, is whether some species go to extinct or not. Here, extinction means approaching asymptotically zero. Finding criteria for the long term coexistence of multiple species, referred to as permanence or uniform persistence, is thus an important issue.

Predator-prey system is a typical model of permanence. Orbits near stable equilibrium exhibit also the permanence in general. Coexistence criteria for arbitrary initial data, however, have not yet been made sufficiently comprehensive, particularly for systems with huge numbers of components. Under this situation, one approach to the question of permanence may be finding periodic-in-time orbits. In §4.8 we see that some systems are provided with infinitely many periodic-in-time solutions, which guarantees the spatial homogenization in corresponding to partial differential equation provided with diffusion term, that is, the reaction diffusion system.

In this section we formulate a class of such systems, characterized by an algebraic condition on $A = (a_{jk})$ and $e = (e_j)$. More precisely, we say that the Lotka-Volterra system (5.74) is provided with *property* (P) if it satisfies the following conditions:

$(P1)$ The set of equilibria E is the intersection of L, an affine space of co-dimension 2, and the positive cone $\mathbf{R}_+^N$, that is, $E = L \cap \mathbf{R}_+^N$.

$(P2)$ Any non-stationary solution is periodic-in-time, with the orbit $\mathcal{O} \cong S^1$ contractible to an equilibrium, that is, a point belonging to E, in $\mathbf{R}_+^N \setminus E$.

$(P3)$ Any two distinct orbits $\mathcal{O}_1, \mathcal{O}_2 \cong S^1$ do not link in $\mathbf{R}_+^N$.

Henceforth, we assume the following conditions for the matrix $A = (a_{jk})$:

$(a1)$ A is irreducible.

$(a2)$ A is skew-symmetric, $\quad {}^t A + A = 0.$ $\qquad\qquad\qquad\qquad (5.75)$

$(a3)$ A has both positive and negative components in any row.

We recall that the square matrix A is called reducible if there is a permutation matrix P such that

$$^tPAP = \begin{pmatrix} A_{11} & A_{12} \\ 0 & A_{22} \end{pmatrix},$$

where A_{11} and A_{22} are non-trivial square matrices. Although this assumption is not essential in later arguments, since $A = (a_{jk})$ is skew-symmetric, system (5.74) does not have any non-trivial proper sub-systems under $(a3)$. Our results are divided into two cases, $e = 0$ and $e \neq 0$, where $e = (e_j)$.

Theorem 5.5. *Let* $A = (a_{jk})$ *satisfy (a1), (a2), and (a3). Let* $e = (e_j) = 0$, $N \geq 3$, *and*

$$a_{ij}a_{kl} + a_{il}a_{jk} - a_{ik}a_{jl} = 0, \quad \forall\, i,j,k,l \in \{1,\dots,N\}. \tag{5.76}$$

Then, system (5.74) is provided with property (P).

Condition (5.76) is void if $N = 3$. Therefore, when $N = 3$, $e_1 = e_2 = e_3 = 0$, and

$$A = \begin{pmatrix} 0 & c_3 & -c_2 \\ -c_3 & 0 & c_1 \\ c_2 & -c_1 & 0 \end{pmatrix}, \quad c_1, c_2, c_3 \text{ have the same sign}, \tag{5.77}$$

any non-stationary solution to (5.74) is periodic-in-time. This result is well-known. Here we confirm that the total set of orbits takes the topological feature described in $(P1)$, $(P2)$, and $(P3)$. Hence the set of equilibria E is a half line in $\mathbf{R}_+^3$, any non-stationary orbit $\mathcal{O}$ is homeomorphic to S^1, surrounding E in $\mathbf{R}_+^3$, and any distinct two non-stationary orbits $\mathcal{O}_1$ and $\mathcal{O}_2$ do not link in $\mathbf{R}_+^3$.

Let $N \geq 4$. We first note that the set of skew-symmetric matrices, denoted by X, is identified with $\mathbf{R}^{N(N-1)/2}$. Second, if i, j, k, l are not distinct, equation (5.76) for $A = (a_{ij}) \in X$ is obvious. Third, equation (5.76) is invariant under the change of order on i, j, k, l for $A = (a_{jk}) \in X$. Therefore, system (5.76) of equations for $N \geq 4$ is reduced to that of $_NC_4$ equations,

$$a_{ij}a_{kl} + a_{il}a_{jk} - a_{ik}a_{jl} = 0, \quad 1 \leq i < j < k < l \leq N. \tag{5.78}$$

The set of irreducible skew-symmetric matrices satisfying (5.78), denoted by Y, however, is identified with a dense set in $\mathbf{R}^{2N-3}$. To see this remarkable property, let $a_{12} \neq 0$ without loss of generality. Then the entries

$a_{34}, \ldots, a_{N-1 N}$ of $A = (a_{jk})$ are represented by $a_{12}, \ldots, a_{2N}$ from the first $(N-2)(N-3)/2$ equations of (5.78), that is,

$$a_{kl} = \frac{a_{1k}a_{2l} - a_{1l}a_{2k}}{a_{12}}, \quad 3 \leq k < l \leq N. \tag{5.79}$$

Then the rest of equations of (5.78) is satisfied for a_{kl} defined above. In other words, equations (5.78) for $i = 1$, $j = 2$, $3 \leq k < l \leq N$, assure all the other equations of (5.78), provided that $a_{12} \neq 0$. Since irreducibility is a generic property of matrices, the above Y is identified with a dense set in $\mathbf{R}^{2N-3}$. The proof of the above reduction is elementary.

To see this property, let $N \geq 4$ and $a_{12}, \ldots, a_{2N}$ with $a_{12} \neq 0$ be given. Putting $a_{34}, \ldots, a_{N-1 N}$ by (5.79), we shall show that (5.78) holds. Henceforth, $j << l$ and $j <<< l$ denote $l \geq j+2$ and $l \geq j+3$, respectively. It holds that

$$a_{kl} = \frac{a_{1k}a_{2l} - a_{1l}a_{2k}}{a_{12}}, \quad 4 \leq k < l \leq N$$

$$a_{jk} = \frac{a_{1j}a_{2k} - a_{1k}a_{2j}}{a_{12}}, \quad 3 \leq j < k \leq N-1$$

$$a_{jl} = \frac{a_{1j}a_{2l} - a_{1l}a_{2j}}{a_{12}}, \quad 3 \leq i << j \leq N.$$

First, we have

$$a_{1j}a_{kl} + a_{1l}a_{jk} - a_{1k}a_{jl} = \frac{1}{a_{12}}\{a_{1j}(a_{1k}a_{2l} - a_{1l}a_{2k}) + a_{1l}(a_{1j}a_{2k} - a_{1k}a_{2j})$$
$$-a_{1k}(a_{1j}a_{2l} - a_{1l}a_{2j})\} = 0$$

for $2 \leq j < k < l \leq N$, and hence (5.78), $i = 1$. Second, it holds that

$$a_{2j}a_{kl} + a_{2l}a_{jk} - a_{2k}a_{jl} = \frac{1}{a_{12}}\{a_{2j}(a_{1k}a_{2l} - a_{1l}a_{2k}) + a_{2l}(a_{1j}a_{2k} - a_{1k}a_{2j})$$
$$-a_{2k}(a_{1j}a_{2l} - a_{1l}a_{2j})\} = 0$$

for $3 \leq j < k < l \leq N$ which means (5.78), $i = 2$.

To treat the case $3 \leq i \leq N-3$, finally, we use

$$a_{ij} = \frac{a_{1i}a_{2j} - a_{1j}a_{2i}}{a_{12}}, \quad 3 \leq i < j \leq N-2$$

$$a_{il} = \frac{a_{1i}a_{2l} - a_{1l}a_{2i}}{a_{12}}, \quad 3 \leq i <<< l \leq N$$

$$a_{ik} = \frac{a_{1i}a_{2k} - a_{1k}a_{2ji}}{a_{12}}, \quad 3 \leq i << k \leq N-1$$

derived from (5.79). Then it holds that

$$a_{ij}a_{kl} + a_{il}a_{jk} - a_{ik}a_{jl} = \frac{1}{a_{12}{}^2}\{(a_{1i}a_{2j} - a_{1j}a_{2i})(a_{1k}a_{2l} - a_{1l}a_{2k})$$
$$+(a_{1i}a_{2l} - a_{1l}a_{2i})(a_{1j}a_{2k} - a_{1k}a_{2j})$$
$$-(a_{1i}a_{2k} - a_{1k}a_{2i})(a_{1j}a_{2l} - a_{1l}a_{2j})\} = 0$$

for $i < j < k < l \leq N$ which means (5.78), $3 \leq i \leq N - 3$. The proof is thus complete.

We note also that if $\text{sign}(a_{ij}) = (-1)^{i+j}$ is the case for $i = 1, 2$ and $i < j \leq N$, the matrix $A = (a_{jk}) \in Y$ satisfies the sign condition, $(a3)$. Thus, there is $(2N - 3)$ degree of freedom for system (5.74) with $e = (e_j) = 0$, provided with property (P).

Condition $(a3)$ is associated with the permanence of (5.74). Let $A = (a_{jk}) \in Y$ and $e = (e_j) = 0$ satisfy $(a1)$ and $(a2)$. Then, if property (P) does not arise to (5.74) any non-stationary solution $v = (v_j(t))$ is global-in-time but satisfies

$$\lim_{t \uparrow +\infty} v_j(t) = 0 \tag{5.80}$$

for some $1 \leq \exists j \leq N$. In particular, any $v_* \in E$ is dynamically unstable. A simple example of this case is $N = 3$ of (5.77), with the sign condition of c_1, c_2, c_3 violated.

We turn to the case $e = (e_j) \neq 0$. It is also classical that system (5.74) is provided with property (P) when $N = 2$, $e_1 \cdot e_2 < 0$, and $a_{12} \cdot a_{21} < 0$. Thus, in this case there is a unique equilibrium $v_* \in \mathbf{R}_+^2$, and the total set of orbits denoted by $\mathcal{F}$ is composed of curves homeomorphic to S^1 and the equilibrium v_*. This property has a generalization.

Theorem 5.6. *Let $N \geq 3$ and $A = (a_{jk})$ satisfy (a1), (a2), and (a3). Let $e = (e_j)$ have both positive and negative components in (5.74). Furthermore, assume*

$$a_{jk}e_i - a_{ik}e_j + a_{ij}e_k = 0, \quad \forall i, j, k \in \{1, \ldots, N\}. \tag{5.81}$$

Then, system (5.74) is provided with property (P).

Condition (5.81) concerning $A = (a_{jk}) \in X$ and $e = (e_j) \in \mathbf{R}^N$ has the following reduction similar to (5.76). First, this equation is obvious if i, j, k are not distinct. Second, this equation is invariant under the change of order on i, j, k. Hence it is reduced to the system of $_N C_3$ equations

$$a_{jk}e_i - a_{ik}e_j + a_{ij}e_k = 0, \quad 1 \leq i < j < k \leq N. \tag{5.82}$$

Finally, the degree of freedom of $\{a_{ij}, e_k\}$ satisfying (5.82) is $(2N - 1)$.

To see the last fact, we assume $a_{1N} \cdot e_1 \neq 0$ without loss of generality, and define $a_{23}, \ldots, a_{2N}$ by $a_{12}, \ldots, a_{1N}$ and $e_1, \ldots, e_N$, using the first $(N - 1)(N - 2)/2$ equations of (5.82):

$$a_{jk} = \frac{a_{1k}e_j - a_{1j}e_k}{e_1}, \quad 2 \leq j < k \leq N. \tag{5.83}$$

Then the other equations of (5.82) are satisfied automatically.

In fact, by (5.83) it holds that

$$a_{jk} = \frac{a_{1k}e_j - a_{1j}e_k}{e_1}, \quad 3 \le j < k \le N$$

$$a_{ik} = \frac{a_{1k}e_i - a_{1i}e_k}{e_1}, \quad 2 \le i << k \le N$$

$$a_{ij} = \frac{a_{1j}e_i - a_{1i}e_j}{e_1}, \quad 2 \le i < j \le N - 2$$

which implies

$$\begin{aligned}
&a_{jk}e_i - a_{ik}e_j + a_{ij}e_k \\
&= \frac{a_{1k}e_j - a_{1j}e_k}{e_1}e_i - \frac{a_{1k}e_i - a_{1i}e_k}{e_1}e_j + \frac{a_{1j}e_i - a_{1i}e_j}{e_1}e_k = 0
\end{aligned}$$

for $2 \le i < j < k \le N$. We thus obtain the result.

Let $\hat{X}$ be the set of systems (5.74) with (5.75). This time, since both $A = (a_{ij})$ and $e = (e_k)$ are regarded as parameters, the set $\hat{X}$ is identified with $\mathbf{R}^{N(N+1)/2}$. Let $\hat{Y}$ be the set of $\{a_{ij}, e_k\}$ satisfying $A = (a_{ij}) \in X$ and (5.81). By the above reduction, this $\hat{Y}$ is identified with $\mathbf{R}^{2N-1}$. To detect a class of $\{a_{ij}, e_k\} \in \hat{Y}$ with $A = (a_{ij})$ satisfying $(a3)$, let $\text{sign}(a_{1j}) = (-1)^j$, $2 \le j \le N$, and $\text{sign}(e_k) = (-1)^k$, $3 \le k \le N$ with $e_1, e_2 > 0$. Then it holds that $\text{sign}(a_{2j}) = (-1)^{j+1}$ for $3 \le j \le N$, and therefore, $A = (a_{ij})$ actually satisfies $(a3)$. Thus, there is $(2N - 1)$ degree of freedom of $\{a_{ij}, e_k\}$ such that (5.74) is provided with property (P). Furthermore, if property (P) does not arise to (5.74) for $\{a_{ij}, e_k\} \in \hat{Y}$, any non-stationary solution to (5.74) exhibits extinction, that is, (5.80).

First, we prove Theorem 5.5. Let $N \ge 3$ and $e = (e_j) = 0$ in (5.74):

$$\tau_j \frac{dv_j}{dt} = \sum_{k=1}^{N} a_{jk}v_k v_j, \quad v_j|_{t=0} = v_{j0} > 0, \quad 1 \le j \le N. \tag{5.84}$$

Equality (5.75) implies the total mass conservation

$$\frac{dM}{dt} = 0 \tag{5.85}$$

where $M = \tau \cdot v$. Second, by $v = (v_j) > 0$ and (5.85) we obtain $T = +\infty$ with

$$\sup_{t \ge 0} \max_{1 \le j \le N} v_j(t) < +\infty. \tag{5.86}$$

Here and henceforth, $v = (v_j) > 0$ indicates $v_j > 0$ for any $1 \le j \le N$. Let $\mathcal{O} = \{v(t)\}_{t \ge 0} \subset \mathbf{R}_+^N$ be the orbit.

Boltzmann's H-function or entropy, on the other hand, is defined under the assumption

$$\exists b \in \mathbf{R}^N \setminus \{0\} \quad \text{such that} \quad {}^t\widetilde{A}b = 0 \tag{5.87}$$

for $\widetilde{A} = (\tau_j^{-1} a_{jk})$, that is,

$$H = b \cdot \log v = \sum_{j=1}^N b_j \log v_j$$

where $b = (b_j)$ and $\log v = (\log v_j)$ for $v = (v_j) \in \mathbf{R}^N$. Thus we obtain

$$\frac{dH}{dt} = \sum_{j=1}^N b_j v_j^{-1} \frac{dv_j}{dt} = \sum_{j,k=1}^N b_j \tau_j^{-1} a_{jk} v_k = 0$$

by (5.87) for the solution $v = (v_j(t))$ to (5.74). This H actually induces an increasing quantity for the reaction diffusion system.

Summarizing, we obtain the following lemma.

Lemma 5.10. *Assume (5.75) and the existence of linearly independent r-vectors $b^i \in Ker\ {}^t\widetilde{A}$, $1 \leq i \leq r$, where $r \geq 1$ and $\widetilde{A} = (\tau_j^{-1} a_{ij})$. Put $\beta_0 = \tau \cdot v_0$ and $\beta_i = b^i \cdot \log v_0$, $1 \leq i \leq r$, where $v_0 = (v_{j0}) > 0$. Then it holds that $\mathcal{O} \subset \mathcal{O}_*$, where*

$$\mathcal{O}_* = \{v \in \mathbf{R}_+^N |\ \tau \cdot v = \beta_0,\ b^i \cdot \log v = \beta_i,\ 1 \leq i \leq r\}. \tag{5.88}$$

Now we show the following lemma.

Lemma 5.11. *If $r = N - 2$ and $b^i > 0$ for some $i \in \{1, \cdots, r\}$ in Lemma 5.10, the above $\mathcal{O}_*$ is a Jordan curve or singleton.*

Proof. Under the transformation $\xi = \log v \in \mathbf{R}^N$, the set $\mathcal{O}_*$ is homeomorphic to

$$\widetilde{\mathcal{O}_*} = \{\xi \in \mathbf{R}^N \mid \tau \cdot e^\xi = \beta_0,\ b^i \cdot \xi = \beta_i,\ 1 \leq i \leq N - 2\}. \tag{5.89}$$

Let $M = \{\xi \in \mathbf{R}^N \mid \tau \cdot e^\xi \leq \beta_0\}$. This set is strictly convex, and $\widetilde{\mathcal{O}_*}$ is the intersection of its boundary $\partial M = \{\xi \in \mathbf{R}^N \mid \tau \cdot e^\xi = \beta_0\}$ and linearly independent $(N - 2)$-hyperplanes

$$P_i = \{\xi \in \mathbf{R}^N \mid b^i \cdot \xi = \beta_i\}, \quad 1 \leq i \leq N - 2.$$

Hence $\widetilde{\mathcal{O}_*} \subset \mathbf{R}^N$ is a Jordan curve or singleton.

Now we show the boundedness of $\widetilde{\mathcal{O}_*}$. First, there is $C_1 > 0$ such that

$$\sup_{1 \leq j \leq N} \xi_j \leq C_1, \quad \forall \xi = (\xi_j) \in M.$$

Second, since $b^i > 0$ for some $1 \leq i \leq N - 2$ we have $C_2 > 0$ such that

$$\inf_{1 \leq j \leq N} \xi_j \geq -C_2, \quad \forall \xi = (\xi_j) \in M \cap P_i.$$

Therefore, $\widetilde{\mathcal{O}_*} \subset \mathbf{R}^N$ is bounded. Then we see that $\mathcal{O}_*$ is a Jordan curve or singleton by the inverse transformation $v_j = e^{\xi_j}$, $1 \leq j \leq N$. $\qquad \square$

If $\widetilde{\mathcal{O}_*}$ is not bounded, the set $\mathcal{O}_*$ is an open curve in $\mathbf{R}_+^N$. Then any non-stationary solution $v = (v_j(t)) > 0$ satisfies (5.80) by (5.86), because any stationary solution does not lie on $\mathcal{O}_*$ in this case as is shown in Lemma 5.14 below.

To confirm the assumption of Lemma 5.11, we require the following elementary lemma.

Lemma 5.12. *If $N \geq 4$, one of the following cases arises to $A = (a_{jk})$ satisfying (a2) and (a3):*

Case 1. *There are (distinct) $i, j, k \in \{1, 2, \cdots, N\}$ such that $a_{ij} > 0$, $a_{ik} < 0$, $a_{jk} > 0$.*

Case 2. *There are distinct $i, j, k, l \in \{1, 2, \cdots, N\}$ such that $a_{ij} > 0$, $a_{ik} < 0$, $a_{il} = 0$, $a_{jk} = 0$, $a_{jl} > 0$, $a_{kl} < 0$.*

Proof. We have (distinct) $n_1, n_2, n_3 \in \{1, 2, \cdots, N\}$ such that $a_{n_1 n_2} > 0$, $a_{n_1 n_3} < 0$. Then we distinguish three cases of the sing of $a_{n_2 n_3}$, that is, (A) $a_{n_2 n_3} > 0$, (B) $a_{n_2 n_3} = 0$, (C) $a_{n_2 n_3} < 0$. If (A) arises, the first case of the lemma holds with $(i, j, k) = (n_1, n_2, n_3)$.

Suppose (B). From the assumption, there is n_4 such that $a_{n_2 n_4} > 0$. In addition, $a_{n_3 n_4} < 0$ holds by (5.76) for $(i, j, k, l) = (n_1, n_2, n_3, n_4)$. Therefore, if $a_{n_1 n_4} > 0$, the first case arises with $(i, j, k) = (n_1, n_4, n_3)$. If $a_{n_1 n_4} < 0$, we obtain the first case with $(i, j, k) = (n_1, n_2, n_4)$. If $a_{n_1 n_4} = 0$, the second case holds for $(i, j, k, l) = (n_1, n_2, n_3, n_4)$.

Suppose (C). From the assumption, there is n_4 such that $a_{n_2 n_4} > 0$. Then, we distinguish three cases, (a) $a_{n_3 n_4} < 0$, (b) $a_{n_3 n_4} = 0$, (c) $a_{n_3 n_4} > 0$. If (a) occurs, then we have the first case of the lemma for $(i, j, k) = (n_2, n_3, n_4)$. If (b) occurs, we have either the first or the second case of the lemma as in (B). If (c) occurs, we take n_5 such that $a_{n_3 n_5} < 0$. In addition, $a_{n_1 n_4} > 0$ holds by (5.76) for $(i, j, k, l) = (n_1, n_2, n_3, n_4)$.

Then we distinguish three cases according to the sign of $a_{n_4 n_5}$. If $a_{n_4 n_5} \geq 0$, we obtain the first or the second case of the lemma, similarly. If $a_{n_4 n_5} < 0$, there is n_6 such that $a_{n_4 n_6} > 0$. In addition,

$a_{n_1 n_5} < 0$ and $a_{n_2 n_5} < 0$ hold by (5.76) for $(i,j,k,l) = (n_1, n_3, n_4, n_5)$ and $(i,j,k,l) = (n_2, n_3, n_4, n_5)$, respectively.

We continue the argument. Let any one of the first or the second cases of the lemma do not arise from n_1 to n_{N-1}. From this fact and assumption (a3), if N is even it holds that $a_{n_{N-2} n_N} > 0 > a_{n_{N-1} n_N}$. Then the first case of the lemma arises to $(i,j,k) = (n_{N-2}, n_N, n_{N-1})$. If N is odd, on the contrary, we obtain $a_{n_{N-2} n_N} < 0 < a_{n_{N-1} n_N}$. Then the first case of the lemma holds for $(i,j,k) = (n_{N-2}, n_{N-1}, n_N)$. The proof is thus complete. $\square$

Now we show the following lemma.

Lemma 5.13. *The requirements of Lemma 5.11 are fulfilled under the assumption of Theorem 5.5.*

Proof. If $N = 3$ the assumption implies (5.77). Since $b = {}^t(b_1, b_2, b_3) \in$ Ker ${}^t\tilde{A}$, $b_1 b_2 b_3 \neq 0$, means

$$\frac{b_3}{b_2} = \frac{c_3}{c_2} \cdot \frac{\tau_3}{\tau_2}, \quad \frac{b_1}{b_3} = \frac{c_1}{c_3} \cdot \frac{\tau_1}{\tau_3}, \quad \frac{b_2}{b_1} = \frac{c_2}{c_1} \cdot \frac{\tau_2}{\tau_1},$$

we can take $0 < b = {}^t(b_1, b_2, b_3) \in$ Ker ${}^t\tilde{A}$.

Let $N \geq 4$. First, we examine the condition for the existence of linearly independent $(N-2)$-vectors, denoted by $b^i = (b_j^i) \in$ Ker ${}^t\tilde{A}$, $1 \leq i \leq N-2$. This condition is equivalent to the linear independence of $B^i = (B_j^i)$, $1 \leq i \leq N-2$, for $B_j^i = \tau_j^{-1} b_j^i$, which means rank $A = 2$.

To confirm this property under the assumption of Theorem 5.5, let $A = [\vec{a_1} \cdots \vec{a_N}]$. Since A is irreducible, we have $\vec{a_k} \neq 0$ for $1 \leq k \leq N$. Also, since A is skew-symmetric, the vectors $\{\vec{a_k}, \vec{a_l}\}$, $k \neq l$, are linearly independent if $a_{kl} \neq 0$. For such k, l, which actually exists, condition (5.76) implies

$$\vec{a_i} - \frac{a_{il}}{a_{kl}}\vec{a_k} - \frac{a_{ik}}{a_{lk}}\vec{a_l} = 0, \quad \forall i \neq k, l, \tag{5.90}$$

and hence rank $A = 2$. Thus we have actually linearly independent $(N-2)$-vectors in Ker ${}^t\tilde{A}$.

Now we shall show the existence of $0 < b \in$ Ker ${}^t\tilde{A}$, or equivalently, that of $0 < B \in$ Ker ${}^tA =$ Ker A. For this purpose, first, assume the first case of Lemma 5.12, that is, the existence of $i,j,k \in \{1,2,\cdots,N\}$ such that $a_{ij} > 0$, $a_{ik} < 0$, and $a_{jk} > 0$. The condition $B = {}^t(B_1, \cdots, B_N) \in$ Ker ${}^tA$ is equivalent to

$$B_k = -\frac{1}{a_{jk}}\sum_{n \neq j,k} a_{jn}B_n, \quad B_j = \frac{1}{a_{jk}}\sum_{n \neq j,k} a_{kn}B_n.$$

Consequently, we obtain Ker $^tA = \langle \vec{c_n} \mid n \neq j, k \rangle$ for $\vec{c_n} = {}^t(c_n^1, \cdots, c_n^N) \in \mathbf{R}^N$ defined by

$$c_n^m = \begin{cases} a_{jk}, & m = n \\ a_{kn}, & m = j \\ -a_{jn}, & m = k \\ 0, & \text{otherwise.} \end{cases}$$

Hence $\vec{c_i}$ is a vector whose i, j, k components are positive and the others are zero. Recalling $c_n^n = a_{jk} > 0$, we obtain

$$0 < \tilde{B} \equiv \sum_{n \neq i,j,k} \vec{c_n} + s\vec{c_i} \in \text{Ker} \, {}^tA, \quad s \gg 1.$$

We turn to the second case of Lemma 5.12. For such a tuple (i, j, k, l) of four (distinct) elements in $\{1, \cdots, N\}$, the condition $B = {}^t(B_1, \cdots, B_N) \in$ Ker ^tA means

$$B_j = -\frac{1}{a_{ij}} \sum_{n \neq i,j} a_{in} B_n, \quad B_i = \frac{1}{a_{ij}} \sum_{n \neq i,j} a_{jn} B_n.$$

Consequently, it holds that Ker $^tA = \langle \vec{c_n} \mid n \neq i, j \rangle$ for $\vec{c_n} = {}^t(c_n^1, \cdots, c_n^N) \in \mathbf{R}^N$ defined by

$$c_n^m = \begin{cases} a_{ij}, & m = n \\ a_{jn}, & m = i \\ -a_{in}, & m = j \\ 0, & \text{otherwise.} \end{cases}$$

We observe that $\vec{c_k}$ is a vector whose j, k components are positive and the others are zero. Also, $\vec{c_l}$ is a vector whose i, l components are positive and the others are zero. Recalling $c_n^n = a_{ij} > 0$, we obtain

$$0 < b \equiv \sum_{n \neq i,j,k,l} \vec{c_n} + s(\vec{c_k} + \vec{c_l}) \in \text{Ker} \, {}^tA, \quad s \gg 1.$$

The proof is complete.   □

We see that in (5.90), the pair $(c_1, c_2) = (a_{il}/a_{kl}, a_{ik}/a_{lk})$ is the unique choice for

$$\vec{a_i} - c_1 \vec{a_k} - c_2 \vec{a_l} = 0$$

to hold, regarding the l-th and the k-th components on the left-hand side. Consequently, the condition

$$i, j, k, l \in \{1, \ldots, N\}, \ a_{kl} \neq 0$$
$$\Rightarrow \ a_{ij}a_{kl} + a_{il}a_{jk} - a_{ik}a_{jl} = 0, \tag{5.91}$$

slightly weaker than (5.76), is sufficient for rank $A = 2$ to hold, if $A = [\vec{a_1} \cdots \vec{a_N}]$ is irreducible and skew-symmetric. Condition (5.91) is also necessary for rank $A = 2$, if, furthermore, $\{\vec{a_k}, \vec{a_l}\}$ are linearly dependent whenever $a_{kl} = 0$. By this reason it is natural that the solution set to (5.76) has $(2N - 3)$ degree of freedom, because two vectors $\vec{a_k}, \vec{a_l}$, $k \neq l$, can be free in $A = [\vec{a_1} \cdots \vec{a_N}]$ if rank $A = 2$.

The following lemma excludes the clinic orbit of (5.74), $e = (e_j) = 0$, when $A = (a_{ij})$ satisfies (a1), (a2), and (5.76).

Lemma 5.14. *Let $A \in X$ satisfy rank $A = 2$. Let $b^i \in Ker\ {}^t\tilde{A}$, $1 \leq i \leq N - 2$, be linearly independent $(N - 2)$-vectors. Then the property $\mathcal{O}_* \cap Ker\ A \neq \emptyset$ arises only if $\sharp\mathcal{O}_* = 1$, where $\mathcal{O}_*$ is the set defined by (5.88), $r = N - 2$.*

Proof. We use the sets $\widetilde{\mathcal{O}}_*$, M, and P_i defined in the proof of Lemma 5.11. Let $\mathcal{M} = \partial M$ and $\tau e^\xi = (\tau_j e^{\xi_j})$ for $\tau = (\tau_j)$ and $\xi = (\xi_j)$. First, since $\mathcal{M} = \{\xi \in \mathbf{R}^N \mid \tau \cdot e^\xi = \beta_0\}$ it holds that

$$T_\xi \mathcal{M} = \langle \tau e^\xi \rangle^\perp \equiv \{\eta \in \mathbf{R}^N \mid \tau e^\xi \cdot \eta = 0\}, \quad \xi \in \mathcal{M}.$$

Second, the space $Ker\ {}^t\tilde{A}$ is generated by b^i, $1 \leq i \leq N - 2$, and therefore, it holds that

$$Ker\ A = Ker\ {}^t A = \langle \tau^{-1} b^i \mid 1 \leq i \leq N - 2 \rangle$$

for $\tau^{-1} b^i = (\tau_j^{-1} b_j^i)$ where $b^i = (b_j^i)$.

If $v \in \mathcal{O}_* \cap Ker\ A$, there is $\xi \in \widetilde{\mathcal{O}}_*$ such that $v = e^\xi \in Ker\ A$. Then, we obtain

$$\tau e^\xi \in \langle b^i \mid 1 \leq i \leq N - 2 \rangle. \tag{5.92}$$

Since

$$P \equiv \{\zeta \in \mathbf{R}^N \mid b^i \cdot \zeta = \beta_i,\ 1 \leq i \leq N - 2\} = \{\xi\} + \bigcap_{i=1}^{N-2} \langle b^i \rangle^\perp, \tag{5.93}$$

condition (5.92) implies

$$\tau e^\xi \in (P - \{\xi\})^\perp.$$

Therefore, it holds that $P \subset T_\xi \mathcal{M} + \{\xi\}$, and we end up with

$$\widetilde{\mathcal{O}}_* = P \cap \mathcal{M} \subset T_\xi \mathcal{M} + \{\xi\}. \tag{5.94}$$

Equation (5.94) implies $\widetilde{\mathcal{O}}_* = \{\xi\}$ because $\mathcal{M}$ is the boundary of the strictly convex set $M = \{\xi \in \mathbf{R}^N \mid \tau \cdot e^\xi \leq \beta_0\}$. $\qquad\square$

We are ready to complete the following proof.

Proof of Theorem 5.5. By Lemmas 5.10, 5.11, and 5.13, it holds that $\mathcal{O} \subset \mathcal{O}_*$, where $\mathcal{O}_*$ is a Jordan curve or singleton. If $\mathcal{O}_*$ is a Jordan curve, we have $\mathcal{O} = \mathcal{O}_*$ by Lemma 5.14, which means that the solution $v = v(t)$ is periodic-in-time. If $\mathcal{O}_*$ is singleton, on the other hand, the orbit $\mathcal{O}$ is composed of an equilibrium. Hence the solution is stationary.

The set of equilibria to (5.84) is given by $E = \mathbf{R}_+^N \cap \mathrm{Ker}\ A$. This set is non-empty because of the existence of $0 < B \in \mathrm{Ker}\ A$. Since rank $A = 2$, it is the intersection of the vector space $\mathrm{Ker}\ A$ of co-dimension 2 and the positive cone $\mathbf{R}_+^N$.

We recall that $\mathcal{M}$ is the boundary of the strictly convex space $M = \{\xi \in \mathbf{R}^N \mid \tau \cdot e^\xi \leq \beta_0\}$ determined by β_0, that is, $\mathcal{M} = \partial M$. As $\beta_0 > 0$ varies, $\mathcal{F} = \{\mathcal{M}\}$ forms a nested family of hyper-surfaces of co-dimension one, covering $\mathbf{R}^N \setminus \{0\}$. Furthermore, each orbit $\mathcal{O}$ of (5.84) is homeomorphic to the intersection of some $\mathcal{M} \in \mathcal{F}$ and an affine space P of dimension 2 defined by (5.93). Since M with $\partial M = \mathcal{M}$ is strictly convex and $\mathcal{O}$ is either a Jordan curve or singleton composed of an equilibrium, it is contractible in $\mathbf{R}_+^N \setminus E$ to an equilibrium point as P moves in $\mathbf{R}^N$. Finally, if two distinct Jordan orbits $\mathcal{O}_1, \mathcal{O}_2 \subset \mathbf{R}_+^N$ are realized on the same hyper-surface $\mathcal{M} = \partial M \in \mathcal{F}$, they do not link because of the strictly convexity of M. Therefore, the same property arises to any distinct two non-stationary orbits in $\mathbf{R}_+^N$. $\qquad\qquad\qquad\qquad\qquad\qquad\qquad\qquad\qquad\qquad\quad$ $\square$

Now we turn to the proof of the Theorem 5.6. We take the case $N \geq 3$ of (5.74). To begin with, we have

$$\frac{d}{dt}\tau \cdot v = -r \cdot v \leq a\tau \cdot v, \quad a = \frac{\max_j(-e_j)}{\min_j \tau_j} \qquad (5.95)$$

and hence $\tau \cdot v \leq (\tau \cdot v_0)e^{at}$, $v_0 = (v_{0j})$. Then it holds that $T = +\infty$ with

$$0 < v_j(t) \leq \tau_j^{-1}(\tau \cdot v_0)e^{at}.$$

Transform (5.74) to

$$\tau_j \frac{d\xi_j}{dt} = -e_j + \sum_{k=1}^N a_{jk}e^{\xi_k}, \quad \xi_j|_{t=0} = \xi_{j0} \in \mathbf{R}, \quad 1 \leq j \leq N, \qquad (5.96)$$

using $\xi_j = \log v_j$, and let $\tau = (\tau_j) > 0$, $e = (e_j)$, $\xi = (\xi_j)$, and $\xi_0 = (\xi_{j0})$. We use the following lemma analogous to Lemma 5.12. The proof of this lemma is more complicated and omitted here.

Lemma 5.15. *Under the assumption of Theorem 5.6, either the first or the second case of Lemma 5.12 arises with $e_i \neq 0$, that is,*

Case 1. *There are (distinct) $i,j,k \in \{1,2,\cdots,N\}$ such that $a_{ij} > 0$, $a_{ik} < 0$, $a_{jk} > 0$, and $e_i \neq 0$.*

Case 2. *There are distinct $i,j,k,l \in \{1,2,\cdots,N\}$ such that $a_{ij} > 0$, $a_{ik} < 0$, $a_{il} = 0$, $a_{jk} = 0$, $a_{jl} > 0$, $a_{kl} < 0$, and $e_i \neq 0$.*

Henceforth, without loss of generality we assume $(i,j,k) = (1,j,N)$, $j \in \{2,\ldots,N-1\}$ and $(i,j,k,l) = (1,j,N,l)$, $j,l \in \{2,\ldots,N-1\}, j \neq l$, in the first and the second cases of Lemma 5.15, respectively. Thus we have $e_1 \neq 0$ and either

$$a_{1j} > 0, \ a_{1N} < 0, \ a_{jN} > 0 \tag{5.97}$$

for some $2 \leq j \leq N-2$ or

$$a_{1j} > 0, \ a_{1l} = 0, \ a_{1N} < 0, \ a_{jN} = 0, \ a_{jl} > 0 \tag{5.98}$$

for some $2 \leq j \neq l \leq N-2$. In both cases it holds that $a_{1N} < 0$.

Lemma 5.16. *Under the above assumption, define $V_j(\xi)$, $1 \leq j \leq N-1$, by*

$$V_1(\xi) = a_{1N} \sum_{k=1}^{N} \tau_k e^{\xi_k} + e_N \tau_1 \xi_1 - e_1 \tau_N \xi_N$$
$$V_j(\xi) = a_{jN} \tau_1 \xi_1 - a_{1N} \tau_j \xi_j + a_{1j} \tau_N \xi_N, \quad 2 \leq j \leq N-1. \tag{5.99}$$

Then $V_j(\xi)$, $1 \leq j \leq N-1$, are invariant with respect to (5.96). Hence it holds that

$$\widetilde{\mathcal{O}} \equiv \{\xi(t) \mid t \geq 0\} \subset \widetilde{\mathcal{O}}_* \equiv \{\xi \in \mathbf{R}^N \mid V_j(\xi) = c_j, \ 1 \leq j \leq N-1\} \tag{5.100}$$

for $c_j = V_j(\xi_0)$. Moreover, $\widetilde{\mathcal{O}}_*$ is a Jordan curve or singleton.

Proof. We shall confirm

$$\frac{d}{dt} V_j(\xi(t)) = 0, \quad 1 \leq j \leq N-1 \tag{5.101}$$

for the solution $\xi = (\xi_j(t))$ to (5.96). First, (5.75) implies

$$\frac{d}{dt} \sum_{k=1}^{N} \tau_k e^{\xi_k} = -\sum_{k=1}^{N} e_k e^{\xi_k},$$

while

$$e_N \tau_1 \frac{d\xi_1}{dt} - e_1 \tau_N \frac{d\xi_N}{dt} = \sum_{k=1}^{N} (e_N a_{1k} - e_1 a_{Nk}) e^{\xi_k}$$

is obvious. Then, equation (5.101) for $j = 1$ is a consequence of (5.81) with $(i, j, k) = (1, N, k)$:

$$\frac{dV_1}{dt} = \sum_{k=1}^{N}(-a_{1N}e_k + e_N a_{1k} - e_1 a_{Nk})e^{\xi_k} = 0.$$

Here we use (5.81) for $(i, j, k) = (1, j, N)$, $(i, j, k) = (1, j, k)$, and $(i, j, k) = (1, N, k)$ to obtain

$$a_{jN}e_1 - a_{1N}e_j + a_{1j}e_N = 0$$
$$a_{jk}e_1 - a_{1k}e_j + a_{1j}e_k = 0$$
$$a_{Nk}e_1 - a_{1k}e_N + a_{1N}e_k = 0.$$

Then it follows that

$$(a_{jN}a_{1k} + a_{1j}a_{Nk} - a_{1N}a_{jk})e_1 = 0$$

and hence (5.76) for $(i, j, k, l) = (1, j, k, N)$, that is,

$$a_{jN}a_{1k} + a_{1j}a_{Nk} - a_{1N}a_{jk} = 0, \quad 2 \le j \ne k \le N - 1. \qquad (5.102)$$

By (5.102) we obtain

$$\frac{dV_j}{dt} = \sum_{k=1}^{N}(a_{jN}a_{1k} - a_{1N}a_{jk} + a_{1j}a_{Nk})e^{\xi_k} + (-a_{jN}e_1 + a_{1N}e_j - a_{1j}e_N)$$

$$= -a_{jN}e_1 + a_{1N}e_j - a_{1j}e_N, \quad 2 \le j \le N - 1 \qquad (5.103)$$

which implies (5.101) for $2 \le j \le N - 1$ by (5.81), $(i, j, k) = (1, j, N)$.

We show that $\widetilde{\mathcal{O}}_*$ is a Jordan curve or singleton. In fact, it is the intersection of the boundary of

$$M = \{\xi \in \mathbf{R}^N \mid V_1(\xi) \ge c_1\},$$

which is strictly convex by $a_{1N} < 0$, and the $N - 2$ linearly independent hyper-planes

$$P_j = \{\xi \in \mathbf{R}^N \mid V_j(\xi) = c_j\}, \quad 2 \le j \le N - 1.$$

Thus we have only to show that $\widetilde{\mathcal{O}}_*$ defined by (5.100) is bounded. In fact, since $a_{1N} < 0$ implies

$$c_1 = V_1(\xi) \le a_{1N}(\tau_N e^{\xi_N} + \tau_1 e^{\xi_1}) + e_N \tau_1 \xi_1 - e_N \tau_N \xi_N$$

the components ξ_1, ξ_N are bounded above on $\widetilde{\mathcal{O}}_*$. From these bounds inequality (5.100) implies

$$-C \le a_{1N} \sum_{k=2}^{N-1} \tau_k e^{\xi_k}$$

with a constant $C > 0$. Therefore, the components $\xi_2, \cdots, \xi_{N-1}$ are also bounded above on $\widetilde{\mathcal{O}}_*$ by $a_{1N} < 0$.

Thus we have only to derive the boundedness below of $\xi_1, \cdots, \xi_N$ for $\xi = (\xi_1, \cdots, \xi_N) \in \widetilde{\mathcal{O}}_*$. For this purpose, we distinguish two cases of Lemma 5.15.

Case 1. We have (5.97). Hence it holds that $a_{1j} > 0$, $a_{1N} < 0$, $a_{jN} > 0$, and $e_1 \neq 0$ for some $2 \leq j \leq N - 1$. Then, V_j is a linear function of (ξ_1, ξ_j, ξ_N), of which coefficients are positive. Since $\widetilde{\mathcal{O}}_*$ is bounded above, the coordinates ξ_1, ξ_j, ξ_N are bounded below on $\widetilde{\mathcal{O}}_*$. From this property and $V_i(\xi) = c_i$, so is ξ_i for $i \neq 1, j, N$. Hence $\widetilde{\mathcal{O}}_*$ is bounded below.

Case 2. We have (5.98). Hence there are $j \neq l$ satisfying $2 \leq j, l \leq N - 1$, $a_{1j} > 0$, $a_{1N} < 0$, $a_{1l} = 0$, $a_{jN} = 0$, $a_{jl} > 0$, $a_{Nl} < 0$, and $e_1 \neq 0$. Then, V_j and V_l are linear functions of (ξ_j, ξ_N) and (ξ_1, ξ_l), respectively, and all of their coefficients are positive. Hence the cordinates $\xi_1, \xi_j, \xi_N, \xi_l$ are bounded below on $\widetilde{\mathcal{O}}_*$ because they are bounded above. From this property and $V_i(\xi) = c_i$, so is ξ_i for $i \neq 1, j, N, l$. Hence $\widetilde{\mathcal{O}}_*$ is bounded below. $\square$

Lemma 5.17. *Under the above assumption, the equilibrium set of (5.74) is given by $E = \mathbf{R}_+^N \cap E_*$ for*

$$E_* = \{v_*\} + \langle \vec{b_j} \mid 2 \leq j \leq N - 1 \rangle,$$

where $v_* = \frac{1}{a_{1N}}{}^t(-e_N, 0, \cdots, 0, e_1)$, $\vec{b_j} = (b_j^m)$, $2 \leq j \leq N - 1$, and

$$
b_j^m = \begin{cases}
a_{jN}, & m = 1 \\
a_{1j}, & m = N \\
-a_{1N}, & m = j \\
0, & otherwise.
\end{cases}
$$

Proof. We have $v = (v_k) \in E$ if and only if $v > 0$ and

$$I_j \equiv -e_j + \sum_{k=1}^{N} a_{jk}v_k = 0, \quad 1 \leq j \leq N. \tag{5.104}$$

Equalities (5.104) and (5.102) imply

$$a_{1j}I_N - a_{Nj}I_1 = a_{1N}I_j, \quad 2 \leq j \leq N - 1$$

by (5.81) with $(i, j, k) = (1, j, N)$.

Since $a_{1N} \neq 0$, therefore, system of equations (5.104) is reduced to $I_1 = I_N = 0$, that is,

$$v_1 = \frac{1}{a_{1N}}\left(-e_N - \sum_{k=2}^{N-1} a_{kN}v_k\right)$$

$$v_N = \frac{1}{a_{1N}}\left(e_1 - \sum_{k=2}^{N-1} a_{1k}v_k\right). \tag{5.105}$$

Then the result follows from (5.105). $\square$

Let

$$\widetilde{E} = \{\xi \in \mathbf{R}^N \mid \xi = \log v, \ v \in E\}.$$

Lemma 5.18. *Under the above assumption, the property $\widetilde{\mathcal{O}}_* \cap \widetilde{E} \neq \emptyset$ arises only if $\sharp \widetilde{\mathcal{O}}_* = 1$.*

Proof. The hyper-surface $\mathcal{M} = \{\xi \in \mathbf{R}^N \mid V_1(\xi) = c_1\}$ is the boundary of a strictly convex set $M = \{\xi \in \mathbf{R}^N \mid V_1(\xi) \geq c_1\}$, and $P = \{\xi \in \mathbf{R}^N \mid V_j(\xi) = c_j, \ 2 \leq j \leq N-1\}$ is a two-dimensional affine space.

Given $\xi \in \mathcal{M}$, we have

$$T_\xi \mathcal{M} = \{\eta = (\eta_j) \in \mathbf{R}^N \mid$$

$$(a_{1N}\tau_1 e^{\xi_1} + e_N \tau_1)\eta_1 + \sum_{i=2}^{N-1} a_{1N}\tau_i e^{\xi_i}\eta_i + (a_{1N}\tau_N e^{\xi_N} - e_1\tau_N)\eta_N = 0\}.$$

In addition, for $\xi \in \widetilde{\mathcal{O}}_* = \mathcal{M} \cap P$, the plane P is parallel to $T_\xi \mathcal{M}$ if and only if

$$a_{1N}\tau_1 e^{\xi_1} + e_N \tau_1 = \sum_{i=2}^{N-1} k_i a_{iN}\tau_1$$

$$a_{1N}\tau_j e^{\xi_j} = -k_j a_{1N}\tau_j, \quad 2 \leq j \leq N-1$$

$$a_{1N}\tau_N e^{\xi_N} - e_1\tau_N = \sum_{i=2}^{N-1} e_i a_{1i}\tau_N$$

for some $k_2, \cdots, k_{N-1} \in \mathbf{R}$, which is equivalent to

$$^t(e^{\xi_1}, e^{\xi_2}, e^{\xi_3}, \cdots, e^{\xi_{N-1}}, e^{\xi_N}) \in E,$$

or $\xi \in \widetilde{E}$. Therefore, if $\xi \in \widetilde{\mathcal{O}}_* \cap \widetilde{E}$, the plane P is parallel to $T_\xi \mathcal{M}$. Then $\widetilde{\mathcal{O}}_* = \{\xi\}$ follows because M is strictly convex. $\qquad \square$

We are ready to complete the following proof.

Proof of Theorem 5.6. To begin with, we show the existence of $v = (v_k) > 0$ such that (5.105). In fact, in the case (5.97) we have $j \in \{2, \cdots, N-1\}$ such that $a_{1j} > 0$, $a_{jN} > 0$. Then the requirement $v = (v_k) > 0$ holds in (5.105) for $v_j > 0$ sufficiently large by $a_{1N} < 0$. In the case (5.98), on the other hand, we have $j \in \{2, \cdots, N-1\}$ such that $a_{1j} > 0 = a_{jN}$. Here we find $i \in \{2, \cdots, N-1\}$ with $a_{iN} > 0$, using (a1), (a3), and $a_{1N} < 0$. Then $v = (v_k) > 0$ holds in (5.105) for sufficiently large $v_j > 0$ and $v_i > 0$.

Since (5.104) is equivalent to (5.105), now we have $E \neq \emptyset$. Then the rest the proof is similar to that of Theorem 5.5, using Lemmas 5.16, 5.17, and 5.18. $\qquad \square$

5.7 Dynamical Equilibrium and Its Break Down

Biological phenomena are maintained through metabolism. Usually this process is under the dynamical equilibrium. For example, bone metabolism is achieved under the balance of two kinds of cell, osteoblast and osteoclast, associated with bone formation and bone resorption, respectively. Break down of this balance, therefore, makes the individual unstable, and sometimes causes diseases, that is, osteopetrosis and osteoporosis if osteoblast dominates osteoclast and if osteoclast dominates osteoblast, respectively.

Here we study this break down of dynamical equilibrium, using mathematical modeling. The principal part of this model is composed of two pathways of maturation, that is, from pre-osteoblast to osteoblast and from pre-osteoclast to osteoclast. There is also a pathway of acceleration to the formation of pre-osteoclast by pre-osteoblast. This pathway is evoked by a cytokine, called RANKL. Experimental data, on the other hand, suggest a differentiation annihilation factor, to the maturation pathways above. Total mathematical modeling on these positive and negative feedback loops induces an insight, how the dynamical equilibrium of this metabolism breaks down, via mathematical analysis and numerical simulations.

More precisely, first, these cells, osteoblast and osteoclast are formed by differentiations of pre-osteoblast and pre-osteoclast, respectively. Here, hematopoiesis stem cell maturates to pre-cell of pre-osteoclast. Then there occurs proliferation of this pre-cells. There is also an acceleration in the differentiation of the above pre-cell to pre-osteoclast, from the pre-osteoblast through a cytokine, called, RANKL. In the process of differentiation, finally, pre-osteoclasts form cluster, called MN osteoclast, and this cluster matures to PN osteoclast. Recent experimental data, however, strongly suggest the production of differentiation annihilation factor (DAF) by MN osteoclast. This DAF annihilates both maturations, from pre-osteoblast to osteoblast and from MN osteoclast to PN osteoclast.

Here we examine the above hypothesis of DAF, using mathematical modeling. We apply two methods for this purpose, that is, multi-scale modeling and break down of dynamical equilibrium, to predict what should be observed in experimental data and also evaluate the drug effect. We formulate these feedback loops as a system of ordinary differential equations, pick up dynamical equilibria, and study their break down. First, we apply multi-scale modeling on DAF. The event is on tissue level, where each cell is regarded as a point, and the densities of four kinds of cell are counted, that is, pre-osteoblast, osteoblast, pre-osteoclast, and osteoclast. Here we identify pre-osteoclast and MN osteoclast, while DAF is on the molecular level. We assume three functions of DAF, that is, production by MN osteoclast, decay by itself, and annihilation of two pathway of differentiation, from pre-osteoblast to osteoblast and pre-osteoclast to osteoclast. These effects on the molecular level are modeled as functional relations. Then dynamical equilibrium is formulated, and dependence on the parameter is examined in connection with its break down. Finally, transit to osteoporosis is suggested by mathematical analysis and numerical simulations at the occasion of break down of dynamical equilibrium.

Densities of the four kinds of cell are defined on tissue level. Hence, pre-osteoblast, osteoblast, pre-osteoclast, and osteoclast, are denoted by X_1, X_2, X_3, and X_4, respectively. Then it holds that

$$\frac{dX_1}{dt} = -\ell_1 X_1 + m_1$$

$$\frac{dX_2}{dt} = \ell_1 X_1$$

$$\frac{dX_3}{dt} = -\ell_2 X_3 + m_2$$

$$\frac{dX_4}{dt} = \ell_2 X_3 \tag{5.106}$$

where m_1 and m_2 denote the amounts of supply per unit time of pre-osteoblast and pre-osteoclast, respectively, and ℓ_1 and ℓ_2 denote the rates of differentiations, from pre-osteoblast to ostepblast and from pre-osteoclast to osteoclast, respectively. These differentiations are annihilated by a factor, which we call DAF. It lies on the molecular level, produced by MN osteoclast, identified with pre-osteoclast. Hence DAF density, denoted by X_5, is subject to

$$\frac{dX_5}{dt} = \gamma X_3 - \delta X_5 \tag{5.107}$$

where γ and δ denote the rates of production and self-inhibition, respectively, which are assume to be positive constants. We call (5.106)–(5.107) the top down model, totally.

Positive and negative feedback loops, on the other hand, arise in the molecular level. Below, a, b, c, d, e, f, g, and h denote positive constants. First, pre-osteoblast accelerates the production of pre-osteoclast through the activation of RANKL, which is identified with the pre-osteoblast in this model. Thus we take

$$m_2 = m_2(X_1) = aX_1 + b. \tag{5.108}$$

Since DAF annihilates the maturations of osteoblast and osteoclast, we assume

$$\ell_1 = \ell_1(X_5) = \frac{c}{dX_5 + e}$$

$$\ell_2 = \ell_2(X_5) = \frac{f}{gX_5 + h}. \tag{5.109}$$

We call (5.108)–(5.109) the bottom up model totally, under the agreement that m_1 is a positive constant. The precise forms of the bottom up model, however, are not essential. For the moment it is sufficient to assume the strict convexity of the continuous mapping $x \in [0, \infty) \mapsto \varphi(x) \in (0, \infty)$, where

$$\varphi(x) = m_2 \left(\frac{m_1}{\ell_1(x)} \right) \cdot \frac{1}{\ell_2(x)}. \tag{5.110}$$

In fact we have

$$\varphi(x) = \frac{1}{f} \left(\frac{am_1}{c}(dx + e) + b \right) (gx + h)$$

in the case of (5.108)–(5.109).

From the above description, dynamical equilibrium in bone metabolism is formulated by

$$\frac{dX_1}{dt} = \frac{dX_3}{dt} = \frac{dX_5}{dt} = 0$$

which is equivalent to

$$\ell_1 X_1 = m_1, \quad \ell_2 X_3 = m_2, \quad \gamma X_3 = \delta X_5. \tag{5.111}$$

System of equations (5.111) is reduced to

$$\frac{\delta}{\gamma} X_5 = \varphi(X_5) \tag{5.112}$$

and then the other variables are determined by

$$X_1 = \frac{m_1}{\ell_1(X_5)}, \quad X_3 = \varphi(X_5). \tag{5.113}$$

From the strict convexity of $y = \varphi(x) > 0$, $x \geq 0$, there is a critical value $\overline{\lambda} > 0$ of $\lambda = \delta/\gamma$ concerning the number of solutions to (5.112). This number is acutally 2, 1, and 0, according to $\lambda > \overline{\lambda}$, $\lambda = \overline{\lambda}$, and $0 < \lambda < \overline{\lambda}$, respectively. Assume $\lambda > \overline{\lambda}$, let $X_5^+ = X_5^+(\lambda) > X_5^- = X_5^-(\lambda) > 0$ be the solutions to (5.112), and put

$$X_1^{\pm} = \frac{m_1}{\ell_1(X_5^{\pm})}, \quad X_3^{\pm} = \varphi(X_5^{\pm}).$$

Then we obtain linearly non-degenerate equilibria of the system (4.213), (3.81), and (5.107), that is,

$$(X_1^\pm, X_3^\pm, X_5^\pm) = (X_1^\pm(\lambda), X_3^\pm(\lambda), X_5^\pm(\lambda)).$$

Dynamics of this system around $(X_1^\pm, X_3^\pm, X_5^\pm)$, on the other hand, is reduced to that of

$$\frac{dX_5}{dt} = \gamma X_3 - \delta X_5 \approx \gamma\varphi(X_5) - \delta X_5 \qquad (5.114)$$

around $X_5 = X_5^\pm$.

By the strict convexity of $y = \varphi(x) > 0$, $x \geq 0$, therefore, the only stable dynamical equilibrium arises to $\lambda > \overline{\lambda}$, that is,

$$(X_1, X_3, X_5) = (X_1^-(\lambda), X_3^-(\lambda), X_5^-(\lambda)).$$

Then the other variables $(X_2, X_4) = (X_2(t), X_4(t))$ exhibit linear growth for t large.

This dynamical equilibrium $(X_1^-, X_3^-, X_5^-) = (X_1^-(\lambda), X_3^-(\lambda), X_5^-(\lambda))$ breaks down as λ decreases to exceed $\overline{\lambda}$. At this occasion it arises the increase of the value $X_5^-(\lambda)$. Although $\lim_{\lambda \downarrow \overline{\lambda}} X_5^-(\lambda)$ exists, its increasing rate becomes extremely high, if the malignancy proceeds on time.

A natural question is what happens if the dynamical equilibrium breaks down. To approach this problem we introduce the notion of near from dynamical equilibrium. Here we assume that m_1, γ, and δ are positive constants, $m_2 = m_2(X_1)$, $\ell_1 = \ell_1(X_5)$, and $\ell_2 = \ell_2(X_5)$, with the strict convexity of $y = \varphi(x) > 0$, $x \geq 0$, defined by (5.110). Then we say that the solution $(X_1, X_2, X_3, X_4, X_5)$ to (4.213)–(5.107) lies on near from dynamical equilibrium if it is in the region where the approximation

$$X_1 \approx \frac{m_1}{\ell_1(X_5)}, \quad X_3 \approx \varphi(X_5) \qquad (5.115)$$

is valid, recalling (5.113).

This is the region where the dynamics of (X_1, X_2, X_3, X_4) is controlled by that of X_5. In particular, it holds that

$$\frac{dX_4}{dX_2} = \frac{\ell_2 X_3}{\ell_1 X_1} \approx \frac{\ell_2(X_5)}{m_1} \cdot \varphi(X_5) = \frac{1}{m_1} \cdot m_2\left(\frac{m_1}{\ell_1(X_5)}\right)$$

and hence

$$\frac{d}{dt}\left(\frac{dX_4}{dX_2}\right) \approx \psi'(X_5)\frac{dX_5}{dt}, \quad \psi(x) = \frac{1}{m_1} \cdot m_2\left(\frac{m_1}{\ell_1(x)}\right). \qquad (5.116)$$

From the positive and negative feedback loops underlying this model, it holds that $\psi'(x) > 0$, $x \geq 0$. We can actually confirm this property for the case of (5.108)–(5.109). Relation (5.116) means that break down of dynamical equilibrium, near from dynamical equilibrium, arises in accordance with the velocity of DAF density. More precisely, this break down leads to osteoclast and osteopetrosis in the cases of $\frac{dX_5}{dt} > 0$ and $\frac{dX_5}{dt} < 0$, respectively.

Even in the case of $\lambda > \bar{\lambda}$, the orbit can stay around the unstable dynamical equilibrium $(X_1^+(\lambda), X_3^+(\lambda), X_5^+(\lambda))$ in a relatively long time. Since this is the region near from dynamical equilibrium, such transient experience may cause a serious damage to the individual, although the variables

$$(X_1, X_3, X_5) = (X_1(t), X_3(t), X_5(t))$$

eventually approach the stable dynamical equilibrium

$$(X_1^-(\lambda), X_3^-(\lambda), X_5^-(\lambda))$$

as $t \uparrow +\infty$.

If

$$a = 1,\ b = 1,\ c = 1,\ d = 1,\ e = 3,\ f = 10,\ g = 1,\ h = 1$$

in the bottom up model (5.108)–(5.109), for example, we obtain two solutions to (5.112), denoted by $X_5^+(\lambda) = 4$ and $X_5^-(\lambda) = 1$, for $\lambda = \frac{\delta}{\gamma} = 10$. The unstable dynamical equilibrium is detected as $(X_1^+(\lambda), X_3^+(\lambda), X_5^+(\lambda)) = (7, 4, 4)$. We can see that the Morse index of this unstable equilibrium is 1, and hence it is associated with a stable manifold of codimension 1. Consequently, generic orbit stays near this unstable equilibrium in a relatively long time. Numerical simulation shows that the orbit stays still around there at $t = 100$. Then $\frac{dX_5}{dt} > 0$ is kept, and consequently, we observe saturation of X_2 after $t \geq 60$, recalling (5.116). Here, increase of X_5 is due to that of the supply of X_3, which matures to X_4, and therefore, X_2 relatively saturates in spite of two pathways of annihilation by X_5. The orbit stays near from dynamical equilibrium at least initially if the initial value is taken there. Then, as we have mentioned, even transient saturation of either X_2 or X_4 takes an important clinical role.

At the break down of dynamical equilibrium the value X_5 takes $X_5^+(\bar{\lambda}) = X_5^-(\bar{\lambda})$, denoted by X_5^*. Since it holds that $\gamma\varphi(X_5^*) - \delta X_5^* > 0$ for $\lambda = \frac{\delta}{\gamma} < \bar{\lambda}$, we have $\frac{dX_5}{dt} > 0$ near this point, recalling (5.114). Hence break down of dynamical equilibrium occurs with the saturation of osteoblast by (5.116), which will be a driving force to osteoporosis. The other near from dynamical equilibrium is achieved around the unstable dynamical equilibrium

$$(X_1^+(\lambda), X_3^+(\lambda), X_5^+(\lambda))$$

for $\lambda > \overline{\lambda}$. Then the conditions $\frac{dX_5}{dt} > 0$ and $\frac{dX_5}{dt} < 0$ arise initially if $X_5(0) > X_5^+(\lambda)$ and $X_5(0) < X_5^+(\lambda)$, respectively. Such initial dynamics that $X_5(0)$ is close to $X_5^+(\lambda)$ may occur often if $0 < \lambda - \overline{\lambda} \ll 1$. In other words, if bone metabolism is close to the break down of dynamical equilibrium and the concentration of DAF becomes higher in some reason, saturation is induced to the production of either osteoblast or osteoclast, associated with increase and decrease of DAF, respectively.

Let us call the curve in $X_1 X_3 X_5$ space,

$$X_1 = \frac{m_1}{\ell_1(X_5)}, \quad X_3 = \varphi(X_5),$$

the quasi-dynamical equilibrium. Then near from dynamical equilibrium is formulated as a tubuler neighbourhood of this curve, denoted by $\mathcal{T}$. The principal dynamics of near from dynamical equilibrium is thus governed by

$$\frac{dX_5}{dt} = \gamma\varphi(X_5) - \delta X_5. \tag{5.117}$$

For $\lambda > \overline{\lambda}$ there should be a hetero-clinic orbit $\mathcal{O}_\lambda$ connecting $(X_1^+(\lambda), X_3^+(\lambda), X_5^+(\lambda))$ at $t = -\infty$ and $(X_1^-(\lambda), X_3^-(\lambda), X_5^-(\lambda))$ at $t = +\infty$. Hence $\frac{dX_5}{dt} < 0$ occurs mostly along $\mathcal{O}_\lambda$. As $\lambda \downarrow \overline{\lambda}$, this $\mathcal{O}_\lambda$ shrinks to a point denoted by P_*, which is a conditionary stable equilibrium of (5.117). More precisely, this P_* has a stable manifold $\mathcal{M} \subset \mathbf{R}^3$ of codimension 1, and except for this $\mathcal{M}$ it always holds that $\frac{dX_5}{dt} > 0$ near P_*. This tendency will be kept even after the break down of dynamical equilibrium in $\lambda < \overline{\lambda}$. Thus our conclusion is that the break down of dynamical equilibrium arises with osteoclast.

5.8 Transient Oscillation

Cells are constantly sensing external stimuli through various receptors that recognize their specific ligands. Signals triggered by the ligand-receptor interaction are then properly processed, thereby regulating differentiation, proliferation and apoptosis etc. These series of processes are called signal transduction. Since various diseases are caused by abnormality in signal transduction, elucidation of molecular mechanisms of signal transduction pathways is in great demand.

Transcription factor $NF\text{-}\kappa B$ is responsible for various biological cell functions such as proliferation, differentiation and blocking apoptosis. $NF\text{-}\kappa B$ induces expression of multiple genes by shuttling between cytoplasm and nucleus. The mechanism of this nuclear-cytoplasmic $(N-C)$ oscillation, a characteristic of canonical $NF\text{-}\kappa B$ pathway is not fully understood. It has been intensively investigated that the transcription of $NF\text{-}\kappa B$ target genes is regulated by phosphorylation of $I\kappa B\alpha$ and $RelA$ subunit of $NF\text{-}\kappa B$, suggesting that these phosphorylation events are crucial for the oscillation. In this study, we constructed a new mathematical model considering how the phosphorylation of $I\kappa B\alpha$ and $RelA$ modulates the oscillation phenomena. The new model considering $I\kappa B\alpha$ phosphorylation explained an appearance of a stable periodic orbit, which appeared in a transitional manner in response to the attenuation of an external stimulus. Because the $NF\text{-}\kappa B$ oscillation is caused by the periodic orbit, the amplitude and period of $NF\text{-}\kappa B$ oscillation in phosphorylation model was constant regardless of changing initial condition: we defined this property as reproducibility of oscillation. In addition, the amplitude and period of the oscillation depend on the parameter related to $RelA$ phosphorylation. Therefore, it is suggested that the oscillation period is regulated by the phosphorylation of $RelA$. Furthermore, by comparing the models with and without adding phosphorylation component, we concluded that phosphorylation conferred the robustness of oscillation on $NF\text{-}\kappa B$ signaling module.

$NF\text{-}\kappa B$ is a transcription factor that regulates variety of biological functions such as cell proliferation, differentiation, inflammation and cell death. By receiving the stimuli of cytokines, growth factors and DNA damages, $NF\text{-}\kappa B$ is activated and regulates numerous gene expression. There are two signaling pathways for $NF\text{-}\kappa B$ activation: the canonical pathway and the noncanonical pathway. Although both pathways are biologically significant, most of the mathematical studies has been focused on the canonical pathway. $NF\text{-}\kappa B$ is sequestered in the cytoplasm of unstimulated cells by mainly binding to the inhibitor of $NF\text{-}\kappa B$, that is, $I\kappa B\alpha$. First, an $I\kappa B$ kinase (IKK), activated by $TNF\alpha$ stimulation, phosphorylates $I\kappa B\alpha$, then the phosphorylated $I\kappa B\alpha$ is ubiquitinated and degraded. $NF\text{-}\kappa B$, released from $I\kappa B\alpha$ is translocated to the nucleus and regulates expression of its target genes. Since $I\kappa B\alpha$ is also one of the $NF\text{-}\kappa B$ target genes, $I\kappa B\alpha mRNA$ is induced by $NF\text{-}\kappa B$ activation and transferred to the cytoplasm, where the $I\kappa B\alpha$ protein becomes translated. Newly synthesized free $I\kappa B\alpha$ goes to nucleus and binds to nuclear $NF\text{-}\kappa B$, leading to export of the $NF\text{-}\kappa B/I\kappa B\alpha$ complex to the cytoplasm. Thus, $I\kappa B\alpha$ induction by $NF\text{-}\kappa B$

forms a negative feedback loop. If IKK is still activated, the above process repeats because the $I\kappa B\alpha/NF\text{-}\kappa B$ complex is the target for $I\kappa B$ phosphorylation by IKK. A few min pulse stimulation of $TNF\alpha$ causes transient nuclear localization of $NF\text{-}\kappa B$, on the other hand, continuous $TNF\alpha$ stimulation induces the long term $N-C$ oscillation of $NF\text{-}\kappa B$. Nuclear localization of $NF\text{-}\kappa B$ immediately starts by $TNF\alpha$ stimulation. The nuclear concentration of $NF\text{-}\kappa B$ reaches the maximum at 30 min after stimulation and then reaches minimum at about 60 min after stimulation. After that, $NF\text{-}\kappa B$ oscillates between nucleus and cytoplasm in a period of 90 to 120 min. We think that the oscillation with a period of 90 to 120 min can be obtained because the $NF\text{-}\kappa B$ signaling system has the property of obtaining constant oscillation regardless of the initial conditions: we defined this property as reproducibility of oscillation.

Among a number of critical findings about $NF\text{-}\kappa B$ pathways, one of important components is phosphorylation of $I\kappa B\alpha$ and $RelA$: in the $NF\text{-}\kappa B$ signaling pathway, phosphorylation of $I\kappa B\alpha$ at $Ser32$ and 36 plays an important part in inducing $I\kappa B\alpha$ degradation, and phosphorylation of $RelA$ regulates transcriptional activity. After $I\kappa B\alpha$ is phosphorylated by $IKKb$, the phosphorylated $I\kappa B\alpha$ becomes ubiquitinated followed by its proteasomal degradation. $RelA$ subunit of the $NF\text{-}\kappa B$, released from $I\kappa B\alpha$, is subsequently phosphorylated by $IKKb$ or other kinases and moves into nucleus. It has been reported that the influence on transcriptional activity is different depending on the phosphorylation site of $RelA$: Phosphorylation of $Ser276$, 205, 281, or 311 upregulates its transcriptional activity and that of $Ser468$ suppresses transcriptional activity in vivo. In addition, although phosphorylation of $Ser536$ inhibits its transcriptional activity in vivo while the same phosphorylation does not affect transcriptional activity in vitro. Therefore, considering the effect of phosphorylation is essential for understanding the $NF\text{-}\kappa B$ signaling.

From the model analysis and experimental data, the oscillation does not seem a damped oscillation nor sustained oscillation but some sort of transitional phenomenon. In particular, we analyze the stability of the $NF\text{-}\kappa B$ signaling, focusing on the concentration of activate $IKKb$ rather than the intensity of $TNF\text{-}\alpha$ stimulation. Here, to understand the function of $I\kappa B\alpha$ and $RelA$ phosphorylation in $NF\text{-}\kappa B$ signaling, we construct a phosphorylation conscious mathematical model. In particular, we focus on the $I\kappa B\alpha$ phosphorylation leading to its proteasomal degradation and the $RelA$ phosphorylation at $Ser-539$ leading to the inhibition transcriptional activity of $RelA$. Here, we first construct a mathematical model of $NF\text{-}\kappa B$

signaling, considering phosphorylation using ordinary differential equations. Then, we confirm what kind of properties of the oscillation phenomenon is caused by time evolution through the stability analysis of the equilibrium point.

To construct a mathematical model of the $NF\text{-}\kappa B$ signaling pathway considering phosphorylation of $I\kappa B\alpha$ and $RelA$ (phosphorylation model), the processes related phosphorylation are assumed below. Henceforth, phosphorylation of $I\kappa B\alpha$ means phosphorylation of $Ser32$ and 36, and phosphorylation of $NF\text{-}\kappa B$ means phosphorylation of $Ser536$ of $RelA$.

(1) IKK affects only $I\kappa B\alpha/NF\text{-}\kappa B$ to induce $I\kappa B\alpha$ phosphorylation.
(2) This phosphorylation of $I\kappa B\alpha$ results in proteasomal degradation of $I\kappa B\alpha$.
(3) Both phosphorylated $NF\text{-}\kappa B$ and unphosphorylated $NF\text{-}\kappa B$ exists in cells.
(4) Cytoplasmic $NF\text{-}\kappa B$ is phosphorylated at a constant rate.
(5) Nuclear transport of $NF\text{-}\kappa B$ is independent of $NF\text{-}\kappa B$ phosphorylation.
(6) Transcription of target gene is activated by only unphosphorylated $NF\text{-}\kappa B$.
(7) Nuclear phosphorylated $NF\text{-}\kappa B$ is dephosphorylated at a constant rate.
(8) $I\kappa B\alpha$ binds $NF\text{-}\kappa B$ irrespective of the phosphorylation status of $NF\text{-}\kappa B$.

The first two assumptions relate with the proteasomal degradation of $I\kappa B\alpha$, and others relate with phosphorylation of $NF\text{-}\kappa B$. In the reaction about $I\kappa B$ phosphorylation, we assume;

(1) concentration of active $IKKb$ does not change before and after the reaction, because $I\kappa B\alpha$ is immediately phosphorylated by $IKKb$.
(2) phosphorylated $I\kappa B\alpha$ is degraded at constant rate.

To verify the effect of phosphorylation on oscillation, phosphorylated $NF\text{-}\kappa B$ and unphosphorylated $NF\text{-}\kappa B$ must be distinguished. Here we assume;

(1) free $NF\text{-}\kappa B$ is phosphorylated at a constant rate in cytoplasm only.
(2) regarding the nuclear import of $NF\text{-}\kappa B$, there is no effect of phosphorylation.
(3) $NF\text{-}\kappa B$ moves into the nucleus with the same ratio regardless of phosphorylation.

(4) since phosphorylation of NF-κB attenuates transcription and dephosphorylation enhances it, only unphosphorylated NF-κB is involved in the transcription of target gene, and dephosphorylation occurs at a constant rate in the nucleus.

(5) $I\kappa B$ binds irrespective of the phosphorylation of NF-κB and inhibits the activity of NF-κB.

Other differences between our phosphorylation model and the standard model are;

(1) NF-κB simplex does not export from nucleus.
(2) The dissociation of $I\kappa B\alpha/NF$-κB does not occur.
(3) $I\kappa B\alpha$ of $I\kappa B\alpha/NF$-κB complex is degraded only after phosphorylated by $IKKb$.

These differences are based on experimental knowledge and are natural assumptions.

To investigate the influence of proteasome degradation and phosphorylation of NF-κB, we analyze two models;

(1) the model assuming proteasome degradation only, $I\kappa B\alpha$ phosphrylation model.
(2) the model assuming proteasome degradation and phosphorylation of NF-κB, full phosphorylation model.

We applied stability analysis around the equilibrium to this $I\kappa B\alpha$ phosphorylation model numerically. In this model, we find only one equilibrium point, and the result of stability analysis showed that all eigenvalues of linearized matrix around the equilibrium point were real negative numbers. This result showed that the stability of the equilibrium point is the same as that of classical model without phosphorylation. However, the amplitude and period of oscillation are almost constant regardless of initial values. To analyze the details of the stable structure of our $I\kappa B\alpha$ phosphorylation model, we focus on the result of simulations and the ordinary differential equation of our $I\kappa B\alpha$ phosphorylation model. According to the simulation results, the attenuation of the oscillation seems to be related to the decrease of X_3 ($IKKb$. In fact, in this model X_3 is independent from other values. We then obtain the solution

$$X_3(t) = X_3(0)e^{-d2t}$$

by integration. Furthermore, the changing of X_3 is slower than the oscillation period of other values, and activation of $IKKb$ is dependent of outer stimuli. Hence we regard X_3 as a parameter constant for a long time. In order to investigate how the concentration of X_3 affects oscillation phenomenon,we analyze the stability of the equilibrium point under the situation that X_3 is constant. As a result of the stability analysis of the equilibrium point, the stability of the equilibrium point changed with the $IKKb$ concentration. In the high $IKKb$ concentration, the real part of all eigenvalues was negative and there arise complex eigenvalues. As the $IKKb$ concentration decreases, the real part of a set of conjugated complex eigenvalues became positive. Then the real part of all the eigenvalues became negative again, and finally all the eigenvalues became negative real numbers. This means that in our $I\kappa B\alpha$ phosphorylation model the stability of the equilibrium is transiently changing with the decreasing of $IKKb$. Especially, when the equilibrium is an unstable focul, the simulation indicates that the periodic solution becomes stable. Therefore, Hopf bifurcation is expected when changing from a stable focul to an unstable focul. Therefore, it turns out that the oscillation in the $I\kappa B\alpha$ phosphorylation model is caused by a stable limit cycle, and this model has reproducibility of oscillation, that is, a robust transient oscillation is obtained regardless of the initial value.

Next, we analyze the influence of assumption three to eight on the oscillation of NF-κB. In order to compare with the $I\kappa B\alpha$ phosphorylation model, we analyze the stability around the equilibrium point using our full phosphorylation model. The result of stability analysis shows the same result as in $I\kappa B\alpha$ phosphorylation model. That is, this full phosphorylation model has a globally stable transient equilibrium and stable transient periodic orbit. Hence this full phosphorylation model has reproducibility of the oscillation. In addition, we analyze the sensitivity to coefficients related to phosphorylation of NF-κB, that is, $p1$ and $p2$. Numerical simulations show that the coefficients related to phosphorylation and dephosphorylation affect the period of oscillation. The period of oscillation is important to regulate expression gene; some gene expressions do not occur unless NF-κB oscillate with a period of about 100 minutes. This result suggests that phosphorylation of NF-κB relates the regulation of early expression and late expression.

We also analyze the model when phosphorylation of NF-κB promotes transcriptional activity as well. That is, we constructed the model in which the pathway

$$2NF - \kappa B \rightarrow 2NF - \kappa B + I\kappa BmRNA \ (tr2)$$

is replaced by

$$2NF - \kappa B^* \to 2NF - \kappa B^* + I\kappa BmRNA \ (tr2).$$

However, as in the case where phosphorylation of NF-κB suppresses transcription, the reproducibility of oscillation is obtained.

Mass conservation law concerning NF-κB holds in this phosphorylation models, and total NF-κB is taken as one parameter. The amplitude and period of this limit cycle are also determined by the total concentration of NF-κB. In addition, depending on the values of the phosphorylation coefficients $p1$ and $p2$, the range of $IKKb$ for existence of the stable periodic orbit changes. We confirm that when the $IKKb$ concentration is fixed, the amplitude of the limit cycle becomes large with increasing the value of total NF-κB. In other words, the expression level of NF-κB is a parameter affecting the period and amplitude of oscillation.

5.9 Notes

The first and the second cases of Theorem 5.2 are due to [Horn (1972)] and [Horn and Jackson (1972)], respectively. Then, theory of chemical reaction network is developed by [Feinberg (1987); Feinberg and Horn (1974)]. The complex balanced equilibrium $c_\infty > 0$ in the second case of Theorem 5.2 is locally stable. In the global attractor conjecture this equilibrium is suspected to be a global attractor.

Section 5.2 is due to [Pierre, Suzuki, and Zou (2017)]. Global-in-time existence is also proved for (5.4) when the space-dimension n is small enough with respect to the degree of the polynomial f or when the diffusion coefficients d_k are close enough to each other. See the discussion in [Pierre (2010)]. In the non-separative type, different from (5.3),

$$\alpha_1 A_1 + \cdots + \alpha_n A_n \rightleftharpoons \alpha_1 A_1 + \cdots + \alpha_n A_n,$$

there may arise a boundary spatially homogeneous equilibrium. Even in this case, we have a renormalized solution global-in-time with pre-compact orbit in $L^1(\Omega)$, of which ω-limit set is spatially homogeneous. Furthermore, there arises finiteness of spatially homogeneous equilibria with prescribed masses for A_i, $1 \le i \le n$, and consequently, convergence to a spatially homogeneous equilibrium as $t \uparrow +\infty$ in $L^1(\Omega)$. See [Pierre, Suzuki, and Umakoshi (2018)].

The following list may be useful to overview the study for general space-dimension n and general positive $d_k \in (0, \infty)$. Here we assume $c_1 = c_2 = 1$ for simplicity.

(1) If $m = 1, N = 2$, that is, $f(u) = u_1^{\alpha_1} - u_1^{\alpha_2}$, then global existence of uniformly bounded, and therefore classical, solutions easily follows from the invariance of the rectangles

$$\{(u_1, u_2); 0 \leq u_1 \leq M_1, 0 \leq u_2 \leq M_2\} \text{ where } M_1^{\alpha_1} = M_2^{\alpha_2}.$$

(2) If $N = m + 1$, $\alpha_N = 1$, i.e. $f(u) = \prod_{k=1}^m u_k^{\alpha_k} - u_N$, global-in-time classical solutions exist. See [Bothe, Fisher, Pierre, and Rolland (2017)]. The same symmetrically holds if $m = 1, \alpha_1 = 1$, that is, $f(u) = u_1 - \prod_{k=2}^N u_k^{\alpha_k}$.

(3) If $m = 2$, $N = 3$, i.e. $f(u) = u_1^{\alpha_1} u_2^{\alpha_2} - u_3^{\alpha_3}$ and $\alpha_3 > \alpha_1 + \alpha_2$, then global-in-time existence of classical solutions holds by [Laamri (2011)].

(4) If again $m = 2$, $N = 4$, that is, $(f(u) = u_1 u_2 - u_3 u_2)$, then global-in-time weak solutions are proved to exist by [Pierre (2003); Desvillettes, Fellner, Pierre, and Vovelle (2007)]. See Section 4.5 for the definition of the weak solution.

(5) More generally, if for some reason, the nonlinearity $f(u)$ is a priori bounded in $L^1((0,T) \times \Omega)$ for all $T > 0$, then global *weak solutions* do exist. See [Pierre (2003, 2010)]. Thanks to quadratic a priori estimates valid for these systems, this is for instance the case if

$$N = m + 1, \ f(u) = \prod_{k=1}^m u_k^{\alpha_k} - u_N^2;$$
$$N = m + 2, \ f(u) = \prod_{k=1}^m u_k^{\alpha_k} - u_{m+1} u_{m+2}$$

(6) For sub-quadratic reaction-diffusion systems, global smooth solutions are proved to exist. See [Caputo and Vasseur (2009)]. For super-quadratic systems, the existence of global classical or weak solutions are verified if the diffusion coefficients are quasi-uniform. See [Fellner, Latos, and Suzuki (2016)].

(7) In the general situation of system (5.4), global-in-time existence of the renormalized solution is proved in [Fisher (2015)].

Theorem 5.3 is due to [Fisher (2015)]. When $\tau_k = 1$ for all k, the limit u is a weak solution to system (5.4) as soon as $f(u) \in L^1_{loc}([0, \infty); L^1(\Omega))$. It is only a renormalized solution in the sense of [Fisher (2015)] in general.

To check that the results of [Fisher (2015)] do apply here, let $U_k^\varepsilon := \tau_k u_k^\varepsilon$. Then system (5.6) may be rewritten as

$$\frac{\partial U_k^\varepsilon}{\partial t} - \frac{d_k}{\tau_k} \Delta U_k^\varepsilon = \chi_k \frac{F(U^\varepsilon)}{1 + \epsilon |F(U^\varepsilon)|} \quad \text{in } Q_T = \Omega \times (0, T)$$

$$\left.\frac{\partial U_k^\varepsilon}{\partial \nu}\right|_{\partial\Omega} = 0, \quad U_k^\varepsilon|_{t=0} = \tau_k u_{k0}^\varepsilon(x) \geq 0 \tag{5.118}$$

for $1 \leq k \leq N$, where, for $U \in bfR^N$,

$$F(U) = C_1 \prod_{i=1}^{m} U_i^{\alpha_i} - C_2 \prod_{j=m+1}^{N} U_j^{\alpha_j}$$

$$C_1 = c_1 \prod_{i=1}^{m} (\tau_i)^{-\alpha_i}, \quad C_2 = c_2 \prod_{j=m+1}^{N} (\tau_j)^{-\alpha_j}.$$

For this new system, the entropy inequality required in [Fisher (2015)] holds, namely

$$\sum_{k=1}^{N} \chi_k F(U^\epsilon)[\mu_k + \log U_k^\varepsilon]$$

$$= -F(U^\epsilon) \left[\log \left(C_1 \prod_{i=1}^{m} (U_i^\varepsilon)^{\alpha_i} \right) - \log \left(C_2 \prod_{j=m+1}^{N} (U_j^\varepsilon)^{\alpha_j} \right) \right] \leq 0$$

with $\mu_k = \log(C_2/C_1)/(N\chi_k)$ for $1 \leq k \leq N$. The a.e. convergence of U^ϵ, up to a subsequence, is stated in Lemma 7 of [Fisher (2015)]. It implies the a.e. convergence of u^ϵ. Together with the estimate of $U_k^\epsilon \log U_k^\epsilon$ in $L_{loc}^\infty([0,\infty); L^1(\Omega))$, it also implies the convergence of U_k^ϵ and therefore of u_k^ϵ in $L_{loc}^1([0,\infty); L^1(\Omega))$. Moreover, this implies

$$u \in L^\infty([0,\infty); L^1(\Omega)), \quad u_k \log u_k \in L_{loc}^\infty([0,\infty); L^1(\Omega))$$

for any k. From the proof, the same result of Theorem 5.3 holds for quite more general approximations f_ϵ of f. For instance, we can choose

$$f_\epsilon(s) = f(s)G_\epsilon(s), \ 0 \leq G_\epsilon(s) \leq M, \ |f_\epsilon(s)| \leq 1/\epsilon, \quad s \in [0,\infty)^N$$

with $f_\epsilon(s) \to f(s)$ as $\epsilon \downarrow 0$. Then any pointwise limit of the corresponding approximate solution will satisfy the conclusion of Lemma 5.5 and of Theorem 5.4 as well.

Theorem 5.4 does not handle the interesting case when the chemical species are not separated, contrary to the reversible reaction (5.3). This is the case for instance with the typical following reaction

$$A_1 + 2A_2 \rightleftharpoons 2A_1 + A_2.$$

The corresponding system writes

$$\frac{\partial u_1}{\partial t} - d_1 \Delta u_1 = u_1 u_2^2 - u_1^2 u_2 = - \left[\frac{\partial u_2}{\partial t} - d_2 \Delta u_2 \right]$$

$$\frac{\partial u_1}{\partial \nu} = 0 = \frac{\partial u_2}{\partial \nu}, \quad (u,v)|_{t=0} = (u_0(x), v_0(x)) \geq 0. \qquad (5.119)$$

Here, the only positive solution of system (5.24), namely of

$$z_1 z_2^2 = z_1^2 z_2, \quad z_1 + z_2 = \overline{u}_{10} + \overline{u}_{20} \equiv U_{12}$$

is given by $z = (U_{12}/2, U_{12}/2)$. The situation, however, is quite different from Theorem 5.4. Indeed if $U_{12} > 0$, the solution does not always converge to this z. If we chose for instance, $u_{10} \equiv 0, u_{20} \equiv a > 0$, then, by uniqueness, the solution is independent of the space variable x and is given by $(u_1(t), u_2(t)) = (0, a)$. Actually, the solution of the spatially homogeneous part of this system is given by $(u_1(t), u_2(t)) = (v(t), a - v(t))$ where v is solution of

$$v' = v(a - v)(a - 2v).$$

This equation has three stationary states, $0, m_0/2, m_0$. The second one is stable, while the first and the third ones are unstable. Such a behavior probably holds for system (5.119).

Now we show a reduction of system (5.6) to the case $c_1 = c_2 = 1, \alpha_k = 1$, $1 \le k \le N$. Namely, we may only consider these particular values without loss of generality. Let us check that system (5.6) is actually a particular case of the following system (5.120) whose solutions are exactly α_k copies of $u_k, 1 \le k \le N$:

$$l_0 = 0, \ l_k = \sum_{j=1}^{k} \alpha_j, \ 1 \le k \le N; \quad \lambda^{-l_m} = c_1, \ \mu^{l_m - l_N} = c_2$$

$$D_l = \lambda d_k / \alpha_k, \quad \tau_l = \lambda \tau_k / \alpha_k, \ l_{k-1} < l \le l_k, \ 1 \le k \le m$$

$$D_l = \mu d_k / \alpha_k, \quad \tau_l = \mu \tau_k / \alpha_k, \ l_{k-1} < l \le l_k, \ m+1 \le k \le N.$$

Then we put

$$v_l^\epsilon = u_k^\epsilon / \lambda, \quad l_{k-1} < l \le l_k, \ m+1 \le k \le N,$$

to obtain

$$f(u^\epsilon) = \lambda^{-l_m} \prod_{l=1}^{l_m} (v_l^\epsilon \lambda) - \mu^{l_m - l_N} \prod_{l=l_m+1}^{l_N} (v_l^\epsilon \mu) = \prod_{l=1}^{l_m} v_l^\epsilon - \prod_{l=l_m+1}^{l_N} v_l^\epsilon.$$

Now we consider the extended system

$$\tau_l \frac{\partial v_l^\epsilon}{\partial t} - D_l \Delta v_l^\epsilon = \chi_l g(v^\epsilon) / [1 + \epsilon |g(v^\epsilon)|] \quad \text{in } Q_T = \Omega \times (0, T)$$

$$\left. \frac{\partial v_l}{\partial \nu} \right|_{\partial\Omega} = 0, \quad g(v^\epsilon) = \prod_{l=1}^{l_m} v_l^\epsilon - \prod_{l=l_m+1}^{l_N} v_l^\epsilon, \ v^\epsilon = (v_l^\epsilon)_{1 \le l \le l_N}$$

$$\chi_l = -1, \ v_l|_{t=0} = u_{k0} / \lambda, \quad 1 \le l \le l_m$$

$$\chi_l = 1, \ v_l|_{t=0} = u_{k0} / \mu, \quad l_m < l \le l_N. \tag{5.120}$$

By uniqueness, we have

$$v_l^\epsilon = v_{l_k}^\epsilon, \quad l_{k-1} < l \le l_k, \ 1 \le k \le N.$$

Then, we see that u^ϵ is the solution of system (5.6) if v^ϵ is the solution of the extended system (5.120).

The uniqueness of z in Lemma 5.3 is classical. See [Horn and Jackson (1972); Mielke, Haskovec, and Markowich (2015); Fellner and Tang (2017)].

A main contribution of [Pierre, Suzuki, and Zou (2017)] in the proof of Theorem 5.4 is to simplify rather significantly the proof Lemma 5.6 and consequently to be able to reach the general case (5.4). In the particular cases already known by [Desvillettes and Fellner (2008, 2006, 2014); Fellner, Latos, and Suzuki (2016)], namely for 3 or 4 systems, this part of the proof is rather involved and requires much technicality. For instance, we compare the variation of $\sqrt{u}$ with the square root $\sqrt{\overline{u}}$ of its average rather than with the average of the square root. The corresponding computation turns out to be quite simpler and sufficient for the expected estimate of Lemma 5.9. We also simplify the proof of the estimate from below of $f(\sqrt{\overline{u}})$ in Lemma 5.8. In fact, entropy decay estimate is exploited for some particular 3×3 and 4×4 systems. See [Desvillettes and Fellner (2006, 2008, 2014); Fellner, Latos, and Suzuki (2016)]. Here we adapt and extend to any number of components, following the entropy method developed there. This method is based on the use of relative entropy, entropy dissipation, treated by Csizár-Kullback inequality (4.207) valid to (4.208), and the logarithmic Sobolev inequality derived by [Gross (1975)],

$$\int_{\mathbf{R}^n} u \log u \ dx \le 2\|\nabla\sqrt{u}\|_2^2 + A(\|u\|_1) \tag{5.121}$$

for $0 \le u = u(x) \in L^1(\mathbf{R}^n)$ with $\nabla\sqrt{u} \in L^2(\mathbf{R}^n)$ in the sense of distributions, where $A(s) = n + s\log(s/(2\pi)^{n/2})$. See Theorem 17 of [Carrillo, Jüngel, Markowich, Toscani, Unterriter (2001)] for (5.121).

Inequality (5.37) follows from Lemma 4 of [Pierre and Rolland (2016)]. See Proposition 6.1 of [Pierre (2010)] when $(u_{k0}) \in L^2(\Omega)^N$ in which case we may take $\tau = 0$. The adaptation of the Cziszár-Kullback inequality for the proof of Lemma 5.6 is taken by [Desvillettes and Fellner (2008, 2006, 2014); Fellner, Latos, and Suzuki (2016); Arnlod, Markowich, Toscani, and Unterreiter (2000)]. Inequality (5.40) is due to Theorem 31 in [Carrillo, Jüngel, Markowich, Toscani, Unterriter (2001)]. See also [Csiszár (1963)]).

Section 5.4 is due to [Pierre, Suzuki, and Zou (2017)]. This proof is inspired from those given in [Desvillettes and Fellner (2006, 2008, 2014); Fellner and Laamri (2016)] for the 4×4 systems, with some significant improvements and simplifying modifications. For the logarithmic Sobolev inequality in the form of (5.56), see Theorem 17 in [Carrillo, Jüngel, Markowich, Toscani, Unterriter (2001)].

So far, the integrability of what is called the N-system is known. See [Itano and Suzuki (2016)]. Biological theory on MMP2 activation is illustrated in [Seiki (1994)]. Mathematical model (5.58) is used in [Hoshino (2012)] for the study of cell biology. See Figure 1.3 of [Suzuki (2017)] for this reaction network. See [Hoshino (2012)] for the parameters used for simulation, following experimental data. The explicit formula of solutions to (5.58) is given in [Kawasaki, Minerva, Suzuki, and Itano (2017)].

By the total mass conservations of a, b, and c, the quantities

$$X_1 + X_4 + X_7 + X_9 + X_{11} + 2X_{12}$$
$$X_2 + X_4 + X_5 + X_7 + X_8 + X_9 + 2X_{10} + 2X_{11} + 2X_{12}$$
$$X_3 + X_5 + 2X_6 + X_7 + 2X_8 + 2X_9 + 2X_{10} + 2X_{11} + 2X_{12}$$

are invariant in time. Thus it follows that

$$\frac{d}{dt}(X_1 + X_4 + X_7 + X_9 + X_{11} + 2X_{12}) = 0$$
$$\frac{d}{dt}(X_2 + X_4 + X_5 + X_7 + X_8 + X_9 + 2X_{10} + 2X_{11} + 2X_{12}) = 0$$
$$\frac{d}{dt}(X_3 + X_5 + 2X_6 + X_7 + 2X_8 + 2X_9 + 2X_{10} + 2X_{11} + 2X_{12}) = 0$$

from (5.58). By (5.59) we obtain

$$\frac{d}{dt}X_1 = -k_1 X_1(X_2 + X_5 + X_8 + X_{10} + X_{11})$$
$$\frac{d}{dt}(X_2 + X_5 + X_8 + 2X_{10} + X_{11}) = -k_1 X_1(X_2 + X_5 + X_8$$
$$+X_{10} + X_{11}). \tag{5.122}$$

The first and the second equations of (5.122) govern the conservations of mass of a and b, respectively, in this reaction. Since two b's of X_{10} are active in (5.59) we put X_{10} twice on the left-hand side of the second equation of (5.122). In fact the conservation of mass of b now reads

$$\frac{dX_1}{dt} = \frac{d}{dt}(X_2 + X_5 + X_8 + 2X_{10} + X_{11}).$$

By (5.60), next, we obtain

$$\frac{d}{dt}(X_2 + X_4) = -k_2(X_2 + X_4)(X_3 + X_6 + X_8 + X_9)$$
$$+\ell_2(X_5 + X_7 + X_8 + X_9 + X_{10} + 2X_{11} + X_{12}),$$
$$\frac{d}{dt}(X_3 + 2X_6 + X_8 + X_9) = -k_2(X_2 + X_4)(X_3 + X_6 + X_8 + X_9)$$
$$+\ell_2(X_5 + X_7 + X_8 + X_9 + X_{10} + 2X_{11} + X_{12}). \tag{5.123}$$

The first and the second equations of (5.123) govern the conservation of mass of b and c, respectively, in this reaction. Here the term $2X_6$ on the left-hand side of the second equation indicates that two c's of X_6 are active in the attachment of (5.60). The presence of the term $2X_{11}$ on the right-hand sides of both equations of (5.123) is due to the fact that X_{11} is involved by the reactions of both X_2 and X_4 in (5.60).

Section 5.6 is taken from [Kobayashi, Suzuki, and Yamada (2017)]. Model (5.74) is introduced by [Lotka (1925); Volterra (1926)]. It is used in several areas including ecology, economics, and chemistry, and has been studied extensively (see, e.g., [Cheon (2003); Gao (2000)]). See [Gopalsamy (1984); Huston and Schmitt (1992); Kona and Hofbauer (2003)]) for permanence or uniform persistence. We note that the conclusion of Theorem 1.4. of [Suzuki and Yamada (2015)] is valid to $A = (a_{jk}) \in Y$ satisfying ($a3$). Hence there is at least $(2N - 1)$ degree of freedom of $A = (a_{jk})$, such that the ω-limit set of the solution to the associated reaction diffusion system of (5.74) is a Jordan curve or singleton contained in $\mathbf{R}_+^N$. See the original paper [Kobayashi, Suzuki, and Yamada (2017)] for the proof of Lemma 5.15. Permanence of (5.74) means that any solution $v = (v_j(t))$ exists global-in-time and it holds that $\liminf_{t\uparrow+\infty} v_j(t) > 0$ for any j. From the above method we can show the following theorem.

Theorem 5.7. *System (5.74), $N \geq 3$, exhibits permanence if $^t A + A \leq 0$ and there exists $i, j, l \in \{1, \cdots, N\}$ such that $a_{jl}, a_{ij} > 0 > a_{il}$ and*

$$a_{ij}a_{kl} + a_{il}a_{jk} - a_{ik}a_{jl} \leq 0, \quad a_{jk}e_i - a_{ik}e_j + a_{ij}e_k \leq 0 \qquad (5.124)$$

for $1 \leq k \leq N$.

Proof. Since $^t A + A \leq 0$ we have

$$\frac{d}{dt}\tau \cdot v \leq -r \cdot v$$

for (5.125). Then it follows that $T = +\infty$ from $v = (v_k(t)) > 0$. We assume $i = 1, l = N$ without loss of generality, and take $V = V_j(\xi)$, $2 \leq j \leq N-1$, in (5.99), where $\xi = (\log v_j)$. By the first equality of (5.103) and the assumption (5.124), it holds that

$$\frac{dV}{dt} \geq 0.$$

Then the set $\hat{\mathcal{O}}_* = \{\xi \in \mathbf{R}^N \mid V(\xi) \geq V(\xi_0)\}$ is bounded below for $\xi_0 = (\log v_{j0})$, because of the assumption $a_{jl}, a_{ij} > 0 > a_{il}$ with $i = 1, l = N$. $\square$

Some part of Section 5.7 is taken from [Suzuki, Itano, Zou, Iwamoto, and Mekada (2016)]. For maturation pathways of osteoblast and osteoclast, see [Harada and Rodan (2003); Teitelbaum and Ross (2003)]. Several mathematical models used for biological phenomena are collected in [Murray (2003a,b)]. A numerical simulation is illustrated in [Suzuki, Itano, Zou, Iwamoto, and Mekada (2016)] for the case of $a = 1$, $b = 1$, $c = 1$, $d = 1$, $e = 3$, $f = 10$, $g = 1$, and $h = 1$ in (5.108)–(5.109) and $X_1(0) = 7(1+0.01)$, $X_3(0) = 4(1 + 0.01)$, $X_5(0) = 4(1 + 0.01)$, $X_2(0) = 0$, $X_4(0) = 0$ as an initial value near from the unstable dynamical equilibrium.

To identify the cell molecule casting DAF, knock down technique may be used. Here we specify what should be observed under this operation, modifying the bottom up model (5.108)–(5.109). We take three modifications, cutting the effects of annihilation, that is, annihilation of differentiation of pre-osteoblast to osteoblast, that of pre-osteoclast to osteoclast, and both. We thus obtain three bottom up models, that is,

$$m_2 = aX_1 + b, \quad \ell_1 = c, \quad \ell_2 = \frac{f}{gX_5 + h}, \tag{5.125}$$

$$m_2 = aX_1 + b, \quad \ell_1 = \frac{c}{dX_5 + e}, \quad \ell_2 = f, \tag{5.126}$$

and

$$m_2 = aX_1 + b, \quad \ell_1 = c, \quad \ell_2 = f. \tag{5.127}$$

Their dynamical equilibria are reduced to

$$\lambda X_5 = \frac{1}{f}\left(\frac{am_1}{c} + b\right)(gX_5 + h)$$
$$\lambda X_5 = \frac{1}{f}\left(\frac{am_1}{c}(dX_5 + e) + b\right)$$
$$\lambda X_5 = \frac{1}{f}\left(\frac{am_1}{c} + b\right) \tag{5.128}$$

with $\lambda = \frac{\delta}{\gamma}$. In the first and second cases of (5.128), there is $\overline{\lambda} > 0$ such that the unique solution $X_5 = X_5(\lambda)$ exists for $\lambda > \overline{\lambda}$, while there is no solution for $0 < \lambda \leq \overline{\lambda}$. This unique solution is stable and the break down of dynamical equilibrium arises with $\lim_{\lambda \downarrow \overline{\lambda}} X_5(\lambda) = +\infty$. This property has a strong contrast with that of the original model, (5.108)–(5.109). Then it is easy to suspect that saturation of X_2 and X_4 arises at this occasion in models (5.125) and (5.126), respectively. In the third case, finally, there is a unique stable dynamical equilibrium for any $\lambda > 0$.

Break down of dynamical equilibrium may be caused several ways. Increase of a or b in (5.108) may be achieved by injecting RANKL. This technique is standard in cell biology. Since (5.112) takes the form

$$\frac{\delta f}{\gamma} X_5 = \left(\frac{am_1}{c} (dX_5 + e) + b \right) (gX_5 + h) \qquad (5.129)$$

we can create a break down of dynamical equilibrium by making a or b large. Since X_2 saturates at this occasion, osteoporosis will be observed for wild type, which will not appear to the control, knock down mise of DAF. In fact, the function $\psi(x)$ defined by (5.116) is constant and any solution will approach the unique dynamical equilibrium in the case of (5.94). We can present several medical insights based on the argument above. First, break down of dynamical equilibrium may be predicted by the increase of DAF. Second, the recovery of dynamical equilibrium plays an essential role against both diseases, osteopetrosis and osteoporosis. Since (5.129), good manipulations are increase of δ, f, c and decrease of γ, a, m_1, d, e, b, g, h. Generally, inhibition of DAF or that of differentiation to pre-osteoblast and pre-osteoclast are efficient to recover dynamical equilibrium. Third, rapid increase of DAF density can be a trigger of osteoclast.

Section 5.8 is based on [Hatanaka, Seki, Inoue, Tero, and Suzuki (2017)]. Here, NF-κB is a transcription factor that regulates numerous genes, and the signal transduction pathway that activates NF-κB is defined by chemical reactions. However, molecular mechanisms of the NF-κB pathway have not been fully elucidated. Here we apply mathematical analysis to the NF-κB pathway as follows:

(1) constructing a new mathematical model which considers phosphorylation of $I\kappa B\alpha$ and $RelA$ for the NF-κB canonical pathway.
(2) analyzing the characteristics of oscillation of NF-κB pathway.

A mathematical model of NF-κB signaling pathway has been originally constructed by [Hoffmann (2002)]. $TNF\alpha$ stimulation is modeled as an addition of active IKK, and the model is composed of three main parts;

(1) the binding and dissociation of $I\kappa B$, NF-κB, and IKK.
(2) $N - C$ shuttling of $I\kappa B$ and NF-κB.
(3) transcription and translation of $I\kappa B$.

In this model, phosphorylation of $I\kappa B$ is represented by binding to IKK, and ubiquitination and proteolysis of $I\kappa B$ are assumed to follow immediately after formation of $I\kappa B/IKK$ complex. The transport between nucleus and cytoplasm is modeled as a primary reaction, and five pathway of $I\kappa Bi \rightarrow I\kappa B$, $NF - \kappa Bi \rightarrow NF$-$\kappa B$, $I\kappa B/NF - \kappa Bi \rightarrow I\kappa B/NF$-$\kappa B$ are assumed. The transcription of $I\kappa B$ $mRNA$ is modeled as secondary reaction, that is, $I\kappa B$ $mRNA$ is generated in proportion to the square of the concentration of nuclear NF-κB. Various mathematical models have been reported after this work.

In [Nelson (2004)] it is mentioned that transcription process should be modeled as a primary reaction in order to maintain consistency with experimental data in single cells. Also, a model in which the transcription reaction is represented as a differential equation with time delay is reported in [Hoffmann (2013)]. On the other hand, there is a model representing transcription reaction using Hill function instead of law of mass action [Nelson (2009)]. However, since the transcription reaction has a complicated structure, it is necessary to properly select mathematical models according to the purpose. In [Lipniacki (2004)] it is reported that transport modeling as a primary reaction is not appropriate because the membrane transport is in proportion to the nuclear volume. Subsequently, 3D simulation NF-κB by [Ohshima and Ichikawa (2014)] shows that spatial factors such as volume ratio of nucleus and cytoplasm, and distribution of proteins affect oscillation. It is confirmed that the oscillation pattern changes particularly with the diffusion coefficient.

Regarding degradation of $I\kappa B$, the following models have been reported in [Nelson (2009); Hoffmann (2013, 2016)]: a model that includes the phosphorylated $I\kappa B$, represents more simply as $IKK + I\kappa B/NF$-$\kappa B \rightarrow IKK + NF - \kappa B$, and express $I\kappa B$ degradation by IKK using Michaelis-Menten equation. For each model, mathematical analysis by numerical simulation and dynamical system theory is performed. In [Hoffmann (2013)] a model with $I\kappa B$-mediated negative feedback is reported. In the model, cytoplasmic and nuclear compartments were not modeled, and the model included the time delay in the transcription/translation process. The result of model simulation showed that the delay time τ affects the magnitude of oscillation. That is, the two feedback loops may be tuned to provide for rapid activation and inactivation capabilities. In [Hoffmann (2016)] it is analyzed mathematically the NF-κB oscillation and switch-like response, characteristic phenomenon in which the amplitude of oscillation rapidly increases when the intensity of stimulation exceeds a certain threshold. There, core models based on three components IKK, $I\kappa B$, and NF-κB are used, and cytoplasmic and nuclear compartments are not considered. The results of the stability analysis of the equilibrium point have implied that the NF-κB signaling pathway always has one stable equilibrium point, and the stable node point changes to the stable focus point depending on the strength of stimulation and the total concentration of IKK, that is, the damped oscillation is caused by the pitch fork bifurcation. More detailed model are developed in [Nelson (2009)] to analyze the effect of pulsatile stimulation on NF-κB dependent transcription. The model includes two negative feedback loops: one in which $I\kappa B$ inhibits transcriptional activation by NF-κB, and another in which it inhibits liberating NF-κB from $I\kappa B/NF$-κB complexes by IKK. Moreover, the regulation of gene expression by NF-κB is modeled by using Hill function, and $I\kappa B$ phosphorylation by IKK is considered. From this model, a reduced model containing 14 ordinary differential equations to capture essential mechanisms of oscillation is constructed in [Wang (2012)]. This oscillation detail was explained as Hopf bifurcation. The model has two steady state depending on the bifurcation parameter TR representing the intensity of $TNF\alpha$ stimulation.

As a result, standard Hoffmann's model has a global stable equilibrium, and the oscillation phenomenon is caused by the unstable limit cycle transiently until the orbit reaches this equilibrium. Our phosphorylation model, on the other hand, demonstrates that a transitional oscillation phenomenon occurs due to the appearance of a transitional stable periodic orbit in the process of decay of the activated $IKKb$. Since the solution obtained from this phosphorylation model is attracted by this stable periodic orbit, it is possible to obtain an oscillation with a constant amplitude and a period regardless of the initial states.

In our model with phosphorylation, the same values as in Hoffmann's model are used for most parameters, but some parameters are unknown. The unknown reaction rates, pa, $p1$, $p2$, $deg11$, in the pathways

$$I\kappa BNF - \kappa B + IKK \rightarrow I\kappa B^*NF - \kappa B + IKK \ (pa)$$
$$I\kappa B^*NF - \kappa B \rightarrow NF - \kappa B \ (deg11)$$
$$NF\kappa B \rightarrow NF - \kappa B \ (p1), \quad NF - \kappa B^* \rightarrow NF - \kappa B \ (p2)$$

are estimated by data fitting. There, parameters are set so as to have a vibration period of 90 minutes to 120 minutes. Compared to the experimental results, we see that our phosphorylation model closely approximates the experimental values. See the original [Hatanaka, Seki, Inoue, Tero, and Suzuki (2017)].

The phosphorylation model constructed in this work is based on the law of mass action and has five main processes:

(1) binding $a + b \to ab$.

(2) degradation $ab \to a$, $a \to 0$.

(3) nuclear-cytoplasmic translocation $a \to ai$.

(4) transcription and translation $a \to a + b$, $2a \to 2a + b$.

(5) phosphorylation and dephosphorylation $ab + d \to a * b + d$, $b^* \to b$.

Transcription reaction is explained as secondary reaction based on Hoffmann's model:

$$2NF\kappa B \to 2NF\kappa B + i\kappa B \ (tr_2).$$

The concentration of $NF\text{-}\kappa B$ does not change before and after the reaction. Therefore, the reaction related to $NF\text{-}\kappa B$ needs not be considered, and the reaction rate with transcription is expressed as

$$\frac{d[i\kappa B]}{dt} = tr_1[NF\kappa B]^2.$$

Since the translation reaction,

$$i\kappa B \to i\kappa B + I\kappa B \ (tr1)$$

is also assumed that the RNA does not change before and after the reaction like the transcription reaction, it is described by

$$\frac{d[I\kappa B]}{dt} = tr_1[i\kappa B].$$

In addition, since the reaction related to phosphorylation does not change IKK before and after the reaction, only the reaction on $I\kappa B$ may be regarded. Moreover, as the reaction related to dephosphorylation dephosphorylation of $NF\text{-}\kappa B$ occurs in a constant rate. Therefore, the equations related to phosphorylation and dephosphorylation

$$I\kappa BNF - \kappa B + IKK \to I\kappa B^* NF - \kappa B + IKK \ (pa)$$
$$NF - \kappa B^* \to NF - \kappa B \ (p2)$$

are given by

$$\frac{d[a^*b]}{dt} = -p_1[a^*b][d]$$
$$\frac{d[b_i^*]}{dt} = -p_2[b^*], \quad \frac{d[b_i]}{dt} = p_2[b_i].$$

By these factors, $I\kappa B$ phosphorylation model and full phosphorylation model are given by

$$\frac{dX_1}{dt} = -(s_1 + d_1)X_1 + u_1X_7 + tr_1X_9 - k_{1b}X_1X_2$$

$$\frac{dX_2}{dt} = -s_2X_2 + deg_{11}X_{11} - k_{1b}X_1X_2$$

$$\frac{dX_3}{dt} = -d_2X_3$$

$$\frac{dX_4}{dt} = u_4X_{10} + k_{1b}X_1X_2 - p_aX_3X_4$$

$$\frac{dX_7}{dt} = s_1X_1 - u_1X_7 - k_{1b}X_7X_8$$

$$\frac{dX_8}{dt} = s_2X_2 - k_{1b}X_7X_8$$

$$\frac{dX_9}{dt} = e - d_3X_9 + tr_2X_8^2$$

$$\frac{dX_{10}}{dt} = -u_4X_{10} + k_{1b}X_7X_8$$

$$\frac{dX_{11}}{dt} = -deg_{11}X_{11} + p_aX_3X_4$$

and

$$\frac{dX_1}{dt} = -(s_1 + d_1)X_1 + u_1X_7 + tr_1X_9 - k_{1b}X_1(X_2 + X_{12})$$

$$\frac{dX_2}{dt} = -(p_1 + s_2)X_2 + deg_{11}X_{11} - k_{1b}X_1X_2$$

$$\frac{dX_3}{dt} = -d_2X_3$$

$$\frac{dX_4}{dt} = u_4X_{10} + k_{1b}X_1(X_2 + X_{12}) - p_aX_3X_4$$

$$\frac{dX_7}{dt} = s_1X_1 - u_1X_7 - k_{1b}X_7(X_8 + X_{13})$$

$$\frac{dX_8}{dt} = s_2X_2 + p_2X_{13} - k_{1b}X_7X_8$$

$$\frac{dX_9}{dt} = e - d_3X_9 + tr_2X_8^2 + tr_3X_{13}^2$$

$$\frac{dX_{10}}{dt} = -u_4X_{10} + k_{1b}X_7(X_8 + X_{13})$$

$$\frac{dX_{11}}{dt} = -deg_{11}X_{11} + p_aX_3X_4$$

$$\frac{dX_{12}}{dt} = p_1X_2 - s_2X_{12} - k_{1b}X_7X_{13}$$

$$\frac{dX_{13}}{dt} = s_2X_{12} - p_2X_{13} - k_{1b}X_7X_{13},$$

respectively, where $X_1 = [I\kappa B\alpha] = [a]$, $X_2 = [NF\text{-}\kappa B] = [b]$, $X_3 = [IKKb] = [d]$, $X_4 = [I\kappa B\alpha/NF\text{-}\kappa B] = [ab]$, $X5 = [IKKb/I\kappa B\alpha] = [da]$, $X6 = [IKKb/I\kappa B\alpha/NF\text{-}\kappa B] = [dab]$, $X7 = [I\kappa B\alpha_i] = [a_i]$, $X_8 = [NF\text{-}\kappa B_i] = [b_i]$, $X9 = [I\kappa B mRNA] = [c]$, $X_{10} = [I\kappa B\alpha/NF\text{-}\kappa B_i] = [ab_i]$, $X_{11} = [I\kappa B\alpha^*/NF\text{-}\kappa B] = [a^*b]$, $X_{12} = [NF\text{-}\kappa B^*] = [b^*]$, $X_{13} = [NF\text{-}\kappa B_i^*] = [b_i^*]$.

Bibliography

Alikakos, N.D. (1979), *An application of the invariance principle to reaction-diffusion equations*, J. Differential Equations, **4**, 827–868.

Arnold, A., Markowich, P., Toscani, G., and Unterreiter, A. (2000), *On generalized Csiszar-Kullback inequalies*, Monatsh. Math., **131**, 235–253.

Ashall, L., Horton, C.A., Nelson, D.E., Paszek, P., Harper, C.V., Sillitoe, K., Ryan, S., Spiller, D.G., Unitt, J.F., Broomhead, D.S., Kell, D.B., Rand, D.A., See, V., and White M.R.H. (2013), *Pulsatile stimulation determines timing and specificity of NF − κB-dependent transcription*, Science **324**, 242–246.

Bandle, C. (1971), *Konstruktion isoperimetrischer Ungleichungen der matehmatischen Physik aus solchen der Geometrie*, Comment. Math. Hlelv., **46**, 182–213.

Bandle, C. (1976), *Isoperimetric inequalities for a nonlinear eigenvalue problem*, Proc. Amer. Math. Soc., **56**, 243–246.

Bandle, C. (1980), *Isoperimetric Inequalities and Applications*, Pitmann, London.

Baras, P. (1978), *Compacité de l'opérateur f → u solution d'une équation non linéaire (du/dt) + Au ∋ f*, C.R. Acad. Sci. Paris A, **286**, 1113–1116.

Baras, P. and Pierre, M. (1984), *Problème paraboliqus semi-linéaires avec donées mesures*, Appl. Anal., **18**, 111–149.

Bartolucchi, D. and Lin, C.-S. (2014), *Existence and uniqueness for mean field equation on multiply connected domain at the critical parameter*, Math. Ann., **359**, 1–44.

Berestycki, H. and Brezis, H. (1980), *On a free boundary problem arising in plasma physics*, Nonlinear Analysis, **4**, 415–436.

Biler, P. (1998), *Local and global stability of some parabolic systems modelling chemotaxis*, Adv. Math. Sci. Appl. **8**, 715–743.

Biler, P., Hilhorst, D., and Nadzieja, T. (1994), *Existence and nonexistence of solutions for a model of gravitational interaction of particles. II.*, Colloq. Math. **67**, 297–308.

Blanchet, A., Carrillo, J.A., and Laurençot, P. (2009), *Critical mass for a Patlak-Keller-Segel model with degenerate diffusion in higher dimension*, Calc. Vari. PDE **35**, 133–168.

Bonhoeffer, S., May, R.M., Shaw, G.M., and Nowak, M.A. (1997), *Virus dynamics and drug therapy*, Proc. Natl. Acad. Sci. USA **94**, 6971–6976.

Bothe, D., Fisher, A., Pierre, M., and Rolland, G. (2017), *Global wellposedness for a class of reaction-advection-anisotropic-diffusion systems*, J. Evolution Equations **17**, 101–130.

Brezis, H. (1983), *Analyse Fonctionnelle, Théorie et Applications, Masson*, Paris.

Brezis, H., Li, Y.Y., and Shafrir, I. (1993). *A sup + inf inequality for some nonlinear elliptic equations involving exponential nonlinearities*, J. Funct. Anal. **116**, 344–358.

Brezis, H. and Merle, F. (1991), *Uniform estimates and blow-up behavior for solutions of −Δu = V(x)eᵘ in two dimensions*, Comm. Partial Differential Equations **16**, 1223–1253.

Brezis, H. and Peletier, L.A. (1989), *Asymptotics for elliptic equations involving critical growth*, Partial Differential Equations and Calculus of Variations (Colombini, I.F., Marino, A., Modica, L., and Spagnolo, S. ed.), Birkhäuser, Boston, 149–192.

Brezis, H. and Strauss, W. (1973), *Semi-linear second-order elliptic equations in L¹*, J. Math. Soc. Japan **25**, 565–590.

Burago, Yu.D. and Zalgaller, V.A. (1988), *Geometric Inequalities*, Springer Verlag, Berlin.

Caffarelli, L.A. and Friedmann, A. (1989), *Convexity of solutions of semilinear elliptic equations*, Duke Math. J. **52**, 431–456.

Caffarelli, L.A., Gidas, B., and Spruck, J. (1989), *Asymptotic symmetry and local behavior of semilinear elliptic equations with critical Sobolev growth*, Comm. Pure Appl. Math. **42**, 271–297.

Caglioti, E., Lions, P.-L., Marchioro, C., and Privilenti, M. (1992), *A special class of stationary flows for two-dimensional Euler equations: a statistical mechanics description*, Comm. Math. Phys. **143**, 501–525.

Cãnizo, J.A., Desvilletes, J., and Fellner, K. (2014), *Improved duality estimates and applications to reaction-diffusion equations*, Comm. Partial Differential Equations **39**, 1185–1204.

Caputo, C., Goudon, T., and Vasseur, A. (2017), *Solutions of the 4-species quadratic reaction-diffusion system are bounded and C∞-smooth, in any space dimension*, preprint.

Caputo, C. and Vasseur, A. (2002), *Global regularity of solutions to systems of reaction-diffusion with sub-quadratic growth in any dimensions*, Comm. Partial Differential Equations **34**, 1228–1250.

Cardaliaguet, P. and Tajrapio, R. (2002), *On the strict concavity of the harmonic radius in dimension N ≥ 3*, J. Math. Pure Appl. **81**, 223–240.

Carrillo, J.A., Hittmeir, S., Volzone, B., and Yao, Y. (2017), *Nonlinear aggregation-diffusion equations: radial symmetry and long term asymptotics*, arXiv:1603.07767v1.

Carrillo, J.A., Jüngel, A., Markowich, P.A., Toscani, G., and Unterriter, A. (2001), *Entropy dissipation methods for degenerate parabolic problems and generalized Sobolev inequalities*, Monatsh Math. **133**, 1–82.

Chandrasekhar, S. (1939), *An Introduction to the Study of Steller Structure*,

University Chicago Press, Chicago.

Chan, S.Y.A., Chen, C.C., and Lin, C.S. (2003), *Extremal functions for a mean field equations in two dimensions*, In; Lectures on Partial Differential Equations (S.-Y.A. Chang, C.-S. Lin, and S.-T. Yau, eds.), International Press, New York, pp. 61–93.

Chanillo, S. and Li, Y.Y. (1992), *Continuity of solutions of uniformly elliptic equations in* $\mathbf{R}^2$, *Manus. Math.* **77**, 415–433.

Chang, S.-Y.A. and Yang, P.-C. (1987), *Prescribing Gaussian curvature on* S^2, *Acta Math.* **159**, 215–259.

Chavanis, P.H. (2004), *Generalized kinetic equations and collapse of self-gravitating Langevin particles in D dimensions*, *Banach Center Publ.* **66**, 79–101.

Chavanis, P.H. (2008), *Two-dimensional Brownian vortices*, *Physica A* **387**, 6917–6942.

Chavanis, P.H. and Sire, C. (2004), *Anomalous diffusion and collapse of self-gravitating Langevin particles in D simensions*, *Phys. Rev. E* **69**, 016116.

Chavanis, P.H., Sommeria, J., and Robert, R. (1996), *Statistical mechanics of two-dimensional vortices and collisionless stellar systems*, *Astrophys. J.* **1** **(471)**, 385–399.

Chen, C.C. and Lin, C.S. (1998), *A sharp* sup + inf *inequality for a semilinear elliptic equation in* $\mathbf{R}^2$, *Comm. Anal. Geom.* **6**, 1–19.

Chen, W. and Li, C. (1991), *Prescribing Gaussian curvatures on surfaces with conical singularities*, *J. Geom. Anal.* **1**, 615–622.

Cheng, K.S. and Lin, C.S. (1997), *On the asymptotic behavior of solutions to the conformal Gaussian curvature equations in* $\mathbf{R}^2$, *Math. Anal.* **308**, 119–139.

Cheon, T. (2003), *Evolutionary stability of echological hierarchy*, *Phys. Rev. Lett.* **4**, 58–105.

Crandall, M.G. and Rabinowitz, P.H. (1975), *Some continuation and variation methods for positive solutions of nonlinear elliptic eigenvalue problems*, *Arch. Rational Mech. Anal.* **58**, 207–218.

Csiszár, I. (1963), *Eine informationstheoretische Ungleichung und ihre Anwendung auf den Beweis von Markoschen Ketten*, *Magyar Tud. Akad. Mat. Kutató Int. Közl.* **8**, 85–108.

Damlamian, A. (1978), *Application de la dualité non convexe à une problème non lineéaire à frontiére libre*, *C.R. Acad. Sci. Paris* **286A**, 153–55.

de Figueiredo, D.G., Lions, P.L. and Nussbaum, R.D. (1982), *A priori estimates and existence of positive solutions of semilinear elliptic equations*, *J. Math. Pure Appl.* **61**, 41–63.

del Pino, M., Kowalczyk, M., and Musso, M. (2005), *Singular limits in Liouville-type equations*, *Calc. Vari. PDE* **24**, 47–81.

Devillettes, L. and Fellner, K. (2006), *Exponential decay toward equilibrium via entropy methods for reaction-diffusion equations*, *J. Math. Anal. Appl.* **319**, 157–176.

Devillettes, L. and Fellner, K. (2008), *Entropy methods for reaction-diffusion equations: slowly growing a priori bounds*, *Rev. Mat. Ibero Amer.* **24**, 407–431.

Devillettes, L. and Fellner, K. (2014), *Duality-entropy methods for reaction-diffusion equations arising in reversible chemistry*, System. Mod. Optm. *IFIP AICT* **443**, 96–104.

Devillettes, L., Fellner, K., Pierre, M., and Vovelle, J. (2014), *Global existence for quadratic systems of reaction diffusion*, Adv. Nonlinear Stud. **7**, 491–511.

Doi, M. and Edwars, S.F. (1986), *The Theory of Polymer Dynamics*, Oxford Science Publications, Oxford.

Ekeland, I. and Temam, R. (1976), *Convex Analysis and Variational Problems*, North-Holland, Amsterdam.

Evans, L.C. (1998), *Partial Differential Equations*, Amer. Math. Soc. Providence.

Evans, L.C. and Gariepy, R.F. (1992), *Measure Theory and Fine Properties of Functions*, CRC Press Roca Raton.

Eyink, G.L., Spohn H., and Chen, W. (1993), *Negative-temperature states and large-scale, long-lived vortices in two-dimensional turbulence*, J. Stat. Phys. **70**, 833–886.

Feinberg, M. (1987), *Chemical reaction network structure and the stability of complex isothermal reactors, I. The deficiency zero and deficiency one theorems*, Chem. Eng. Sci. **42**, 2229–2268.

Feinberg, M. and Horn, F.J.M. (1974), *Dynamics of open chemical systems and the algebraic structure of the underlying reaction network*, Chem. Eng. Sci. **29**, 775–787.

Fellner, K. and Laamri, E.H. (2016), *Exponential decay towards equilibrium and global classical solutions for nonlinear reaction-diffusion systems*, J. Evolution Equations **16**, 681–704.

Fellner, K., Latos, E., and Suzuki, T. (2016), *Global classical solutions for mass-conserving, (super)-quadratic reaction-diffusion systems in three and higher space dimensions*, Discrete Cont. Dyn. Syst. B **21**, 3441–3462.

Fellner, K. and Tang, B.Q., *Explicit exponential convergence to equilibrium for nonlinear reaction-diffusion systems with detailed balance condition*, Nonlinear Analysis **159**, 145–180.

Fisher, J. (2015), *Global existence of renormalized solutions to entropy-dissipating reaction-diffusion systems*, Arch. Rational Meth. Anal. **218**, 553–587.

Friedman, A. (1982), *Variational Principles and Free-boundary Problems*, Wiley, New York.

Gao, P.Y. (2000), *Hamilton structure and first integrals for the Lotka-Voterra systems*, Phys. Lett. A. **273**, 85–96.

Gajewski, H. and Zacharias, K. (1998), *Global behaviour of a reaction-diffusion system modelling chemotaxis*, Math. Nachr. **195**, 77–114.

Gidas, B., Ni, W.-M., and Nirenberg, L. (1979), *Symmetry and related properties via the maximum principle*, Comm. Math. Phys. **68**, 209–598.

Gidas, B. and Spruck, J. (1981), *Global and local behavior of positive solutions of nonlinear elliptic equations*, Comm. Pure Appl. Math. **34**, 525–598.

Gidas, B. and Spruck, J. (1981), *A priori bounds for positive solutions of nonlinear elliptic equations*, Comm. Partial Differential Equations **6**, 883–901.

Gilbarg, D. and Trudinger, N.S. (1983), *Elliptic Partial Differential Equations of Second Order*, Springer Verlag, Berlin.

Gladiali, F., Grossi, M., Ohtsuka, H., and Suzuki, T. (2014), *Morse index of multiple blowup solution to the two-dimensional Gel'fand problem*, Comm. *Partial Differential Equations* **39**, 2028–2063.

Gopalsamy, K. (1984), *Persistence in periodic and almost periodic Lotka-Volterra systems*, J. Math. Biol. **21**, 145–148.

Goudon, T. and Vasseur, A. (2010), *Regularity analysis for systems of reaction-diffusion equations*, Ann. Sci. Ec. Norm. Super. *(4)* **43**, 117–141.

Gross, L. (1975), *Logarithmic Sobolev inequalities*, Amer. J. Math. **97**, 1061–1083.

Grossi, M. and Takahashi, F. (2010), *Nonexistence of multi-bubble solutions to some elliptic equations on convex domains*, J. Funct. Anal. **259**, 904–917.

Gui, C., Ni, W.M., and Wang, X. (1992), *On the stability and instability of positive steady states of a semilinear heat equation*, Comm. Pure Appl. Math. **45**, 1153–1181.

Gustaffson, B. (1979), *On the motion of a vortex in two-dimensional flow of an ideal fluid in simply connected domains*, Tech. Report, Department of Math. Royal Insititute of Technology, Stockholm.

Haegi, H.R. (1951), *Extremalprobleme und Ungleichungen Konformer Gebietsgrössen*, Commposito Math. **8**, 81-111.

Hale, J.K. (1988), *Asymptotic Behavior of Dissipative Systems*, Amer. Math. Soc., Providence.

Han, Z.C. (1991), *Asymptotic approach to singular solutions for nonlinear elliptic equations involving critical Sobolev exponent*, Ann. Inst. Henri Poincaré, Analyse Non Linéaire **8**, 159-174.

Harada, S. and Rodan, G.A. (2003), *Control of osteoblast function and regulation of bone mass*, Nature **423**, 349–355.

Harrabi, A. and Rebhi, S. (1998), *Solutions for superlinear elliptic equations and their Morse indices, I.*, Duke Math. J. **94**, 141-157.

Harrabi, A. and Rebhi, S. (1998), *Solutions for superlinear elliptic equations and their Morse indices, II.*, Duke Math. J. **94**, 159-179.

Hatanaka, N., Seki, T. Inoue, J. Tero, A., and Suzuki, T. (2017), *Critical roles of $I\kappa B\alpha$ and RelA phosphorylation in transitional oscillation in $NF - \kappa B$ signaling module*, PLoS Comput. Biol., submitted.

Henry, D. (1981), *Geometric Theory of Semilinear Parabolic Equations*, Lecture Notes in Math. **840**, Springer-Verlag.

Herrero, M.A. and Velázquez, J.J.L. (1996), *Singularity patterns in a chemotaxis model*, Math. Ann. **306**, 583–623.

Hoffmann, H., Levchenko, A., Scott, M.L., Baltimore, D. (2002), *The $I\kappa B$-$NF\kappa B$ signaling module: temporal control and selective gene activation*, Science **298**, 1241–1245.

Horn, F.J.M. (1972), *Necessary and sufficient conditions for complex balancing in chemical kinetics*, Arch. Rational Mech. Anal. **49**, 172–186.

Horn, F.J.M. and Jackson, R. (1972), *General mass action kinetics*, Arch. Rational Mech. Anal. **47**, 81–116.

Hoshino, D., Koshikawa, N., Suzuki, T., Quaranta, V., Seiki, M., and Ichikawa, K. (2012), *Establishment of computational model for MT1-MMP dependent ECM degradation and intervention strategies*, PLoS Comput. Biol. **8 (4)**, e1002479.

Huang, S.Z. (2006), *Gradient Inequalities*, Amer. Math. Soc., Providence.

Huston, V. and Schmitt, K. (1992), *Permanence and the dynamics of biological systems*, Math. Biosc. **111**, 1-71.

Ikehata, R. and Suzuki, T. (2000), *Semilinear parabolic equations involving critical Sobolev exponent: local and asymptotic behavior of solutions*, Differential Integral Equations **13**, 869–901.

Inoue, K., Shinohara, H., Behar, M., Yumoto, N., Tanaka, G., Hoffmann, A., Aihara, K., and Okada-Hatameyama, M. (2016), *Oscillation dynamics underline functional switching of $NF - \kappa B$ for B-cell activation*, Systems Biology and Applications **2**, 16024.

Itano K. and Suzuki, T. (2016), *Mathematical modeling and mathematical analysis for a pathway network on N molecules (in Japanese)*, Bulletin of Japan Soc. Indu. Appl. Math. Sci. **26**, 44–83.

Itoh, T. (1989), *Blow-up of solutions for semilinear parabolic equations*, Kokyuroku RIMS **679**, 127–139.

Jäger, W. and Luckhaus, S. (1992), *On explosions of solution to a system of partial differential equations modelling chemotaxis*, Trans. Amer. Math. Soc. **329**, 819–824.

Joyce, G and Montgomery, D. (1973), *Negative temperature states for the two-dimensional guiding-centre plasma*, J. Plasma Phys. **10**, 107–121.

Karali, G., Suzuki, T., and Yamada, Y. (2013), *Global-in-time behavior of the solution to a Gierer-Meinhardt system*, Discrete Contin. Dyn. Syst. Ser. A **33**, 2885–2900.

Kan, T. (2013), *Global structure of the solution set for a semilinear elliptic problem related to the Liouville quation on an annulus*, Geometric Properties for Parabolic and Elliptic PDE's (Magnanini, R. and Sakaguchi, S. ed.), Springer Verlag Italia, Roma.

Kato, T. (1985), *Abstract Differential Equations and Nonlinear Mixed Problems*, Academia Nazionale Dei Lincei, Scoula Normale Superiore, Pisa.

Kawasaki, S., Minerva, D., Suzuki, T., and Itano, K. (2017), *Finding solvable units of variables in nonlinear ODEs of ECM degradation pathway network*, Comp. Math. Mech. Medicine **2017** 1-15.

Kazdan, J.L. and Warner, F.W. (1975), *Remarks on some quasilinear elliptic equations*, Comm. Pure Appl. Math. **28**, 567–597.

Keller, E.F. and Segel, L.A. (1970), *Initiation of slime mold aggregation viewed as an instability*, J. Theor. Biol. **30**, 235–248.

Kobayashi, M., Suzuki, T., and Yamada, Y., *Lotka-Volterra systems with periodic orbits*, Funkcial. Ekvac., to appear.

Kon, R. and Hofbauer, J., *Structure of n-dimensional Lotka-Volterra systems for qualitative permanence*, Kokyuroku RIMS, Kyoto Univ. **1309**, 146–153.

Korobeinikov, A. (2004), *Global properties of basic virus dynamics models*, Bull. Math. Biol. **66**, 879–883.

Kurokiba, M. and Ogawa, T. (2003), *Finite time blow-up of the solution for a nonlinear parabolic equation of drift-diffusion type*, Differential Integral Equations **16**, 427–452.

Ladyzĕnskaya, O.A., Solonikov, V.A., and Ural'ceva, N.N. (1968), *Linear and Quasi-linear Equations of Parabolic Type, Amer. Math. Soc.*, Providence.

Laamri, E.H. (2011), *Global existence of classical solutions for a class of reaction-diffusion systems, Acta Appl. Math.* **115**, 153–165.

Latos, E. Suzuki, T., and Yamada, Y. (2012), *Transient and asymptotic dynamics of a prey-predator system with diffusion, Math. Meth. Appl. Sci.* **35**, 1101–1109.

Lepoutre, T., Pierre, M., and Rolland, G. (2012), *Global well-poseness of a conservative relaxed cross diffusion system, SIAM J. Math. Anal.* **44**, 1041–1053.

Li, D. and Zhang, X. (2010), *On a nonlocal aggregation model with nonlinear diffusion, Discrete Contin. Dyn. Syst.* **27**, 1–23.

Li, Y.Y. and Shafrir, I. (1994), *Blow-up analysis for solutions of* $-\Delta u = Ve^u$ *in dimension two, Indiana Univ. Math. J.* **43**, 1255–1270.

Lin, S.S. (1989), *On non-radially symmetric bifurcation in the annulus, J. Differential Equations* **80**, 251–279.

Lions, P.L. (1984), *The concentration compactness principle in the calculus of variation. The locally compact case. Part I., Ann. Inst. H. Poincaré, Anal. Non Linéaire* **1**, 109–145.

Lions, P.L. (1997), *On Euler Equations and Statistical Physics, Cattedra Galileiana*, Pisa.

Lojasiewicz, S. (1963), *Une propriétè topologique des sous ensembles analytiques réels, Coloques du CNRS, Les Équations aue Dèrivées Partielles*, 117.

Lipniacki, T., Paszek, P., Brasier, A.R., Luxon, B., and Kimmel, M. (2004), *Mathematical model of* $NF - \kappa B$ *regularoty module, J. Theor. Biol.* **228**, 195–145.

Longo, D.M., Selimkhanov, J., Kearns, J.D., Hasty, J., Hoffmann, A., Tsimring, L.S. (2013), *Dual delayed feedback provides sensitivity and robustness to the* $NF - \kappa B$ *signaling module, PLoS Comput. Biol.* **9** (6), e1003112.

Lotka, A.J. (1925), *Elements of Physical Biology, William and Wilkins*, London.

Luckhaus, S., Sugiyama, Y., and Velázquez, J.J.L. (2012), *Measure valued solutions of the 2D Keller-Segel system, Arch. Rational Mech. Anal.* **26**, 31–80.

Ma, L and Wei, J.C. (2001), *Convergence of Liouville equation, Comment. Math. Helv.* **76**, 506–511.

Marchioro, C. and Pulvirenti, M. (1994), *Mathematical Theory of Incompressible Nonviscous Fluids, Springer Verlag*, New York.

Masuda, K. and Takahashi, K. (1994), *Asymptotic behavior of solutions of reaction-diffusion systems of Lotka-Volterra type, Differential Integral Equations* **7**, 1041–1053.

Mielke, A., Haskovec, J., and Markowich, P.A. (2015), *On uniform decay of the entropy for reaction-diffusion systems, J. Dyn. Differential Equations* **27**, 897–928.

Moser, J. (1971), *A sharp form of an inequality of N. Trudinger, Indiana Univ. Math. J.* **20**, 1077–1092.

Murakami, M., Nishihara, K., and Hanawa, T. (2004), *Self-gravitational collapse of radially cooling spheres, Astrophys. J.* **607**, 879–889.

Murray, J.D. (2003), *Mathematical Biology, I: An Introduction*, third edition,

Springer-Verlag, New York.

Murray, J.D. (2003), *Mathematical Biology, II: Spatial Models and Biomedical Applications*, third edition, Springer-Verlag, New York.

Nagai, T. (1994), *Blow-up of radially symmetric solutions to a chemotaxis system*, Adv. Math. Sci. Appl. **5**, 581–601.

Nagai, T., Senba, T., and Yoshida, K. (1997), *Applications of the Trudinger-Moser inequality to a parabolic system of chemotaxis*, Funkcial. Ekvac. **20**, 411–433.

Nagasaki, K. and Suzuki, T. (1990a), *Asymptotic analysis for two-dimensional elliptic eigenvalue problem with an exponential non-linearity*, Asymptotic Analysis **3**, 173–188.

Nagasaki, K. and Suzuki, T. (1990b), *Radial and nonradial solutions for nonlinear eigenvalue problem $\Delta u + \lambda e^u = 0$ on annului in $\mathbf{R}^2$*, J. Differential Equations **87**, 144–168.

Naito, Y. and Suzuki, T. (2008), *Self-similarity in chemotaxis systems*, Colloquium Mathematics **111**, 11–34.

Nehari, Z. (1958), *On the principal frequency of a membrane*, Pacific J. Math. **8**, 285–293.

Nehari, Z. (1960), *On a class of nonlinear second-order differential equations*, Trans. Amer. Math. Soc. **95**, 101–123.

Nehari, Z. (1961), *Characteristic values associated with a class of nonlinear second-order differential equations*, Acta Math. **105**, 141–175.

Nelson, D.E., Ihekwaba, A.E.C., Elliott, M., Johnson, J.R., Gibney, C.A., Foreman, B.E., et. al. (2004), *Oscillations in $NF - \kappa B$ signaling control the dynamics of gene expression*, Science **306**, 704–708.

Ni, W.M. (2005), *Qualitative properties of solutions to elliptic problems*, Handbook of Differential Equations, Stationary Partial Differential Equations (Chipot, M. Quittner, P. ed.) **1**, Elsevier, Amsterdam, 157–233.

Ni, W.M., Suzuki, K., and Takagi, I. (2006), *The dynamics of a kinetic activator inhibitor system*, J. Differential Equations **229**, 426–465.

Nowak, M.A. and Bangham, C.R.M. (1996), *Population dynamics of immune responses to persistent viruses*, Science **272**, 74–79.

Obata, M. (1971), *The conjecture on conformal transformations of Riemannian manifolds*, J. Differential Geometry **6**, 247–253.

Ohshima, D., Ichikawa, K. (2014), *Regulation of nuclear $NF - \kappa B$ oscillation by a diffusion coefficient and its biological implications*, PLoS One **9 (10)**, e109895.

Ohtsuka, H. and Suzuki, T. (2003), *Palais-Smale sequence relative to the Trudinger-Moser inequality*, Calc. Vari. PDE **17**, 235–255.

Ohtsuka, H. Senba, T., and Suzuki, T. (2007), *Blowup in infinite time in the simplified system of chemotaxis*, Adv. Math. Sci. Appl. **17**, 445–472.

Onsager, L. (1949), *Statistical hydrodynamics*, Suppl. Nuovo Cimento **6**, 279–287.

Ôtani, M. (2004), *L^∞-energy method and its applications*, Nonlinear Partial Differential Equations and Applications, Gakkotosho, pp. 505–516.

Pierre, M. (2003), *Weak solutions and supersolutions in L^1 for reaction diffusion systems*, J. Evolution Equations **3**, 153–168.

Pierre, M. (2010), *Global existence in reaction-diffusion systems with control of mass: a survey*, Milan J. Math. **78**, 417–455.

Pierre, M. and Rolland, G. (2016), *Global existence for a class of quadratic reaction-diffusion system with nonlinear diffusion and L^1 initial data*, Nonlear Analysis **138**, 369–387.

Pierre, M., Suzuki, T., and Umakoshi, H., *Asymptotic behavior in chemical reaction-diffusion systems with boundary equilibria*, J. Appl. Anal. Comp. **8**, 836–858.

Pierre, M., Suzuki, T., and Yamada, Y., *Dissipative reaction diffusion systems with quadratic growth*, Indiana Univ. Math. J., to appear.

Pierre, M., Suzuki, T., and Zou, R., *Asymptotic behavior of renormalized solution to chemical reaction diffusion sysetms*, J. Math. Anal. Appl. **450**, 152–168.

Pohozaev, S.I. (1965), *Eigenfunction of the equation $\Delta u + \lambda f(u) = 0$*, Soviet Math. Dokl. **6**, 1408–1411.

Pointin, Y.B. and Lundgren, T.S. (1976), *Statistical mechanics of two-dimensional vortices in a bounded container*, Phys. Fluid **19**, 1459–1470.

Prüss, J. (1993). *Evolutionary Integral Equations and Applications*, Birkhäuser, Boston.

Prüss, J., Zacher, R., and Schnaubelt, R. (2008), *Global asymptotic stability of equilibria in models for virus dynamics*, Phys. Fluid **19**, 1459–1470.

Robert, R. (1991), *A maximum-entropy principle for two-dimensional perfect fluid dynamics*, J. Stat. Phys. **65**, 531–553.

Robert, R. and Rosier, C. (1997), *The modeling of small scales in two-dimensional turbulent flows: A statistical mechanics approach*, J. Stat. Phys. **86** 481–515.

Robert, R. and Sommeria, J. (1991), *Statistical equilibrium states for two-dimensional flows*, J. Fluid Mech. **229**, 291–310.

Robert, R. and Sommeria, J. (1992), *Relaxation towards a statistical equilibrium state in two-dimensional perfect fluid dynamics*, Phys. Review Letters **69** 2776–2779.

Rothe, F. (1984), *Global Solutions of Reaction-Diffusion Systems*, Lecture Notes in Math. *Springer Verlag* **1072**, Berlin.

Sasaki, T. and Suzuki, T. (2018), *Asymptotic behaviour of the solution to a virus dynamics model with diffusion* Discrete Cont. Dyn. Syst. Ser. B **23**, 525–541.

Sato, H., Takino, T., Okada, Y., Cao, J., Shinagawa, A., Yamamoto, E., and Seiki, M. (1994), *A matrix metalloproteinase expressed on the surface of invasive tumor cells* Nature **370**, 61–65.

Sawada, S. and Suzuki, T. (2008), *Derivation of the Equilibrium Mean Field Equations of Point Vortex System and Vortex Filament System* Theor. Appl. Mechanics Japan **56**, 285–290.

Sawada, S. and Suzuki, T. (2017), *Relaxation theory for point vortices*, Vortex Structures in Fluid Dynamic Propblems (Perez-de-Tejada, H. ed.) INTECH 2017, Chapter 11, pp. 205–224.

Senba, T. (2007), *Type II blowup of solutions to a simplified Keller-Segel system in two dimensions*, Nonlinear Analysis **66**, 1817–1839.

Senba, T. and Suzuki, T. (2000), *Some structures of the solution set for a stationary system of chemotaxis*, Adv. Math. Sci. Appl. **10**, 191–224.

Senba, T. and Suzuki, T. (2001), *Chemotactic collapse in a parabolic-elliptic system of chemotaxis*, Adv. Differential Equations **6**, 21–50.

Senba, T. and Suzuki, T. (2002a), *Weak solutions to a parabolic-elliptic system of chemotaxis*, J. Funct. Anal. **191**, 17–51.

Senba, T. and Suzuki, T. (2002b), *Time global solutions to a parabolic-elliptic system modelling chemotaxis*, Asymptotic Analysis **32**, 63–89.

Senba, T. and Suzuki, T. (2003), *Blowup behavior of solutions to rescaled Jäger-Luckhaus system*, Adv. Differential Equations **8**, 787–820.

Serrin, J. and Weinberger, H. (1960), *Isolated singularities of solutions of quasilinear equations*, Amer. J. Math. **88**, 258–272.

Shafrir, I. (1992), *A sup + inf inequality for the equation* $-\Delta u = V e^u$, C.R. Acad. Sci. Paris, Série I **315**, 159–164.

Simon, L. (1983), *Asymptotics for a class of nonlinear evolution equations with applications to geometric problems*, Ann. Math. **118**, 525–571.

Sire, C. and Chavanis, P.-H. (2002), *Thermodynamics and collapse of self-gravitating Brownian particles in D dimensions*, Phys. Rev. E **66**, 046133.

Smoller, J. (1986), *Shock Waves and Reaction Diffusion Systems*, Springer Veralag, New York.

Spruck, J. (1988), *The elliptic Sinh Gordon equation and construction fo toroidal soap bubbles*, Calculus of Variations and Partial Differential Equations, Lecture Notes in Math. Springer Verlag **1340**, 275–301.

Stampacchia, G. (1965), *Le problème de Dirichlet pour les équations elliptiques de second ordre à coefficients discontinous*, Ann. Inst. Fourier **15**, 189–258.

Straumann, N. (1995), *General Relativity and Relativistic Astropysics*, Springer Verlag, Berlin.

Sugiyama, Y. (2006), *Global existence in sub-critical cases and finite time blow-up in super-critical cases to degenerate Keller-Segel systems*, Differential Integral Equations **19**, 841–876.

Sugiyama, Y. (2015), *Partial regularity and blow-up asymptotics of weak solutions to degenerate parabolic system of porous medium type*, Manus. Math. **147**, 311–363.

Sugiyama, Y. and Kunii, H. (2006), *Global existence and decay properties for a degenerate Keller-Segel model with a power factor in drift term*, J. Differential Equations **227**, 333–364.

Suzuki, T. (1990), *Introduction to geometric potential theory*, Functional Analytic Methods in Partial Differential Equations (Fujita, H., Ikebe, T. and Kuroda, S.T. eds), Lecture Notes in Math. Springer Verlag **1450**, 83–103.

Suzuki, T. (1992), *Global analysis for a two-dimensional elliptic eigenvalue problem with exponential nonlinearity*, Ann. Inst. Henri Poincaré, Analyse Non Linéaire **9**, 367–398.

Suzuki, T. (1993), *Some remarks about singular perturbed solutions for Emden-Fowler equation with exponential non-linearity*, Functional Analysis and Related Topics (Fujita, H., Komatsu, H. and Kuroda, S.T. eds), Lecture Notes in Math. Springer Verlag **1540**, 341–360.

Suzuki, T. (2005), *Free Energy and Self-Interacting Particles*, Birkhäuser, Boston.

Suzuki, T. (2013), *Exclusion of boundary blowup for 2D chemotaxis system provided with Dirichlet boundary condition, J. Math. Pure Appl.* **100**, 347–367.

Suzuki, T. (2014), *Brownian point vortices and DD-model. Discrete Cont. Dyn. Sys. S.* **7**, 161–176.

Suzuki, T. (2015). *Mean Field Theories and Dual Variation*, second edition, Atlantis Press, Paris, France.

Suzuki, T. (2015a), *Almost collapse mass quantization in 2D Smoluchowski-Poisson equation, Math. Meth. Appl. Sci.* **38**, 3587–3600.

Suzuki, T. (2015b), *Blowup in infinite time for 2D Smoluchowski-Poisson equation, Differential Integral Equations* **28**, 601–630.

Suzuki, T. (2015c), *Residual vanishing for blowup solutions to 2D Smoluchowski-Poisson equation*, arXiv:1502.01795.

Suzuki, T. (2017). *Mathematical Methods for Cancer Evolution*, Springer, Singapore, 2017.

Suzuki, T., Itano, K., Zou, R., Iwamoto, R., and Mekada, E. (2016), *Mathematical modeling for break down of dynamical equilibrium in bone metabolism*, The Role and Importance of Mathematics in Innovation, Proceedings of Forum Mathematics for Industry 2015, Springer Verlag, 25–34.

Suzuki, T. and Senba, T. (2011). *Applied Analysis, Mathematical Methods in Natural Science*, second edition, Imperial College Press, London, UK.

Suzuki, T. and Takahashi, F. (2008), *Nonlinear eigenvalue problem with quantization*, Handbook of Differential Equations, Stationary Partial Differential Equations (Chipot, M. ed.) **5**, Elsevier, Amsterdam, 277–370.

Suzuki, T. and Takahashi, R. (2009), *Degenerate parabolic equation with critical exponent derived from the kinetic theory, I. Generation of the weak solution, Adv. Differential Equations* **14**, 503–524.

Suzuki, T. and Takahashi, R. (2009), *Degenerate parabolic equation with critical exponent derived from the kinetic theory, II. Blowup threshold, Differential Integral Equations* **14**, 1153–1172.

Suzuki, T. and Takahashi, R. (2010), *Degenerate parabolic equation with critical exponent derived from the kinetic theory, IV. Structure of the blowup set, Differential Integral Equations* **15**, 223–250.

Suzuki, T. and Takahashi, R. (2012), *Degenerate parabolic equation with critical exponent derived from the kinetic theory, III. ε-regularity, Differential Integral Equations* **25**, 223–250.

Suzuki, T. and Takahashi, R. (2017), *Critical exponent to a class of semilinear elliptic equations with constraints in higher dimensions - local properties, Ann. Mate. Pura Appl.* DOI:10.1007/s10231-015-0508-9.

Suzuki, T. and Tasaki, S. (2010), *Stationary solutions to a thermoelastic system on shape memory materials, Nonlinearity* **23**, 2623-2656.

Suzuki, T. and Yamada, Y. (2013), *A Lotka-Volterra system with diffusion*, Nonlinear Analysis and Interdisciplinary Sciences (Aiki, T., Fukao, T., Kenmochi, N., Niezgódka, M., and Ôtani, M. eds.), Gakuto International Series on Mathematical Sciences and Applications, *Gakkotosyo* **36**, 215-236.

Suzuki, T. and Yamada, Y. (2015), *Global-in-time behavior of Lotka-Volterra*

314   Chemotaxis, Reaction, Network

system with diffusion - skew symmetric case, Indiana Univ. Math. J. **64**, 110-126.

Tanabe, H. (1979), *Equations of Evolution*, Pitman, London.

Tarantello, G. (2008), *Selfdual Gauge Field Vortices*, Birkhäuser, Boston.

Talenti, G. (1976), *Elliptic equations and rearrangements*, Ann. Scoula Norm. Sup. Pisa **IV 3**, 697–718.

Teitelbaum, S.L. and Ross, F.P. (2003), *Genetic regulation of osteoclast development and function*, Nature Reviews Genetics **4**, 638–649.

Temam, R. (1975), *A nonlinear eigenvalue problem: the shape at equilibrium of a confined plasma*, Arch. Rational Mech. Anal. **60**, 51–73.

Temam, R. (1977), *Remarks on a free boundary value problem arising in plasma physics*, Comm. Partial Differential Equations **2**, 563–585.

Toland, J.F. (1978), *Duality in nonconvex optimization*, J. Math. Anal. Appl. **66**, 399–415.

Toland, J.F. (1979), *A duality principle for non-convex optimzation and the calculus of variations*, Arch. Rational meth. Anal. **71**, 41–61.

Volterra, V. (1926), *Variazioni e fluttuazioni del numero d'individui in specie animali conviventi*, Mem. Acad. Licen. **2**, 31–113.

Wang, G. and Ye, D. (2003), *On a nonlinear elliptic equation arising in a free boundary problem*, Math. Z. **244**, 531–548.

Wang, J., Yang, J., and Kuniya, T. (2016), *Dynamics of a PDE viral infection model incorporating cell-to-cell transmission*, J. Math. Anal. Appl. **444**, 1542–1564.

Wang, Y., Paszek, P., Horton, C.A., Yue, H., White, M.R.H., Kell, D.B., Muldoon, M.R., and Broomhead, D.S. (2003), *Systematic survey of the response of model NF-κB signaling pathways to $TNF\alpha$ stimulation*, J. Theor. Biol. **297**, 137–147.

Weissler, F.B. (1980), *Local existence and nonexistence for semilinear parabolic equations in L^p*, Indiana Univ. Math. J. **29**, 79–102.

Weissler, F.B. (1985), *An L^∞ blow-up estimate for a nonlinear heat equation*, Comm. Pure Appl. Math. **38**, 291–295.

Winkler, M. (2016), *Finite-time blow-up in the higher-dimensional parabolic-parabolic Keller-Segel system*, J. Math. Pure Appl. **100**, 748-767.

Yagi, A. (2010), *Abstract Parabolic Evolution Equations and their Applications*, Springer Verlag, Berlin.

Yang, Y. (2001), *Solitons in Field Theory and Nonlinear Analysis*, Springer Verlag, New York.

Ye, D. (1997), *Une remarque sur le comportement asymptotique des solutions de $-\Delta u = \lambda f(u)$*, C. R. Acad. Scii. Paris, Série I **325**, 1279–1282.

Index

Printed in the United States
By Bookmasters